T0257953

Encyclopedia of Herbicides: Weed Control

Volume III

Encyclopedia of Herbicides: Weed Control
Volume III

Edited by **Molly Ismay**

New York

Published by Callisto Reference,
106 Park Avenue, Suite 200,
New York, NY 10016, USA
www.callistoreference.com

Encyclopedia of Herbicides: Weed Control
Volume III
Edited by Molly Ismay

International Standard Book Number: 978-1-63239-257-2 (Hardback)

Printed in the United States of America.

Contents

Preface   VII

Control of Weeds   1

Chapter 1   **Evaluation of the Contamination
by Herbicides in Olive Groves**   3
Antonio Ruiz-Medina and Eulogio J. Llorent-Martínez

Chapter 2   **Forty Years with Glyphosate**   23
András Székács and Béla Darvas

Chapter 3   **Prediction of Herbicides Concentration in Streams**   61
Raj Mohan Singh

Chapter 4   **A Critical View of
the Photoinitiated Degradation of Herbicides**   81
Šárka Klementová

Chapter 5   **Oxidative Stress as a Possible
Mechanism of Toxicity of
the Herbicide 2,4-Dichlorophenoxyacetic Acid (2,4-D)**   99
Bettina Bongiovanni, Cintia Konjuh,
Arístides Pochettino and Alejandro Ferri

Chapter 6   **Effects of Herbicide Atrazine
in Experimental Animal Models**   119
Grasiela D.C. Severi-Aguiar and Elaine C.M. Silva-Zacarin

Chapter 7   **Weed Population Dynamics**   131
Aurélio Vaz De Melo, Rubens Ribeiro da Silva,
Hélio Bandeira Barros and Cíntia Ribeiro de Souza

Chapter 8   **Ecological Production Technology
of Phenoxyacetic Herbicides MCPA
and 2,4-D in the Highest World Standard**   143
Wiesław Moszczyński and Arkadiusz Białek

Chapter 9   **Adverse Effects of Herbicides
on Freshwater Zooplankton**   159
Roberto Rico-Martínez, Juan Carlos Arias-Almeida,
Ignacio Alejandro Pérez-Legaspi, Jesús Alvarado-Flores
and José Luis Retes-Pruneda

Chapter 10   **Herbicide Tolerant Food
Legume Crops: Possibilities and Prospects**   189
N.P. Singh and Indu Singh Yadav

Chapter 11   **Vegetative Response
to Weed Control in Forest Restoration**   207
John-Pascal Berrill and Christa M. Dagley

Chapter 12   **Sugar Beet Weeds in Tadla Region (Morocco): Species
Encountered, Interference and Chemical Control**   225
Y. Baye, A. Taleb and M. Bouhache

Chapter 13   **Herbicides in Winter Wheat
of Early Growth Stages Enhance Crop Productivity**   249
Vytautas Pilipavičius

Chapter 14   **Influence of Degree Infestation
with *Echinochloa crus–galli* Species
on Crop Production in Corn**   271
Teodor Rusu and Ileana Bogdan

Permissions

List of Contributors

Preface

This book covers herbicidal control of weeds in an in-depth manner. It discusses numerous techniques used by herbicides for controlling a particular weed population. The objective of this book is to showcase several characteristics and features of herbicides, the physical and chemical properties of specific types of herbicides, and their effects on physical and chemical features of soil and micro-flora. In addition, an assessment of the extent of contamination in soils as well as crops by herbicides has been elaborated along with a research on the performance and photochemistry of herbicides and the impact of excess herbicides in soils and field crops.

The information contained in this book is the result of intensive hard work done by researchers in this field. All due efforts have been made to make this book serve as a complete guiding source for students and researchers. The topics in this book have been comprehensively explained to help readers understand the growing trends in the field.

I would like to thank the entire group of writers who made sincere efforts in this book and my family who supported me in my efforts of working on this book. I take this opportunity to thank all those who have been a guiding force throughout my life.

Editor

Control of Weeds

Evaluation of the Contamination by Herbicides in Olive Groves

Antonio Ruiz-Medina and Eulogio J. Llorent-Martínez
University of Jaén
Spain

1. Introduction

The application of phytosanitary products on the ground is one of the main employed procedures to solve the problems of plagues and diseases in olive groves. As a result, there is an increasing unease related to the possible presence of residues from these products in the olive-derived foods. Recent studies have shown that these residues can be found in olives and olive oil (mainly insecticides and fungicides) due to the lipophillic nature of the used plaguicides. This is also the effect that has been observed with the introduction of herbicides in this cultive. However, the detected levels of all these phytosanitary products are usually below the maximum residue limits (MRLs) established.

Herbicides are used by olive farmers to weed olive plantations with two essential goals: (1) eliminate the competition for water resources between olive trees and weeds, especially in particular moments of the productive cycle; and (2) keep the soil clean around the olive tree (in that way, the harvest of the olive fruits is easier if they fall off the tree after maturation is over). The main herbicides commonly used in olive groves are triazines (simazine, atrazine, trietazine, terbuthylazine and terbutryn), phenylurea diuron and phenylether oxifluorphen. All of them are directly applied on the ground of the plantation and nowadays widely used in intensified-traditional and modern-intensive olive groves. Herbicide residues remain highly concentrated in the top 5-15 cm of soil, even after several months. This fact presents two main consequences. First, herbicide residues are washed into streams, rivers and reservoirs with the soil that is eroded in heavy rains, polluting surface waters. Second, when harvesting is done using various devices to shake the olives off the trees without extending nets under the tree, olives come in contact with the herbicides present in the soil. This constitutes a risk of these herbicides being incorporated in olives and, consequently, olive oil.

Olive oil is a very important commodity in the Mediterranean basin. This product has a great importance in the sustainable economy of important regions from the main olive oil producers in the world: Spain, Italy and Greece. Due to the facts that virgin olive oil production has increased in recent years and that it is being exported to different countries, exhaustive quality controls are required. Different regulations regarding MRLs in olives and olive oil have been established by the European Union and the Codex Alimentarius of the Food and Agriculture Organization of the United Nations. In addition, new regulations will be established in following years with MRLs of 10 µg/kg.

There are alternative methods available for the analysis of herbicides in different kinds of samples. However, the common methods of analysis for their determination are Gas Chromatography (GC) and Liquid Chromatography (LC). GC is usually the chosen technique due to the high separation efficiency and compatibility with a wide range of detection techniques. Our research group has performed thorough investigation over the analysis of herbicides in soils, olive fruits and virgin olive oils. In particular, we have developed different GC analytical methods with mass spectrometric detection for the analysis of all the main herbicides used in olive groves. In this way, sensitive and reliable procedures can be implemented for the quality control of olives and olive oils in the industry. These methodologies, together with the obtained results over real samples from Spain, will be commented in this chapter.

2. Composition and properties of olive oil. Classification

Olive oil is composed mainly of triglycerides (98-99% of the olive oil) and contains small quantities of free fatty acids (FFA), being the proportion of FFA variable and related to the degree of hydrolysis of the triglycerides. The composition of fatty acids in the oil depends on the variety of the olive tree, climatic conditions and geographical localization of the grove. Both the International Olive Council (IOC) and the Codex Alimentarius Comission have established maximum and minimum percentages for each fatty acid in the composition of the olive and pomace-olive oils. Olive oil is basically composed of mono-insaturated fatty acids. Regarding the degree of insaturation, fatty acids composition is as follows: 72% mono-insaturated, 14% poli-insaturated and 14% saturated.

There are also minor components in olive oil, which are specific markers of its physico-chemical authenticity and they also add unique sensory and biological characteristics:

- Squalene is the major olive oil terpenoid hydrocarbon (300-700 mg L^{-1}), whereas β-carotene, biological precursor of vitamin A, is found in small quantities (mg L^{-1}).
- Triterpenoids alcohols (24-methylenecycloartanol, cycloartenol, α-amirine and β-amirine) are especially important from the biological point of view. Eritrodiol is also important from the analytical perspective, because it can be used to detect the presence of pomace-olive oil.
- Sterols can be used to construct a fingerprint in order to authenticate the olive oil, specifically using the β-sitosterol content, which represents around 93% of the total content of sterols.
- Tocopherols, especially α-tocopherol or vitamin E (150-300 mg L^{-1}) are important antioxidants.
- Phenolic compounds, some of them contribute to the characteristic flavor of olive oil, increase the antioxidant properties of the oil.
- Approximately one hundred aromatic compounds are also present, being the exact chemical composition dependent on the variety, climatic conditions and quality of the oil.

Virgin olive oil is the most digestible of the edible fats and it helps to assimilate vitamins A, D and K. It contains essential acids that cannot be produced by our own bodies and slows down the aging process. It also helps bile, liver and intestinal functions. It is noteworthy that olive oil has a beneficial effect in the dietary treatment of diabetes. In addition, it helps to control blood pressure and increases the bone mass. Moreover, olive oil has a favorable effect on the development of the central nervous and vascular systems, in brain development as well as normal child development (Cicerale et al., 2010; El & Karakaya, 2009).

The human body easily absorbs olive oil. This means that the body absorbs the good ingredients such as vitamin E and phenols, which have anti-oxidizing properties and prevent the oxidization of fatty tissue, therefore helping to delay the aging process. In addition, it is not only easy to digest but it also helps the digestion of other fatty substances because it helps the secretions of the peptic system and stimulates the pancreatic enzyme lipase. On the other hand, olive oil consumption has a very positive effect on blood cholesterol (limits the oxidizing of bad cholesterol because it is rich in anti-oxidizing agents, as it was indicated before).

Olive oil, as any fatty substance, deteriorates during the frying process, especially if it is used over and over and if the frying temperature is very high. High temperature destroys the good ingredients of any oil while it creates harmful agents for the liver, the arteries and the heart. However, it is important to take into consideration that these harmful agents are less likely to be created in olive oil than in all other known vegetable oils because of the different composition. Olive oil contains a high percentage of oleic acid, which is much more resistant to oxidization than polyunsaturated acids (found in large amounts in seed oils). As a result, olive oil is the most stable fat and it stands up well to high frying temperatures.

Taking into account all the procedure in order to obtain the oil, edible olive oil is marketed in accordance with the following designations and definitions (International Olive Council, 2010):

1. Virgin olive oil: it is the oil obtained from the fruit of the olive tree solely by mechanical or other physical means under conditions, particularly thermal conditions, that do not lead to alterations in the oil, and which have not undergone any treatment other than washing, decantation, centrifugation and filtration. The virgin olive oils that are fit for consumption are extra virgin olive oil and virgin olive oil, presenting a free acidity, expressed as oleic acid, of not more than 0.8 and 2 grams per 100 grams, respectively. They also have to fulfil different quality criteria that, according to the IOC, include good organoleptic characteristics (taste and aroma) and low peroxide value.
2. Olive oil: it is the oil consisting of a blend (approximately 80:20 v:v) of refined olive oil and virgin olive oils fit for consumption as they are. It has a free acidity, expressed as oleic acid, of not more than 1 gram per 100 grams and its other characteristics correspond to those fixed for this category in this standard.
3. Olive pomace oil: it is the oil obtained by treating olive pomace with solvents or other physical treatments, to the exclusion of oils obtained by re-esterification processes and of any mixture with oils of other kinds. The marketed olive pomace oil is the oil comprising the blend (approximately 80:20 v:v) of refined olive-pomace oil and virgin olive oils fit for consumption as they are. It has a free acidity of not more than 1 gram per 100 grams.

Chemical processing may improve high acidity olive oil and make it edible; however, it takes away some extremely valuable ingredients such as vitamins and phenols. As a result, processed olive oil (refined) lacks the desirable properties and characteristics that can be found in abundance in (extra) virgin olive oil.

3. Processing and elaboration of olives and olive oil

In this section, the steps required for obtaining high-quality olive oil (extra virgin) and table olives will be described. It is important to know the details of the processing in order to understand in which step can the contamination by herbicides take place.

3.1 Olive oil

3.1.1 Harvest of olives

The harvest of the olives could be understood as an independent activity from the elaboration of the oil. However, the characteristics of the oil are highly influenced by the harvest time and method employed.

Olives must be picked at the moment of optimum ripeness, where the fruit presents the maximum content of oil and the best characteristics. Olives reach their ripeness in autumn and picking starts at the end of November, lasting up to February or March. The methods to harvest the olives have not changed much from the ancient time. The methods used should not damage the fruit and should avoid breaking of boughs or shoots. The high-quality olive oil is obtained by "milking" the olives into a sack tied around the harvester's waist (using ladders for the highest boughs) and extending canvases at the foot of the trees, where the olives will fall when the tree is beaten (using flexible poles). Recent methods make use of harvester machines (shakers) that generate the vibration on the tree for the falling of the fruit. This vibration machines can be small vibrators operated by the farmer in the required boughs, of full-equipped tractors, with vibration units for the whole trunk of the tree and an umbrella to pick the olives.

Harvest by hand is impossible in olive trees of 4-5 meters high, even with the use of ladders. In general, the trunks are to wide to allow the use of vibration machines and the olives are picked directly from the ground after they fall down when they reach ripeness. Hence, harvest period lasts longer, even until spring when it is a high-production year. When the olives are picked from the ground, weeds have to be completely eliminated from the zone in order to make it easier the picking. The quality of the olive oil obtained from this olives is poor due to organoleptics flaws such as soil flavor.

3.1.2 Washing

Traditionally, agricultural workers have cleaned the fruit in the fields by means of sieves. But this cleaning is not complete and the fruit goes into the olive-oil mill with a great quantity of impurities, leaves, branches, mud, etc, which are necessary to eliminate. Therefore, for this purpose, cleaners that use air current or shaking sieves are used to eliminate leaves, branches and other impurities lighter than the fruits. In addition, washing devices are employed to eliminate heavier impurities such as stones or dust. In next figures, both cleaning steps, the one with the air for light impurities (Figure 1) and the other one by washing for heavy impurities (Figure 2), are shown.

Once the fruit has been cleaned and weighed, it is stored in hoppers until it can be crushed. The storage period must be as short as possible in order to avoid alterations, which will produce oils with a higher degree of acidity, lower stability and worse flavor.

Olives carried towards the cleaners   Air Cleaners

Washing device

Fig. 1. Cleaning steps for olive fruit.

Fig. 2. Washing step for olive fruit.

3.1.3 Preparation

The process of releasing the oil from the plant tissue begins by milling the olives to tear the flesh cells in order to let the oil run out of the vacuoles. This is followed by stirring the olive paste to permit the formation of large drops of oil and to break up the oil–water emulsion. In so-called "dual-phase decanters" the oil is then separated by direct continuous centrifugation from the pomace, which consists of vegetable matter and water. The yield of oil varies from 80 to 90% of the total oil content of the olives, because the oil in the olive paste is only partially free to escape and part of it remains in the unbroken cells or is trapped in the tissues of the cytoplasm, or is emulsified in the aqueous phase.

After the extraction of virgin olive oil from the olives, the remaining paste is called pomace and still contains a small quantity (2-6%) of oil that can only be extracted with chemical solvents. This is done in specialized chemical plants, not in the oil mills and the obtained oil is called pomace oil.

The FFA limit in olive oil for direct consumption is 2%. Virgin olive oils with higher content are called "lampantes" and must be refined prior to consumption. The components to be removed are all those ones that are detrimental to the flavor, color and stability of the oil, mainly FFA, phosphoacylglycerols, pigments, volatiles and contaminants. The standard processes used are chemical and physical refining. The main difference between both processes is that chemical refining procedure includes caustic soda treatment to neutralize the oil while, following physical refining, FFA are eliminated by distillation during deodorization. Physical refining reduces the loss of neutral oil, minimizes pollution and enables recovery of high quality FFA. Nevertheless, not all oils can be physically refined.

3.2 Table olives

"Table olives" means the product: a) prepared from the sound fruit of varieties of the cultivated olive tree (*Olea europaea L.*) that are chosen for their production of olives whose volume, shape, flesh-to-stone ratio, fine flesh taste, firmness and ease of detachment from the stone make them particularly suitable for processing; b) treated to remove its bitterness and preserved by natural fermentation or by heat treatment, with or without the addition of preservatives; c) packed with or without covering liquid (International Olive Oil Council, 2004).

Table olives are classified in one of the following types according to the degree of ripeness of the fresh fruits:

- Green olives: Fruits harvested during the ripening period, prior to colouring and when they have reached normal size.
- Olives turning color: Fruits harvested before the stage of complete ripeness is attained, at colour change.
- Black olives: Fruits harvested when fully ripe or slightly before full ripeness is reached.

To prevent olive damage, fruits destined for table olives production are picked by hand and carefully placed in special padded basket that are hung by their neck. Olive transportation is carried out in perforated plastic containers, which have plastic netting as walls (supported by an iron structure). The perforated walls permit the aeration of the fruits and the reduced weight also contributes to minimizing the damage of the fruits. Sometimes, olives are also transported in bulk, although this transportation system is not recommended due to the increased risk of damaging the olives.

Once the olives have been transported, their processing takes place. In general, any processing method aims to remove the natural bitterness of the fruit, caused by the glucoside oleuropein. The bitterness may be removed by alkaline treatment, by immersion in a liquid to dilute the bitter compound, or by biological processes. The product so obtained may be preserved in brine according to its specific characteristics, in dry salt, in a modified atmosphere, by heat treatment, my preservatives, or by acidifying agents. The most common trade preparations are (Sánchez et al., 2006):

- Treated olives: Olives that have undergone alkaline treatment, then packed in brine in which they undergo complete or partial fermentation, and preserved or not by the addition of acidifying agents. The most common preparation is "treated green olives in brine".

- Natural olives: Olives placed directly in brine in which they undergo complete or partial fermentation, preserved or not by the addition of acidifying agents. The most common preparation is "natural black olives".
- Olives darkened by oxidation: Olives preserved in brine, fermented or not, darkened by oxidation in an alkaline medium and preserved in hermetically sealed containers subjected to heat sterilisation; they shall be a uniform black color. They are also known as "black olives".

Other trade preparations include dehydrated and/or shrivelling olives and specialities prepared in different forms.

4. Pesticides: definition and classification

The denomination pesticides (or plaguicides) include a wide variety of products that are very different in their chemical composition and characteristics. A plaguicide can be described as a substance (or formulation containing at least one of them) that presents any of the following uses:

- Fight agents that can be harmful for the crops or prevent the potential effects of these agents.
- Control or regulate the vegetable production.
- Protect the vegetable production, including woods.
- Destruct weeds.
- Destruct part of the vegetables or prevent undesirable growths.
- Destruct or prevent the action of potentially harmful organisms different to the ones that attack plants.

Taking into account the specific action of the plaguicides, different classifications can be made, being the decimal classification one of the most frequently used:

- Insecticides are used against insects and they include ovicides and larvicides (against eggs and larvae of insects, respectively). Nearly all insecticides have the potential to significantly alter ecosystems, being many of them toxic to humans.
- Acaricides kill members of the Acari group, which includes ticks and mites.
- Fungicides are chemical compounds or biological organisms used to kill or inhibit fungi or fungal spores.
- Nematocides, disinfectants and fumigants in general, used to kill parasitic nematodes.
- Herbicides are used to kill unwanted plants while leaving the desired crop relatively unharmed.
- Phytoregulators and similar products can be used to improve the potential of the trees.
- Molusquicides and rodenticides are used to control molluscs (slugs and snails) and rodents pests.
- Post-harvest pesticides and seeds.
- Protectors of woods, fibers and derivatives.
- Other specific plaguicides.

Laboratory studies show that pesticides can cause health problems, such as birth defects, nerve damage, cancer, and other effects that might occur over a long period of time. However, these effects depend on how toxic the pesticide is and how much of it is consumed. Some pesticides also pose unique health risks to children. For these reasons, the governments carefully regulate pesticides to ensure that their use does not pose unreasonable risks to human health or the environment.

The mechanisms of action of the plaguicides over the organism are very different depending on the chemical composition. These mechanisms are well-known for some pesticides, even at the molecular level. However, they are completely unknown for some others. Even among pesticides from the same family, some of them can be classified as scarcely toxic while others are very toxic. As a result, it is difficult to establish general rules when dealing with the toxicity of plaguicides.

Herbicides are required to control the growth of weeds. The absence of weeds in the soil around the olive tree presents two major benefits: the weeds do not waste water resources for the olive trees and the harvest of olives from the soil is easier when they naturally fall down from the tree.

Depending on the period of application, herbicides can be classified as:

- Pre-emergents, applied at the starting of autumn.
- Early post-emergents, applied by the middle of autumn, after the early rainings.
- Post-emergents, applied in spring against perennial weeds.

In Table 1, the herbicides that are usually employed in the Spanish olive groves are shown. Simazine, which is forbidden, is also present because it is very persistent and has to be analyzed in order to ensure the absence of residues. The herbicides comprise triazines (simazine, atrazine, trietazine, terbuthylazine and terbutrine), phenylurea diuron and phenylether oxyfluorphen.

The water solubility (W.S.) at 25 °C and the $K_{O/W}$ value are also presented in Table 1. The $K_{O/W}$ value is the partition coefficient between octanol and water, and its logarithm is an indication of fat solubility of the pesticide. A high value of the coefficient $K_{O/W}$ for a particular plaguicide indicates that its solubility in water is low; hence, this plaguicide would be fat-soluble and there would be risks of bio-accumulation in fatty tissues. It has been demonstrated that fat-soluble pesticides tend to concentrate in olive oil during its production and extraction from the olives. For this reason, higher concentration levels of fat-soluble plaguicides are expected in the oil than in the olives from which the oil was produced. On the other hand, polar pesticides do not tend to preconcentrate in olive oil and their concentration is lower.

Monitoring herbicide residues in olive oil and table olives is of great interest to ensure food safety related to their use. The development of multi-residue methods is required in this case in order to determine all the herbicides in the same analysis. These methods need to present a high sensitivity in order to be able to analyze the samples at the legislated MRLs.

The analysis of herbicides in these samples is very challenging, because of the inherent complexity of the matrix. As a result, it is necessary to extract the pesticide fraction from the whole matrix to isolate the compounds that will be analyzed. Taking into account that some herbicides are fat-soluble, it is difficult to completely separate them from the matrix. Hence, a clean-up step is required after the extraction procedure.

Plaguicide	Structure	W.S. (mg L^{-1})	K$_{O/W}$ 25°C
Simazine		1.3	2.10
Atrazine		33	2.50
Trietazine		20	3.34
Terbuthylazine		8.5	3.21
Terbutrine		22	3.65
Diuron		36.4	2.85
Oxyfluorfen		0.12	4.86

Table 1. Herbicides in olive groves.

5. Sample preparation

The first step required in the analysis of herbicides is to separate the analytes from the matrix and other interfering compounds, therefore isolating the herbicides from the rest of the sample. This separation includes extraction, pre-concentration and clean-up steps, although sometimes they can be performed at the same time (Gilbert-López et al., 2009; Mukherjee & Gopal, 1996).

This part of the analysis is critical, because the final results will be completely related to the success of the separation process. The sample has to come into contact with an extractant (solid, liquid or supercritical fluid) in optimized conditions in order to weaken the analyte-matrix interactions and increase the analyte-extractant interactions.

There is a continuous development of sample-treatment procedures for the isolation of herbicides in samples with relatively high fat content, such as olive oil and olives. The preparation of oil samples for their determination by chromatographic techniques requires the removal of the fatty components from the sample. The main problem associated when working with this kind of samples is that dirty extracts may harm the instruments employed. For this reason, proper extraction and clean-up steps are required. In addition, the different nature and physicochemical properties of the classes of herbicides to be analyzed makes it more complex to select the appropriate extraction methodology.

There are numerous extraction and clean-up procedures. However, the most common ones are solid-liquid or liquid-liquid extraction, gel permeation-chromatography (GPC) and solid-phase extraction (SPE) (Gilbert-López et al., 2009; Mukherjee & Gopal, 1996; Walters, 1990).

5.1 Solid-liquid extraction

In this case, the sample comes to contact with an appropriate solvent. After that, different procedures can be applied to homogenize both phases: ultrasonic agitation and microwave extraction.

Ultrasonic agitation: the solvent interacts with the sample simply by a shaking process. Ultrasonic radiation (frequency of 25-40 Hz) causes the vibration of the molecules, increasing the collision between them. Hence, the contact between sample and solvent is enhanced. After the analyte is dissolved in the extractant, a filtration step and the evaporation of the solvent are required.

Microwave extraction: samples are enclosed in high quality Teflon vessels together with the solvent and heated to a controlled temperature with microwave power, being the extract filtered when the process finishes. Electromagnetic irradiation is used to heat the solvent that acts as extractant, which is in most cases a mixture between hexane and acetone.

5.2 Liquid-liquid extraction

Usually, it has been the chosen method for the extraction, preconcentration and clean-up of liquid samples. It is based on the relative solubility of the analyte in two different immiscible liquids; therefore, it is an extraction of a substance from one liquid phase into another liquid phase. The liquid sample comes to contact with an appropriate solvent (immiscible with the sample) and, after a mixing time, both phases (sample and extractant) are separated. During the mixing period, the analyte is distributed between both phases until it reaches the equilibrium.

The $K_{O/W}$ value is critical to decide the appropriate extractant. If this value is high, it indicates that the analyte tends to dissolve in the organic phase instead of in water. As a general rule, the organic solvents selected are volatile substances that present high affinity for the analytes and are immiscible with the sample.

5.3 Gel-permeation chromatography

It is probably the most extensively used technique for the analysis of pesticide residues in olive oil, usually after a liquid-liquid extraction. In this technique, an aliquot of an olive oil extract (obtained from the extraction step) is injected into the GPC system. The selected fraction is collected and, after a solvent-exchange step, the sample is analyzed by GC. A GPC system is composed of a chromatographic pump, a fraction collector and a detector. The columns are made from polymeric porous microspheres that enable the separation of compounds according to their molecular weights (which are related to the size of the compound). Using this principle, the herbicide fraction is separated from the triglyceride fraction, which presents higher molecular weight. In GPC, the compounds are eluted from higher to lower molecular weight.

5.4 Solid-phase extraction

It is a separation process by which compounds that are dissolved or suspended in a liquid mixture are separated from other compounds in the mixture according to their physical and chemical properties. SPE is used to concentrate and purify samples for analysis and to isolate the analytes of interest from a wide variety of matrices.

SPE uses the affinity of the analytes dissolved or suspended in a liquid (known as the mobile phase) for a solid through which the sample is passed (known as the stationary phase) to separate a mixture into desired and undesired components. The result is that either the desired analytes of interest or undesired impurities in the sample are retained on the stationary phase. The portion that passes through the stationary phase is collected or discarded, depending on whether it contains the desired analytes or undesired impurities. If the portion retained on the stationary phase includes the desired analytes, they can then be removed from the stationary phase for collection in an additional step, in which the stationary phase is rinsed with an appropriate eluent.

The stationary phase comes in the form of a packed syringe-shaped cartridge that can be mounted on its specific type of extraction manifold. The manifold allows multiple samples to be processed by holding several SPE media in place and allowing for an equal number of samples to pass through them simultaneously. A typical cartridge SPE manifold can accommodate up to 24 cartridges and is equipped with a vacuum port. Application of vacuum speeds up the extraction process by pulling the liquid sample through the stationary phase. The analytes are collected in sample tubes inside or below the manifold after they pass through the stationary phase.

Solid phase extraction cartridges are available with a variety of stationary phases, each of which can separate analytes according to different chemical properties. Most stationary phases are based on silica that has been bonded to a specific functional group.

6. Detection techniques for the analysis of pesticides. Gas chromatography

Pesticides (herbicides, fungicides or insecticides) are the most abundant environmental pollutants found in soil, water, atmosphere and agricultural products, and may exist in harmful levels and pose an environmental threat. Even low levels of these contaminants can cause adverse effects on humans, plants, animals and ecosystems. As it has been commented in the Introduction, there are several alternatives for the determination of pesticides in different kinds of samples. In this sense, automated systems of analysis with spectroscopic or electrochemical detection have been developed (Llorent-Martínez et al., 2011). In particular, the development of sensors have been specially useful. Electrochemical sensors present the advantages of miniaturization, simplicity, possibility of in-situ measurements and low-cost. In general, the main contribution of these sensors to pesticide analysis has consisted of the development of methods of analysis for determining a whole family of compounds (Du et al., 2009; Halámek et al., 2005; Liu & Lin, 2006). In spectroscopic methods, fluorescence (Calatayud et al., 2006; Mbaye et al., 2011) or chemiluminescence (Catalá-Icardo et al., 2011; López-Paz & Catalá-Icardo, 2011) detections have been usually employed due to their intrinsic sensitivity and selectivity. Among the spectroscopic methods of analysis, the design of flow-through optosensors has also been paid particular attention (Llorent-Martínez et al., 2011). These sensors present enhanced sensitivity and selectivity and have been applied to the determination of a small number of pesticides in different kind of samples (Llorent-Martínez et al., 2005; Llorent-Martínez et al., 2007; López Flores et al., 2007).

In general, electrochemical sensors can provide a useful tool when a whole family of compounds is targeted, while optosensors provide an interesting approach in order to quantify a small number of analytes in a particular sample. However, chromatographic techniques are still the chosen ones when a multi-residue analysis is required, being GC the most common one for the analysis of pesticides.

GC is an analytical technique that is used for separating and analysing compounds that can be vaporized without decomposition (Harris, 2007; Skoog et al., 1996). Typical uses of GC include testing the purity of a particular substance, or separating the different components of a mixture. In GC, the mobile phase is a carrier gas, usually an inert gas such as helium or an unreactive gas such as nitrogen. The stationary phase is a microscopic layer of liquid or polymer on an inert solid support, inside a piece of glass or metal tubing called column. A gas chromatograph uses a flow-through narrow tube known as the column, through which different chemical constituents of a sample pass in a gas stream (carrier gas, mobile phase) at different rates depending on their various chemical and physical properties and their interaction with a specific column filling, called the stationary phase. As the chemicals exit the end of the column, they are detected and electronically identified. The function of the stationary phase in the column is to separate different components, causing each one to exit the column at a different time (retention time). Other parameters that can be used to alter the order or time of retention are the carrier gas flow rate, column length and the temperature.

In a GC analysis, a known volume of gaseous or liquid sample is injected into the head of the column using a micro syringe. As the carrier gas sweeps the analyte molecules through the column, this motion is inhibited by the adsorption of the analyte molecules either onto the column walls or onto packing materials in the column. The rate at which the molecules progress along the column depends on the strength of adsorption, which in turn depends on the type of molecule and on the stationary phase materials. Since each type of molecule has a different rate of progression, the various components of the analyte mixture are separated as they progress along the column and reach the end of the column at different times. A detector is used to monitor the outlet stream from the column; thus, the time at which each component reaches the outlet and the amount of that component can be determined.

Different detectors can be used in GC. The most common ones until recent years were the flame ionization detector (FID), the thermal conductivity detector (TCD) and the electron capture detector (ECD). However, nowadays most of the developed analytical methods use GC coupled to mass spectrometry (MS) detectors. In the analytical methods that will be described later, only ECD, thermoionic specific detector (TSD) and MS have been used in our research.

ECD is used for detecting electron-absorbing compounds. It uses a radioactive beta particle (electron) emitter. The electrons are formed by collision with a nitrogen molecule because it exhibits low excitation energy. The electron is then attracted to a positively charged anode, generating a steady current. Therefore, there is always a background signal present in the chromatogram. As the sample is carried into the detector by the carrier gas, analyte molecules absorb the electrons and reduce the current between the collector anode and a cathode. The analyte concentration is thus proportional to the degree of electron capture. ECD is particularly sensitive to halogens, organometallic compounds, nitriles, or nitro compounds.

TSD is a very sensitive but specific detector that responds almost exclusively to nitrogen and phosphorous compounds. It contains a rubidium or cesium silicate (glass) bead situated in a heater coil, at little distance from the hydrogen flame. The heated bead emits electrons by thermionic emission. These electrons are collected under a potential of few volts by an appropriately placed anode, and provides a background current. When a solute containing nitrogen or phosphorous is eluted from the column, the partially combusted nitrogen and phosphorous materials are adsorbed on the surface of the bead. The adsorbed material reduces the work function of the surface and, as consequence, the emission of electrons is increased, raising the current collected at the electrode.

Mass spectrometry (MS) is an analytical technique that measures the mass-to-charge ratio of charged particles. It is used for determining masses of particles, for determining the elemental composition of a sample or molecule, and for elucidating the chemical structures of molecules. The MS principle consists of ionizing chemical compounds to generate charged molecules or molecule fragments and measuring their mass-to-charge ratios. In MS detection: 1) the analytes undergo vaporization; 2) they are ionized by one of a variety of methods (e.g., by impacting them with an electron beam), which results in the formation of charged particles (ions); 3) the ions are separated according to their mass-to-charge ratio in an analyzer by electromagnetic fields; 4) the ions are detected, usually by a quantitative method; and 5) the ion signal is processed into mass spectra.

7. Analysis of herbicides

Our research has focused mainly on the determination of herbicides in olives destined for production of olive oil and olive oil. However, at this moment, we are continuing with this research, analyzing herbicide residues in table olives. On average, 5 kg of olives are required for the production of 1 L of oil. Taking into account that most pesticides (including herbicides) are lipophilic, a concentration effect could occur when obtaining the olive oil. Thus, MRL for herbicides have been set by the European Union in both olives and olive oil. For this reason, our research group has developed analytical methods that allow the analysis of the selected herbicides in olive oil and olives. In this section we will describe the analytical procedures that have been employed for this purpose as well as the results obtained.

For olive oil samples (Guardia-Rubio et al., 2006b), the preparation procedure included a liquid-liquid extraction followed by GPC clean-up step: 1) Two grams of the olive oil sample were dissolved in n-hexane saturated in acetonitrile. The solution was transferred to a separation funnel where it was extracted three times with acetonitrile saturated in n-hexane. The extracts were combined in a round-bottomed flask and were concentrated to dryness in a rotary evaporator. 2) The residue was dissolved in the GPC mobile phase (ethyl acetate-cyclohexane, 1:1 (v:v)) and injected into the GPC column. The collected eluate fraction was transferred to a round-bottomed flask and concentrated to dryness in a rotary evaporator. 3) The residue was redissolved in cyclohexane and analyzed by GC-MS.

For olive samples (Guardia-Rubio et al., 2007c), approximately 130 g of olives (including the seeds) were first crushed by means of a hammer mill. Afterwards, a 100 g portion was weighed in a glass tube and 50 g of anhydrous sodium sulphate were added. The sample was then extracted twice with light petroleum by homogenization with Ultra-Turrax (a high flow mixing tool) and the extracts evaporated using a vacuum rotary evaporator. The solid residue was dissolved in n-hexane saturated in acetonitrile and the same liquid-liquid extraction and GPC clean-up procedure employed for olive oil were performed before the GC-MS analysis.

The following Table shows the retention times (t_R) and analytical parameters obtained for each selected herbicide, including the detection limit (DL). The procedures previously detailed were applied to the determination of the cited herbicides in olives and olive oil samples.

As it was commented in Section 3, there are different olive harvesting methods. Depending on this, the olive fruits can be grouped into three categories: a) fruits picked directly from the tree without any contact with the soil (flight olives); b) fruits picked from the ground (soil olives); and c) fruits that are not separated before the elaboration process (non-separated olives). The separation of flight and soil olive fruits before olive oil elaboration is critical in order to obtain appropriate results (high-quality virgin olive oil). If both fruits are not separated, the quality of the oil and the percentage of extra virgin olive oil obtained decrease. In addition, the harvesting method may be very important for the presence of herbicide residues in olives, because they remain concentrated in the top 5-15 cm of soil, even after several months since their application. 94 and 33 samples of olives and olive oil, respectively, were analyzed. Diuron and terbuthylazine were found in many of these samples, 79 of olives (specially soil olives) and 31 of olive oil. In four of the soil olives, diuron levels were higher than the MRL established by the European Union (0.2 mg kg^{-1}); however, none of the olive oil samples presented levels higher than the corresponding MRL (0.8 mg kg^{-1}). Terbuthylazine was also quantified in many samples with values over the MRL in some of them, being the established MRLs 0.05 and 0.2 mg kg^{-1} for olives and olive oil, respectively. Although other herbicides were also detected, in most cases the levels were below the quantification limit of the system. In general, the levels of herbicides found in soil olives have been significantly higher than those ones found in flight olives, which have not been in contact with the soil. These results suggest that herbicide residues are mainly caused by the contamination of the olives when they come to contact with the soil after falling down (Guardia-Rubio et al., 2006b; Guardia-Rubio et al., 2006c).

Herbicide	t_R (min)	Olives		Olive oil	
		Linear Range (µg kg^{-1})	DL (µg kg^{-1})	Linear Range (µg kg^{-1})	DL (µg kg^{-1})
Simazine	11.971	0.75-125	0.25	3-2000	1
Atrazine	12.095	1.25-250	0.50	5-1000	2
Trietazine	12.621	2.50-200	1.25	10-800	5
Terbuthylazine	12.666	0.25-250	0.12	1-1000	0.5
Terbutrine	18.125	5.00-250	1.25	20-1000	5
Diuron	6.908	1.25-250	0.12	5-1000	0.5
Oxyfluorfen	25.918	2.50-250	1.25	10-1000	5

Table 2. Analytical parameters for olives and olive oil determination.

The fruit goes into the olive-oil mill with a great quantity of impurities, leaves, branches, mud, etc, which are necessary to eliminate. Therefore, for this purpose, cleaners and washing devices are employed to eliminate impurities (Section 3), especially present in soil olives. It would be interesting to evaluate what fraction of the herbicides could be eliminated from the olives after the washing process in the olive mill, previous selection of the herbicides that frequently appear in these samples. The selected plaguicides were diuron and terbuthylazine. Our research group presented the first exhaustive study of the influence of the olives washing in the mills over the concentration of these herbicides (Guardia Rubio et al., 2006a; Guardia Rubio et al., 2007a; Guardia-Rubio et al., 2007b; Guardia-Rubio et al., 2008). Olive samples were collected before and after the washing process in the mill and were analyzed by GC-MS.

The most outstanding conclusion from the obtained results was the drastic reduction in the levels of herbicides in soil olives after the washing process when compared to the same olive samples before the washing step. The washing process significantly diminished the levels of residues of herbicides in soil olives, while the influence of the washing step was not clearly appreciated in flight olives. In the case of non-separated olives, it is interesting to remark that some of the washed olives presented higher levels of herbicides than the non-washed ones. This can be due to the contamination of herbicides-free olives during the washing process. The washing machines for the olives are cleaned and filled with fresh water at the starting of the day. With this water, up to 160000 kg of olives can be washed before the water is changed next day. The water can be contaminated after the washing of soil olives, therefore contaminating following herbicides-free olives in the washing process. Therefore, the water and mud from the olive washing devices were analyzed to confirm this theory. The procedures employed for both type of analyses follow:

a.   In the case of washing water samples, they were first filtered and then slowly passed through a SPE cartridge packed with C_{18} using a 12-port SPE vacuum manifold. The retained herbicides were then eluted from the solid phase with dichloromethane. The eluate, filtered and dried with anhydrous Na_2SO_4, was evaporated to dryness and the residue was dissolved in cyclohexane for GC-MS analysis.

b.   In the case of mud samples, a solid-liquid extraction was carried out with a mixture of cyclohexane/acetone (3:1) in an ultrasonic water bath. The extracts were then filtered to eliminate particulate material, dried with anhydrous Na_2SO_4 and evaporated to dryness by means of a rotary evaporator. After that, an additional clean-up step was necessary in order to remove any remaining fat in the extract after the extraction procedure. This step involved the use of a chromatographic column that was packed with activated alumina suspended in cyclohexane. Once the extract was applied to the column, a mixture of cyclohexane/acetone (3:1, v:v) followed by dichloromethane was used to elute the herbicides. Once the eluate was evaporated and redissolved in cyclohexane, the sample could be analyzed by GC with ECD or TSD detection.

From the analyses carried out over mud and washing water samples, the following results were obtained: a) regarding water analysis, the waters from an olive mill were collected at different times during the same day, and it was observed that the concentration of herbicides increased continuously along the day. An increase in the amount of washed olives meant an increase in herbicide residues. These results confirmed that the waters were being continuously contaminated with herbicides and a decontamination process would be required in the middle of the day (Guardia-Rubio et al., 2008); b) with respect to mud samples, 18 samples were analyzed. Diuron and terbuthylazine were found in nearly all the analyzed samples. Diuron appeared in all samples at concentration levels that ranged between 2.8 and 401.3 ng g^{-1}. Terbuthylazine was detected in 16 samples at concentration levels between 7 and 1031.4 ng g^{-1}. Simazine, prohibited in olive farming in the European Community but a very persistent pollutant, was detected in four samples, although in three of them the concentration was below the quantification limit (Guardia Rubio et al., 2006a). In general, the analysis of mud and waters from the washing device showed that, although the washing process eliminate a high percentage of the herbicide residues, a control over the re-used washing water needs to be performed.

8. Conclusions

The production of virgin olive oil has increased in recent years and it is being exported to different countries, representing an important part of the economy in some Mediterranean countries. As we described in this chapter, it poses a lot of beneficial properties for the human health. However, it is important to carry out exhaustive quality controls in order to maintain its high standards. One of the most important ones is the analysis of residues of pesticides (including herbicides), which can be very dangerous for the human body, presenting different adverse effects over the organism depending on their toxicity. Here, we have described the methods developed in our research group for the analysis of herbicides in olives and olive oil. It has been shown that diuron and terbuthylazine are commonly found in these samples, although usually at levels below the established MRLs. In addition, the separation of flight and soil olives, together with the washing process in the olive mills, is a critical step to control the levels of herbicides. However, the efficiency of the washing step decreases over the time and the washing waters need to be replaced in order to keep the process being useful along the whole working day. Although the studies here presented have focused on the analysis of olive oil and olives destined to oil production, further research is currently being performed for the analysis of processed table olives.

9. References

Calatayud, J.M., De Ascenção, J.G. & Albert-García, J.R. (2006). FIA-fluorimetric determination of the pesticide 3-indolyl acetic acid. *Journal of Fluorescence*, Vol. 16, No. 1, (January 2006), pp. 61-67, ISSN: 1053-0509

Catalá-Icardo, M., López-Paz, J.L. & Peña-Bádena, A. (2011). FI-photoinduced chemiluminescence method for diuron determination in water samples. *Analytical Sciences*, Vol. 27, No. 3, (March 2011), pp. 291-296, ISSN: 0910-6340

Cicerale, S., Lucas, L. & Keast, R. (2010). Biological activities of phenolic compounds present in virgin olive oil. *International Journal of Molecular Sciences*, Vol. 11, No. 2, (February 2010), pp. 458-479, ISSN: 1422-0067

Du, D., Wang, J., Smith, J.N., Timchalk, C. & Lin, Y. (2009). Biomonitoring of organophosphorus agent exposure by reactivation of cholinesterase enzyme based on carbon nanotube-enhanced flow-injection amperometric detection. *Analytical Chemistry*, Vol. 81, No. 22, (October 2009), pp. 9314-9320, ISSN: 0003-2700

El, S.N. & Karakaya, S. (2009). Olive tree (*Olea europaea*) leaves: Potential beneficial effects on human health. *Nutrition Reviews*, Vol. 67, No. 11, (November 2009), pp. 632-638, ISSN: 0029-6643

Gilbert-López, B., García-Reyes, J.F. & Molina-Díaz, A. (2009). Sample treatment and determination of pesticide residues in fatty vegetable matrices: A review. *Talanta*, Vol. 79, No. 2, (July 2009), pp. 109-128, ISSN: 0039-9140

Guardia Rubio, M., Banegas Font, V., Molina Díaz, A. & Ayora Cañada, M.J. (2006a). Determination of triazine herbicides and diuron in mud from olive washing devices and soils using gas chromatography with selective detectors. *Analytical Letters*, Vol. 39, No. 4, (March 2006), pp. 835-850, ISSN: 0003-2719

Guardia-Rubio, M., Fernández-de Córdova, M.L., Ayora-Cañada, M.J. & Ruiz-Medina, A. (2006b). Simplified pesticide multiresidue analysis in virgin olive oil by gas chromatography with thermoionic specific, electron-capture and mass spectrometric detection. *Journal of Chromatography A*, Vol. 1108, No. 2, (March 2006), pp. 231-239, ISSN: 0021-9673

Guardia-Rubio, M., Ruiz-Medina, A., Molina-Díaz, A. & Ayora-Cañada, M.J. (2006c). Influence of harvesting method and washing on the presence of pesticide residues in olives and olive oil. *Journal of Agricultural and Food Chemistry*, Vol. 54, No. 22, (November 2006), pp. 8538-8544, ISSN: 0021-8561

Guardia Rubio, M., Ruiz Medina, A., Pascual Reguera, M.I. & Fernández de Córdova, M.L. (2007a). Multiresidue analysis of three groups of pesticides in washing waters from olive processing by solid-phase extraction-gas chromatography with electron capture and thermionic specific detection. *Microchemical Journal*, Vol. 85, No. 2, (April 2007), pp. 257-264, ISSN: 0026-265X

Guardia-Rubio, M., Ayora-Cañada, M.J. & Ruiz-Medina, A. (2007b). Effect of washing on pesticide residues in olives. *Journal of Food Science*, Vol. 72, No. 2, (March 2007), pp. C139-C143, ISSN: 1750-3841

Guardia-Rubio, M., Marchal-López, R.M., Ayora-Cañada, M.J. & Ruiz-Medina, A. (2007c). Determination of pesticides in olives by gas chromatography using different detection systems. *Journal of Chromatography A*, Vol. 1145, No. 1-2, (March 2007), pp. 195-203, ISSN: 0021-9673

Guardia-Rubio, M., Ruiz-Medina, A., Molina-Díaz, A. & Ayora-Cañada, M.J. (2008). Pesticide residues in washing water of olive oil mills: Effect on olive washing efficiency and decontamination proposal. *Journal of the Science of Food and Agriculture*, Vol. 88, No. 14, (November 2008), pp. 2467-2473, ISSN: 0022-5142

Halámek, J., Pribyl, J., Makower, A., Skládal, P. & Scheller, F.W. (2005). Sensitive detection of organophosphates in river water by means of a piezoelectric biosensor. *Analytical and Bioanalytical Chemistry*, Vol. 382, No. 8, (August 2005), pp. 1904-1911, ISSN: 1618-2642

Harris, D.C. (2007) *Quantitative Chemical Analysis*, Freeman and Company, ISBN: 0716770415, New York

International Olive Oil Council. (2004). Trade Standard applying to table olives. COI/OT/NC No.1

International Olive Council. (2010). Trade Standard Applying to Olive Oils and Olive-Pomace Oils. *COI/T.15/NC*.

Liu, G. & Lin, Y. (2006). Biosensor based on self-assembling acetylcholinesterase on carbon nanotubes for flow injection/amperometric detection of organophosphate pesticides and nerve agents. *Analytical Chemistry*, Vol. 78, No. 3, (February 2006), pp. 835-843, ISSN: 0003-2700

Llorent-Martínez, E.J., García-Reyes, J.F., Ortega-Barrales, P. & Molina-Díaz, A. (2005). Flow-through fluorescence-based optosensor with on-line solid-phase separation for the simultaneous determination of a ternary pesticide mixture. *Journal of AOAC International*, Vol. 88, No. 3, (May 2005), pp. 860-865, ISSN: 1060-3271

Llorent-Martínez, E.J., García-Reyes, J.F., Ortega-Barrales, P. & Molina-Díaz, A. (2007). Multicommuted fluorescence based optosensor for the screening of bitertanol residues in banana samples. *Food Chemistry*, Vol. 102, No. 3, (July 2007), pp. 676-682, ISSN: 0308-8146

Llorent-Martínez, E.J., Ortega-Barrales, P., Fernández-de Córdova, M.L. & Ruiz-Medina, A. (2011). Trends in flow-based analytical methods applied to pesticide detection: A review. *Analytica Chimica Acta*, Vol. 684, No. 1-2, (January 2011), pp. 21-30, ISSN: 0003-2670

López Flores, J., Molina Díaz, A. & Fernández de Córdova, M.L. (2007). Development of a photochemically induced fluorescence-based optosensor for the determination of imidacloprid in peppers and environmental waters. *Talanta*, Vol. 72, No. 3, (May 2007), pp. 991-997, ISSN: 0039-9140

López-Paz, J.L. & Catalá-Icardo, M. (2011). Analysis of pesticides by flow injection coupled with chemiluminescent detection: A review. *Analytical Letters*, Vol. 44, No. 1-3, (January 2011), pp. 146-175, ISSN: 0003-2719

Mbaye, M., Gaye Seye, M.D., Aaron, J.J., Coly, A. & Tine, A. (2011). Application of flow injection analysis-photo-induced fluorescence (FIA-PIF) for the determination of α-cypermethrin pesticide residues in natural waters. *Analytical and Bioanalytical Chemistry*, Vol. 400, No., 2, (April 2011), pp. 403-410, ISSN: 1618-2642

Mukherjee, I. & Gopal, M. (1996). Chromatographic techniques in the analysis of organochlorine pesticide residues. *Journal of Chromatography A*, Vol. 754, No., 1-2, (November 1996), pp. 33-42, ISSN: 0021-9673

Sánchez, A.H., García, P. & Rejano, L. (2006). Trends in table olives production. *Grasas y aceites*, Vol. 57, No. 1, (January-March 2006), pp. 86-94, ISSN: 0017-3495

Skoog, D.A.,West, D.M. & Holler, F.J. (1996) *Fundamentals of Analytical Chemistry*, Saunders College, ISBN: 0030355230, Philadelphia

Walters, S.M. (1990). Clean-up techniques for pesticides in fatty foods. *Analytica Chimica Acta*, Vol. 236, pp. 77-82, ISSN: 0003-2670

2

Forty Years with Glyphosate

András Székács and Béla Darvas

Department of Ecotoxicology and Environmental Analysis, Plant Protection Institute,
Hungarian Academy of Sciences
Hungary

1. Introduction

If one were to pick the most notified pesticide of the turn of the millennium, the choice would most likely be glyphosate. Although DDT remains to be the all-time star in the Hall of Fame of pesticides, the second most admitted pesticide active ingredient must be the phosphonomethylglycine type compound of Monsanto Company, glyphosate.

Indeed, the two boasted pesticides show certain similarities in their history of discovery and fate. Both were synthesised first several decades prior to the discovery of their pesticide action. DDT and glyphosate were first described as chemical compounds 65 and 21 years before their discovery as pesticides, respectively. Both fulfilled extensive market need, therefore, both burst into mass application right after the discovery of their insecticide/herbicide activity. They both were, to some extent, connected to wars: a great part of the use of DDT was (and remains to be) hygienic, particularly after Word War II, but also the Vietnam War; while glyphosate plays an eminent role in the "drug war" (Plan Colombia) as a defoliant of marijuana fields in Mexico and South America. And last, not least, ecologically unfavourable characteristics of both was applauded as advantageous: the persistence of DDT had been seen initially as a benefit of long lasting activity, and the zwitterionic structure and consequent outstanding water solubility of glyphosate, unusual among pesticides, also used to be praised, before the environmental or ecotoxicological disadvantages of these characteristics were understood.

Yet there are marked differences as well between these two prominent pesticide active ingredients. Meanwhile the career of DDT lasted a little over three decades until becoming banned (mostly) worldwide, the history of glyphosate has gone beyond that by now, since the discovery of its herbicidal action (Baird et al., 1971). And while DDT is the only Nobel prize laureate pesticide, glyphosate was the "first billion dollar product" of the pesticide industry (Franz et al., 1997). Moreover, meanwhile the course of DDT was rather simple: rapid rise into mass utilisation, discovery of environmental persistence, development of pest resistance, loss of efficacy, and subsequent ban; the history of glyphosate is far more diverse: its business success progressed uncumbered, receiving two major boosts. First, the patent protection of glyphosate preparations was renewed in the US in 1991 for another decade on the basis of application advantages due to formulation novelties, and second, its sales were further strengthened outside Europe with the spread of glyphosate-tolerant (GT) genetically modified (GM) crops. This market success has been limited significantly neither by the recognition of the water-polluting feature of the parent compound, nor by the emerging weed resistance worldwide.

It is not a simple task to predict whether glyphosate continues to rise in the near future, or its application will be abating. To facilitate better assessment of these two possibilities, the present work attempts to provide a summary of the utility and the environmental health problems of glyphosate applications.

2. Glyphosate and its biochemistry

2.1 The discovery of glyphosate

The molecule N-(phosphonomethyl)glycine was first synthesised in 1950 by a researcher of the small Swiss pharmaceutical firm Cilag, Henri Martin (Franz et al. 1997). Yet, showing no pharmaceutical perspective, the compound has not been investigated any further. A decade later through the acquisition of the company, it was transferred to the distributor of laboratory research chemicals, Aldrich Chemical Co., along with research samples of Cilag. This is how it came to the attention of Monsanto Company (St. Louis, MO) in the course of its research to develop phosphonic acid type water-softening agents, through testing over 100 chemical substances related to aminomethylphosphonic acid (AMPA). Monsanto later extended the study of these compounds to herbicide activity testing, and observed their potential against perennial weeds (Dill et al., 2010). N-(phosphonomethyl)glycine (later termed glyphosate) was first re-synthesised and tested by Monsanto in 1970. Its herbicidal effect was described by Baird and co-workers in 1971, the subsequent patent (US 3799758), followed by numerous others, was claimed and obtained by Monsanto, and was introduced as a herbicide product Roundup® (formulation of the isopropylamine salt of glyphosate with a surfactant). Upon its introduction in the mid seventies, glyphosate jumped to a leading position on the pesticide market, became the most marketed herbicide active ingredient by the nineties, and more or less holds that position ever since. A great change came about, when the original patent protection expired in many parts of the world outside the United States in 1991. As a result, an almost immediate price decline occurred (by 30% in one year, 40% in two years and about 50% in two decades (Cox, 1998). Upon the expiration of the patent protection also in the United States in 2000, sales of generic preparations intensively expanded (main international producers include Dow, Syngenta, NuFarm, etc.), but the leading preparation producer remained Monsanto (Duke & Powles, 2008).

The current situation of the international active ingredient producers shows a rather different picture. Recently, Chinese chemical factories (e.g., Zhejiang Wynca Chemical Co., Zeijang Jinfanda Biochemical Co. and Hubei Xingta Chemical Group., Nantong Jiangshan Agrochemical and Chemical Co., Sichuan Fuhua Agricultural Investment Group, Jiangsu Yangnong Chemical Group, Jiangshu Good Harvest-Welen, etc.) gained leading parts of this business. At present, the global glyphosate production capacity is 1.1 million tonnes, while the global demand is only 0.5 million tonnes. The overall glyphosate production capacity of Chinese companies rose from 323,400 tonnes in 2007 to 835,900 tonnes in 2010, by a compounded annual growth rate of 37 percent (Yin, 2011). China has enough glyphosate capacity to satisfy the global demand even if all other glyphosate manufacturers cease production. The domestic demand of China is only 30-40 thousand tonnes, about 0.3 million tonnes of glyphosate is produced for export. Presently Chinese glyphosate production facilities have been suspended being limited by the market demand. Extended use of GT plants in the Word would help on this problem, even if Europe is hesitant to allow commercial cultivation of this kind of GM plants. The overall situation has led to continously decreasing glyphosate prices on the Word market, and has significant effects on dispread of GT plants.

2.2 Mode of action

Glyphosate is a phosphonomethyl derivative of the amino acid glycine. It is an amphoteric chemical substance containing a basic secondary amino function in the middle of the molecule and monobasic (carboxylic) and dibasic (phosphonic) acidic sites at both ends (Fig. 1). Containing both hydrogen cation (H[+]) donor (acidic) and acceptor (basic) functional groups, it can form cationic and anionic sites within the small molecule, the dissociation constants (pK$_a$) of these three functional groups are 10.9, 5.9 and 2.3, and therefore, similarly to amino acids, glyphosate can form a zwitterionic structure (Knuuttila & Knuuttila, 1979). This is reflected in excellent water solubility (11.6 g/l at 25 °C). Consequently, its lipophilicity is very low (logP < -3.2 at 20 °C, pH 2-5), and is insoluble in organic solvents e.g., ethanol, acetone or xylene (Tomlin, 2000). To further increase its already good water solubility it is often formulated in form of its ammonium, isopropylammonium, potassium, sodium or trimethylsulphonium (trimesium) salts. The order of water solubility is glyphosate << ammonium salt < sodium salt < potassium salt < isopropylammonium salt < trimesium salt, the solubility of the trimesium salt being two orders of magnitude higher than that of glyphosate.

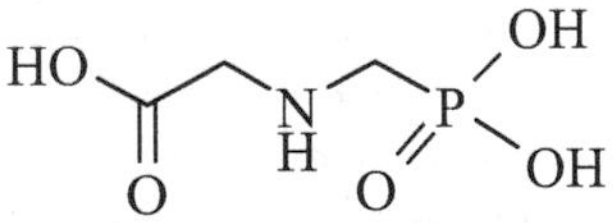

Fig. 1. The chemical structure of N-(phosphonomethyl)glycine, glyphosate, containing a basic function (amine) in the middle of the molecule and two acidic moieties (carboxylic and phosphonic acids) at both ends.

It has been known since the early seventies that glyphosate acts by inhibiting aromatic amino acid biosynthesis in plants (Jaworski, 1972; Amrhein et al., 1980), and elaborate research has revealed that the responsible mechanism is blocking a key step in the so-called shikimate pathway (Herman & Weaver, 1999), responsible for the synthesis of aromatic amino acids and critical plant metabolites. Glyphosate exerts this effect by inhibiting the activity of the enzyme 5-enolpyruvyl shikimate 3-phosphate synthase (EPSPS) catalyzing the transformation of phosphoenol pyruvate (PEP) to shikimate-3-phosphate (S3P) (Amrhein et al., 1980). This metabolic pathway exists in plants, fungi, and bacteria, but not in animals (Kishore & Shah 1988). Although higher order living organisms lack this metabolic route, therefore, are not expected to be directly affected by this herbicide, the environmental consequences of the widespread use of glyphosate have been reported (Cox, 2000; Santillo et al., 1989).

Being an amino acid (glycine) derivative itself, glyphosate inhibits the formation of the main intermediate, by binding as an analogue of the substrate PEP to its catalytic site on the enzyme. The inhibition of this catabolic pathway blocks the synthesis of triptophan, phenylalanine and tyrosine, and in consequence, the synthesis of proteins. The lack of the synthesis of these essential amino acids and the proteins that contain them leads to rapid necrosis of the plant. Because this metabolic pathway is present in all higher order plants, and because the amino acid sequence of the active site of the EPSPS is a very conservative region in higher plants, the herbicidal effect is global among plant species.

Moreover, through its excellent solubility features glyphosate is a systemically active herbicide ingredient. As it is capable to be transported in the plant from the leaves towards the roots, it belongs to the relatively uncommon group of basipetally translocated herbicides (Ashton & Crafts, 1981). Its uptake and translocation is relatively rapid in diverse species (Sprankle et al., 1975).

2.3 Transition state analogue theory of enzyme inhibition

A unique feature of the mechanism of the inhibition of EPSPS by glyphosate is that glyphosate is reported to show close similarity in its structure to the tetrahedral phosphoenolpyruvoyl oxonium ion derivative of PEP, formed during its catalytic conversion to S3P, and the adduct formation with EPSPS has been verified by nuclear magnetic resonance spectroscopy (Christensen & Schaefer, 1993). Therefore, it has been proposed that glyphosate exerts its inhibitory activity as transition-state analogue (TSA) of the putative phosphoenolpyruvoyl oxonium ion derivative of PEP from plants (Anton et al., 1983; Steinrücken & Amrhein, 1984; Kishore & Shah, 1988) and bacteria (Du et al., 2000; Arcuri et al., 2004).

The so-called transition state theory has been advanced by Pauling (1948) to explain the mechanism of enzymatic reactions. Enyzmes are catalysts therefore they accelerate a reaction without influencing its equilibrium constant. One way to achieve that is to diminish the energy barrier of the reaction by lowering the energy of the transition state, transient, unstable intermediate of the reaction. This may be accomplished through stabilizing the transition state by binding to it as soon as it has occurred, and thus facilitating its formation. This results in the enzymatic effect that lowers the activation energy of the catalyzed reaction. Based on this idea, extremely potent inhibitors can be developed for a given enzymatic reaction if one can synthesize "transition state analogues" or "transition state mimics": stable chemical compounds resembling the transition state (Wolfenden, 1969). The TSA theory has therefore been successfully applied to the development of various biologically active substances, including insect control agents (Hammock et al., 1988), sulfonylurea microherbicides (Schloss & Aulabaugh, 1990) or compounds relatd to glyphosate (Marzabadi et al., 1992; Anderson et al., 1995).

The TSA hypothesis as it applies to the mechanism of the inhibition of EPSPS by glyphosate, became widely accepted as it has been evidenced in numerous studies that glyphosate forms a tight ternary complex with EPSPS (Herman & Weaver, 1999). It is easy to understand, however, that a classical TSA inhibitor would cause irreversible inhibition of the enzyme, competeable (although possibly with a low affinity) by the natural substrate of the enzyme. In later studies, it has been evidenced by biochemist researchers of Monsanto that glyphosate was an inhibitor of EPSPS uncompetitive with EPSP, and therefore, the TSA hypothesis has been reconsidered (Sammons et al., 1995; Schönbrunn et al., 2001; Alibhai & Stallings, 2001; Funke et al., 2006). The effects of glyphosate on aromatic amino acid synthesis in *Escherychia coli* have been attributed to chelation of Co^{2+} and Mg^{2+} (Roisch & Lingens, 1980), cofactors for enzymes in this pathway. Moreover, it is interesting, that glyphosate does not inhibit the enzyme UDP-N-acetylglucosamine enolpyruvyl transferase

(Samland et al., 1999), structurally and mechanistically closely related EPSPS, and playing a key role in the biosynthesis of UDP-muramic acid.

2.4 Other biochemical effects of glyphosate

Various biochemical interactions of glyphosate, besides its identified mode of action, in plants and microorganisms were summarised by Hoagland and Duke (1982). The authors refer to numerous secondary or more complex indirect effects of glyphosate, and point out that a compound with such a powerful growth retardant effect or strong phytotoxicity will ultimately affect virtually all biochemical processes in the affected cells.

The effects of glyphosate in the plant possibly include influences on the regulation of hormonal processes. Methionine levels are greatly reduced by glyphosate (Duke et al., 1979), which suggests that this herbicide may alter ethylene biosynthesis. Results of Baur (1979) suggest that glyphosate may inhibit auxin transport by increasing ethylene biosynthesis. Glyphosate may also affect the biosynthesis of non-aromatic amino acids. Nilsson (1977) suggested that the build-up of glutamate and glutamine in glyphosate-treated tissue might be due to blocked transamination reactions.

It has been hypothesised that glyphosate lower phenylalanine and tyrosine pools not only by its primary mode of action, but possibly also by induction of phenylalanine ammonia-lyase (PAL) activity. Indeed, pronounced PAL activity has been detected in glyphosate-treated maize and soy (Duke et al., 1979; Cole et al., 1980), yet not by direct effect according to in vitro tests. Therefore, although glyphosate has been evidenced to cause profound effects on extractable PAL, substrate(s) and end products, increased PAL activity has been evaluated as a secondary effect (Hoagland & Duke, 1982).

Glyphosate did not appear to cause direct effects on photosynthesis, but its possible effect on chlorophyll biosynthesis has been considered, and its strong inhibitory effect on chlorophyll accumulation has been shown (Kitchen et al., 1981). Experimental result indicated that the effect of glyphosate on chlorophyll may be indirect through photobleaching and/or peroxidation of chlorophyll.

Glyphosate has been shown to significantly affect the membrane transport of cellular contents only at very high concentrations (Brecke & Duke, 1980; Fletcher et al., 1980). Phosphorous uptake was retarded (Brecke & Duke, 1980), but loss of membrane integrity, decrease in energy supply or external ion chelation were excluded as causes. Moreover, uptake of amino acids, nucleotides and glucose were also found to be retarded by glyphosate in isolated cells (Brecke & Duke, 1980). Other studies (Cole et al., 1980; Duke & Hoagland, 1981) found inhibition of amino acid uptake by glyphosate not severe. Glyphosate has been reported to uncouple oxidative phosphorylation in plant (Olorunsogo et al., 1979) and mammalian (Olorunsogo & Bababunmi, 1980) mitochondria, the latter is likely to be due to altered membrane transport processes, as glyphosate was found to enhance proton permeability of mitochondrial membranes in a concentration-dependent manner (Olorunsogo, 1990).

3. Pre-emergent application technology of glyphosate

Glyphosate, exerting global herbicidal action, has originally been intended to pre-emergent weed control treatments of field vegetation and weed control of orchards and ruderal areas.

Post-emergent applications are impossible solely with glyphosate-based herbicide formulations due to the phytotoxicity of the compound to the crop as well.

Common first visible phytotoxicity effects of glyphosate include rapid (within 2-10 days upon application) chlorosis, usually followed by necrosis (Suwannamek & Parker, 1975; Putnam, 1976; Campbell et al., 1976; Fernandez & Bayer, 1977; Marriage & Khan, 1978; Segura et al., 1978; Abu-Irmaileh & Jordan, 1978), possibly accompanied with morphological leaf deformities (Marriage & Khan, 1978), root and rhizome damage (Suwannamek & Parker, 1975; Fernandez & Bayer, 1977). Glyphosate accumulation has been reported in the meristems (Haderlie et al., 1978). It is rather surprising that although glyphosate inhibits seedling growth as well, it did not exert significant effect on the germination of various species (Haderlie et al., 1978; Egley & Williams, 1978).

3.1 Formulated glyphosate-based herbicides

Glyphosate-based formulations such as Roundup®, Accord® and Touchdown® represent the most common types used for agricultural purposes (Franz et al., 1997). These formulated herbicides can be used for weed control in agricultural practice, including in no-till agriculture to prepare fields before planting, during crop development and after crop harvest; as well as in silvicultural, urban and, lately, aquatic environments. The main herbicide products currently distributed are listed in Table 1. These preparations contain glyphosate as formulated in form of its ammonium (AMM), dimethylammonium (DMA), isopropylammonium (IPA), potassium (K) or trimesium (TRI) salts. The very first formulations containing IPA, sodium and ammonium salts were patented by Monsanto in 1974. A unique form is the trimesium salt of outstanding water solubility, patented by ICI Agrochemicals (later Zeneca Agricultural Products Inc, then Novartis CP, and after 2000 Syngenta) in 1989 (Tomlin, 2000).

As the actual active ingredients of the formulations are salts, differing from each other in the cation(s) and consequently the molecular mass of the salts, active ingredient concentrations are specified as glyphosate equivalent, in other term acid equivalent (a.i.) referring to the free acid form of glyphosate. This provides instant comparability among various formulations. Moreover, the use of a.i. units is common practice in residue analysis of glyphosate as well.

3.2 Formulating agents

Formulated glyphosate-based herbicides contain various non-ionic surfactants to facilitate their uptake by the plants (Riechers et al., 1995). These components, as all other pesticide additives and diluents, are assumed to be inert, which as it turns out, is not the case for several such ingredients. The most common surfactant applied in combination with glyphosate is polyethyloxylated tallowamine (POEA), which itself has been found to exert ecotoxicity, also in synergy with glyphosate, causing the formulated herbicide (e.g., Roundup) more toxic than its technical grade active ingredient (Folmar et al., 1979; Atkinson, 1985; Wan et al., 1989; Powell et al., 1991; Giesy et al., 2000; Tsui & Chu, 2003; Marc et al., 2005; Benachour et al., 2007; Benachour & Séralini, 2009).

The apparent synergistic toxic effects of the assumedly inert ingredients with glyphosate triggered a legal case between Monsanto and the New York Attorney General's Office in

Manufacturer	a.i. salt [a]	Product [b]
AAKO B.V.	IPA	Akosate
Agriliance LLC	IPA	Cornerstone
Agro-Chemie Ltd.	IPA	Fozát
Albaugh Inc./Agri Star	IPA	Aqua Star, Gly Star Original
Astrachem Ltd.	IPA	Tiller
Barclay Chem. Mfg. Ltd.	IPA	Gallup
Calliope S.A.	IPA	Kapazin
Chemical Products Technologies LLC	IPA	ClearOut; ClearOut Plus
Cheminova	IPA	Glyfos; Glyfos X-tra
Control Solutions Inc.	IPA	Spitfire
Crystal Chem. Inter-America	IPA	Glifonox
Dow AgroSciences	IPA	Dominator; Durango; Glyphomax; Glyphomax Plus; GlyPro; Panzer; Ripper; Rodeo; Vantage
	DMA	Durango DMA; Duramax
Drexel Chem. Co.	IPA	Imitator
	K	DupliKator
FarmerSaver.com LLC	IPA	Glyphosate 4
Griffin LLC	IPA	Glyphosate Original
Growmark Inc.	IPA	FS Glyxphosate Plus
Helena Chemical Co.	IPA	Rattler
	IPA + AMM	Showdown
Helm Agro US Inc.	IPA	Glyphosate 41%; Helosate Plus
Loveland Products Inc.	IPA	Mad Dog; Mirage
Makhteshim-Agan	IPA	Eraser, Gladiator; Glyphogan; Hardflex; Herbolex; Taifun
Micro Flo	IPA	Gly-Flo
Monsanto Co.	IPA	Accord; Aquamaster; Azural; Clinic; Glialka; Honcho; Ranger Pro, Roundup Bioforce / Classic / Original / UltraMAX
	K	Roundup Forte / Mega / PowerMAX / WeatherMAX; VisionMAX
Nufarm	IPA	Amega; Credit; Credit Extra
	IPA + MA	Credit Duo
Oxon Italia S.p.A.	AMM	Buggy
Pinus TKI d.d.	IPA	Boom Efekt
Sinon Corporation	IPA	Glyfozat; Total
Syngenta AG	AMM	Medallon Premium
	DMA	Touchdown IQ
	K	Refuge; Touchdown HiTech / Total; Traxion
	TRI	Coloso; Ouragan
Tenkoz Inc.	IPA	Buccaneer
UAP	IPA	Makaze
Universal Crop Protection Alliance LLC	IPA	Gly-4
Winfield Solutions LLC	IPA	Cornerstone

[a] AMM = ammonium; DMA = dimethylamine; IPA = isopropylamine; K = potassium; TRI = trimesium
[b] Formulations containing only glyphosate salts as active ingredient are listed, herbicide combinations are not included

Table 1. Formulated herbicide preparations containing glyphosate as active ingredient.

1996 (Attorney General of the State of New York, 1996). The toxicological basis of the legal claim was that Monsanto inaccurately implied toxicity data of the active ingredient glyphosate on the formulated product Roundup. As a result of the lawsuit, Monsanto was fined, and agreed to drop description of being "environmentally friendly" and "biodegradable" from the advertisements of the herbicide.

Concerns about application safety, triggered by the above studies and findings on teratogenic effects (see 6.3 Teratogenic activity of glyphosate), have brought re-registration of glyphosate and its formulated products in focus in the European Union, as part of the regular pesticide revision process due to take place in 2012. Nonetheless, the EU Commission dismissed these findings, based on a rebuttal by the EU "rapporteur" member state for glyphosate, Germany, provided by the German Federal Office for Consumer Protection and Food Safety (BVL), and postponed the review of glyphosate and 38 other pesticides until 2015 (European Commission, 2010). To protest against such delay in re-evaluation of these 39 pesticides, the Pesticides Action Network Europe and Greenpeace brought a lawsuit against the EU Commission, and the dismissal of the reported teratogenicity data from the official current evaluation has been judged by several researchers as irresponsible act (Antoniou et al., 2011).

4. Post-emergent application technology of glyphosate

A group so far of the highest financial importance within GM crops has been modified to be tolerant to this active ingredient, outstandingly broadening its application possibilities.

4.1 Glyphosate-tolerant crops

Upon pre-emergent applications of the global herbicide glyphosate, the majority of the weeds decays, perishes, and does not get consumed by wild animals. This situation has been changed tremendously by the appearance of GT crops, leading to increasing environmental herbicide loads due to approved post-emergent treatments (2-3 applications in total). Of these crops, the varieties of Monsanto became most publicised, under the trade mark Roundup Ready® (RR), indicating that these plants can be treated with the herbicide preparation of Monsanto, Roundup® containing glyphosate as active ingredient even, after the emergence of the crop seedlings. Similar varieties by Bayer CropScience, Pioneer Hi-Bred and Syngenta AG are termed Gly-Tol™, Optimum® GAT® and Agrisure® GT, respectively. Two strategies have been followed by plant gene technology in the development of GT varieties: either the genes (*cp4 epsps, mepsps, 2mepsps*) of mutant forms of the target enzyme less sensitive to glyphosate or genes (*gat, gox*) of enzymes metabolizing glyphosate have been transferred into the GM plant varieties (Table 2). The genetically created tolerance to glyphosate does not alter the mode of action of the compound: the molecular mechanism of glyphosate tolerance has been elucidated (Funke et al., 2006), and the sole mechanism of inhibition remains blocking of the shikimate pathway when applied at very high doses on GT soybean and canola (Nandula et al., 2007).

The first GT crop was RR soybean by Monsanto in 1996, followed by GT cotton, GT maize, GT canola, GT alfalfa and GT sugarbeet (Dill et al., 2008). GT crops allow a new form of technology, post-emergent application of glyphosate. The utilizability of post-emergent applications was systematically tested in 2002 and 2003 in field experiments in the United States (Parker et al., 2005). The extensive study involving GT maize and GT soybean sites at

Variety owner	Crop	Genetical event	Transgene introduced [a]
Bayer CropSience (part of Sanofi-Aventis)	Cotton	GHB614	2mepsps
Monsanto Co.	Cotton	MON 1445	cp4 epsps, nptII, aad
	Cotton	MON 88913	cp4 epsps
	Maize	MON 88017	cp4 epsps, cry3Bb1
	Maize	NK603	cp4 epsps
	Rape	GT 73	cp4 epsps, gox
	Soybean	MON40-3-2	cp4 epsps
	Soybean	MON 87705	cp4 epsps, FAD2-1A, FATB1-A
	Soybean	MON 89788	cp4 epsps
	Sugar-beet [b]	A5-15	cp4 epsps, nptII,
	Sugar-beet [c]	H7-1	cp4 epsps
Pioneer Hi-Bred (part of DuPont)	Maize	DP-98140	Gat4601, als
	Soy	DP-356043	gat4601
Syngenta	Maize	GA21	mepsps

[a] *aad* – gene of *Escherichia coli* origin, encoding resistance against aminoglycoside antibotics (streptomycin and spectinomycin); *als* – gene (*zm hra*) of maize origin, enhancing tolerance of ALS inhibiting herbicides (e.g., chlorimuron and thifensulfuron); *cry3Bb1* – gene of *Bacillus thuringiensis* origin, encoding Cry3 toxin; *FAD2-1A* – gene of soy origin, encoding fatty acid desaturease enzyme, silencing of which enhances the proportion of monounsaturated fatty acids; *FATB1-A* – gene of soy origin, encoding medium-chain fatty acid thioesterase, silencing of which reduces the proportion of saturated fatty acids; *cp4 epsps* – *epsps* gene of *Agrobacterium* sp.; *mepsps* – *epsps* gene of maize origin; *2mepsps* – double mutated *epsps* gene of Mexican black, sweet maize origin; *gat4601* – gene of *Bacillus lichiformis* origin, encoding glyphosate acetyltransferase enzyme; *gox* – gene of *Ochrobactrum anthropi* origin, encoding glyphosate oxidase enzyme; *nptII* – gene of *Escherichia coli* K12 origin, encoding neomycin phosphostransferase, causing neomycin and kanamycin resistance.
[b] together with Danisco Seeds and DLF Trifolium as variety owners
[c] together with KWS Saat Ag. as variety owners

Table 2. Glyphosate tolerant crop variety groups under registration process in the European Union.

seven locations, as well as regular or directed post-emergent applications of 10 formulated glyphosate preparations (ClearOut 41 Plus™, Gly Star™, Glyfos®, Glyfos® X-tra, Glyphomax™, Roundup Original™, Roundup UltraMAX®, Roundup WeatherMAX™, Touchdown® and Touchdown Total™) containing isopropylamine or potassium salts of glyphosate found no herbicide efficacy or produce quality differences, no phytotoxicity to maize and medium phytotoxicity to cotton at high doses in some instances, and therefore proposed post-emergent glyphosate applications. As a result, the use of glyphosate has expanded almost 20-fold by 2007 in the United States (Pérez et al., 2011).

Another impact of GT crops on agricultural practices is the spread of no-till agriculture. As the crop tolerates the active ingredient, intensive herbicide treatments are possible to be carried out, instead of former tillage practices, to eradicate vegetation in the field. This has greatly increased herbicide use and consequent chemical pressure on the environment. No-till practice is particularly common in GT crop cultivating areas in South America, including Brazil, Argentina, Paraguay and Uruguay (Altieri & Pengue, 2006).

An interesting detail is that in parallel to industrial development of GT crops, illegal genetic modification projects are also being carried out to achieve "crops" that are resistant to glyphosate e.g, a new marijuana (*Canabis* sp.) hybrid that can be cultivated all year and cannot be controled with herbicides (Anonymous, 2006). The GT marijuana hybrid, first appeared in Mexico in 2004, allows 8-9-times higher yields than "conventional" varieties, and became the plant of choice for drug traffickers in Michoacan.

4.2 The effect of glyphosate-tolerant crops on glyphosate residues

As a result of the combined effect of the expiration of the patent protection of glyphosate (in 2000 in the United States) and the spread of cultivation of GT GM crops (since 1996 in the United States), the use of glyphosate products is again increasing (Woodburn, 2000). Besides GT GM crops, energy crop cultivation is also an and emerging source of glyphosate contamination (Love et al., 2011). Moreover, due to the modified metabolic pool in the GT GM crops, residues of the systemic glyphosate active ingredient are expected to occur in the surviving plants. In case of EPSP-mutant (RR and Agrisure GT) varieties, the residue composition is expected to be similar to those seen at regular glyphosate applications, while in the case of the boosted glyphosate metabolizing (regardless whether *epsps* or *gox* transgene based) varieties, increased amounts of N-acetylglyphosate (NAG) (Optimum GAT variety) or aminomethylphosphonic acid (AMPA) (RR and Agrisure GT varieties) are expected in the plants. In turn, residue patterns not yet seen in food and feed are to be expected. Summarizing the results of their studies in Argentina between 1997 and 1999, Arregui and co-workers (2004) reported glyphosate residue levels after 2-3 glyphosate applications as high as 0.3-5.2 mg glyphosate/kg and 0.3-5.7 mg AMPA/kg in the leaves and stem of RR soy during harvest, and 0.1-1.8 mg glyphosate/kg and 0.4-0.9 mg AMPA/kg in the produce. In turn, glyphosate occurred as surface water, soil and sediment contaminant in a GM soybean cultivating area in Argentina (Peruzzo et al., 2008).

5. The environmental fate of glyphosate

5.1 Residue analysis of glyphosate

Present analytical methods developed for the detection of glyphosate are mostly based on separation by liquid chromatography (LC), as previous methods utilizing gas chromatography (GC) have become of much lesser importance than they used to be (Stalikas & Konidari, 2001). The main obstacle in the GC detection of glyphosate and its main metabolite AMPA is the polaric and zwitterionic structure of these compounds, which required laborious sample preparation steps prior to instrumental analysis. The earliest method accredited for authoritative analytical determination of glyphosate (US FDA, 1977) employed aqueous extraction, anion and cation exchange purification, N-acetylation derivatisation with trifluoroacetic acid and trifluoroacetic anhydride, and subsequent methylation of both the carboxylic acid and phosphonic acid moieties on the parent compound, followed by GC analysis with phosphorous-specific flame ionisation detection. Recoveries above 70% were achieved by the method in plant samples, the limit of detection (LOD) was 0.05 mg/kg. The basis of the protocol was the GC-MS derivatisation method developed by Monsanto (Rueppel et al., 1976). A later method by Alferness and Iwata (1994) also employs aqueous extraction, followed by washing with dichloromethane/chloroform, purification on cation exchange column, derivatisation to trifluoroacetate and heptafluorobutyl ester, followed by GC analysis with mass spectrometry (MS) detection, and a similar methods have also been developed (Tsunoda, 1993; Natangelo et al., 1993;

Royer et al., 2000; Hudzin et al., 2002). Validated LC methods also resulting in similar analytical parameters (Cowell et al., 1986; Winfield et al., 1990; DFG, 1992) utilise washing with chloroform and hydrochloric acid, purification on ion exchange column, and upon neutralisation and derivatisation with o-phthalic aldehyde and mercaptoethanol, determination by high performance liquid chromatography (HPLC) with fluorescence detection. Yet the LOD of the official method (Method 547) established by the U.S. Environmental Protection Agency is as high as 6 μg/l in reagent water and 9 μg/l in surface water (Winfield et al., 1990). Ninhydrin or 9-methlyfluorenyl chloroformiate have also been applied as derivatising agents (Wigfield & Lanquette, 1991; Sancho et al., 1996; Nedelkoska & Low, 2004, Peruzzo et al., 2008). More recent LC procedures with somewhat simplified sample preparation steps offer rapid and more economic analytical methods than GC procedures always requiring complex, often several step derivatisation. As a result, GC methods remain being used solely due to their analytical parameters, including sensitivity. Nonetheless, LODs of LC and ion chromatographic methods were achieved to be lowered (Mallat & Barceló, 1998; Vreeken, 1998; Bauer et al., 1999; Grey et al., 2001; Patsias et al., 2001; Lee et al., 2002a; Nedelkoska & Low, 2004; Ibáñez et al., 2006; Laitinen et al., 2006; Hanke et al., 2008; Popp et al., 2008) to meet the strictening maximal residue levels (MRLs) in environmental and health regulations. The most recent LC-MS methods using electrospray ionisation (Granby et al., 2003; Martins-Júnior et al., 2011) easily meet the MRL by the EU for given pesticide residues in drinking water, 0.1 μg/l, but the instrumentation demand of these methods is substantial.

Among novel innovative analytical methods for the detection of glyphosate, mostly capillary electrophoresis (CE) and immunoanalytical methods are to be mentioned. Initial drawbacks of the CE methods included relatively high LOD and the need for derivatisation or external fluorescent labeling (Cikalo et al., 1996; You et al., 2003; Kodama et al., 2008), later solved by coupling CE with MS (Goodwin et al., 2003) and microextraction techniques (Hsu and Whang 2009; See et al., 2010). Among various immunoanalytical techniques, enzyme-linked immunosorbent assays (ELISAs) gained the highest utility. While in the early nineties we considered yet that effective antibodies are not produced against glyphosate and similar zwitterioninc compounds due to their low immunogeneity (Hammock et al., 1990), difficulties in immunisation have been overcome within a decade, and sensitive ELISAs, also employing derivatisation, were developed (Clegg et al., 1999; Lee et al., 2002b; Rubio et al., 2003; Selvi et al., 2011), proven to be of great utility in environmental analytical studies for glyphosate (Mörtl et al., 2010; Kantiani et al., 2011). On the basis of the immunoassay principle, sensors using glyphosate-sensitive antibodies (González-Martínez et al., 2005) or molecularly imprinted polymers (MIPs) (Zhao et al., 2011) were also developed.

5.2 Glyphosate and its decomposition products

Decomposition of glyphosate takes place mostly by two processes: decarboxylation or dephosphorylation, and the corresponding intermediate metabolites are AMPA or glycine, respectively. The first pathway is catalyzed by oxidoreductases, the second by C–P lyases cleaving the carbon-phosphorous bond. Both pathways occur in environmental matrices (water, soil) and plants, but the main metabolite in all cases is AMPA (Fig. 2). The environmental fate, behaviour and analysis of both AMPA and glyphosate has received considerable attention (Stalikas & Konidari, 2001).

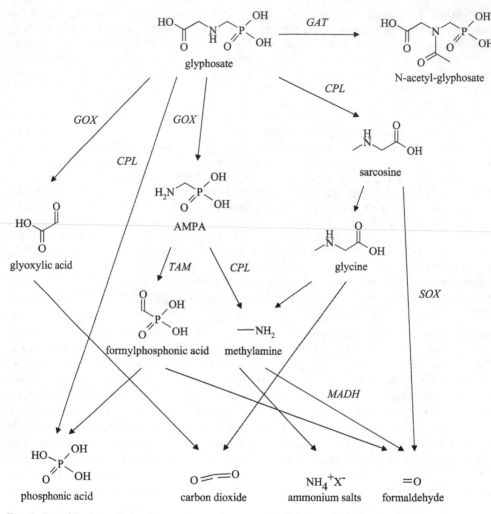

Fig. 2. Possible fate of glyphosate by various metabolizing pathways. Oxidative decomposition (*solid arrows*), non-hydrolytic decomposition (*dashed arrows*), inactivation in plants (*dotted arrow*). Processing enzymes (*Italics letters*) – GOX: glyphosate oxidoreductase, GAT: glyphosate N-acetyltransferase, CPL: C–P-liase, SOX: sarcosin oxidase, TAM: transaminase, MADH: methylamine dehydrogenase.

AMPA has been reported to be rapidly formed microbiologically, but not by chemical action, in water and in various loam soils (Drummer silty clay loam, Norfolk sandy loam, Ray silt loam, Lithonia sandy loam) (Rueppel et al., 1977; Aizawa, 1982; Mallat & Barceló, 1998), and was shown to be degraded subsequently completely to carbon dioxide (Sprankle et al., 1975, Rueppel et al., 1977; Moshier & Penner, 1978). Chemical processes of degradation are ineffective because of the presence of a highly stable carbon-phosphorus bond in the compound (Gimsing et al., 2004). Which pathway is predominant in the microbial degradation depends on bacterial species. The first (AMPA) pathway is commonly seen in mixed soil bacterial cultures (Rueppel et al., 1977) and certain *Flavobacterium* sp. The glycine pathway is characteristic to certain *Pseudomonas* and *Arthrobacter* sp. strains (Jacob et al., 1988). AMPA is further metabolised, providing phosphorus for growth, although the amount eliminated is typically set by the phosphorus requirement of the bacterium in question. Sarcosine and glycine are other possible main degradation products in soils (Rueppel et al., 1977).

As for decomposition in water or soil, the stability of glyphosate depends of a number of parameters. It strongly interacts with soil components by forming tight complexes with numerous metal ions in solution and by being adsorbed on soil particles, including clay minerals. Adsorption is strongly influenced by cations associated with the soil (Carlisle & Trevors, 1988), and it is mainly the phosphonic acid moiety that participates in the process, therefore, phosphate competes with glyphosate in soil adsorption (Gimsing & dos Santos, 2005). As a result of its adsorption on clay particles and organic matter present in the soil, upon application glyphosate remains unchanged in the soil for varying lengths of time (Penaloza-Vazquez et al., 1995). Adsorption of chelating agents by surfaces has been shown to decrease biodegradability. It can be expected that phosphonates with their higher affinity to surfaces are much slower degraded in a heterogeneous compared to a homogeneous system, as seen for glyphosate (Zaranyika & Nyandoro, 1993).

Therefore, differences have been observed between half-lives (DT_{50}) of glyphosate determined in laboratory or field studies. Half-lives were found quite favourable in laboratory, 91 days in water and 47 days in soil. Nonetheless, half-life of the parent compound ranged between a few days to several months or even a year in field studies, depending on soil composition. A reason of such delayed decomposition is partly binding to the soil matrix, through which glyphosate absorbed on soil particles can form complexes with metal (Al, Fe, Mn, Zn) ions (Vereecken, 2005). By the increased solubility of its various alkali metal, ammonium or trimesium salts, the active ingredient can leach into deeper soil layers, in spite of its rapid decomposition and strong complex formation capability under certain conditions (Vereecken, 2005). Its primary metabolite AMPA is more mobile in soil than the parent compound (Duke & Powles, 2008).

Moreover, decomposition dynamics of glyphosate is greatly dependent on the microbial activity of soil, with mostly *Pseudomonas* species as most important microbial components (Borggaard & Gimsing, 2008). If microbial activity is elevated, glyphosate is degraded with reported laboratory and field half-life of < 25 days and 47 days, respectively (Ahrens, 1994). Moreover, glyphosate itself affects the survival of soil microorganisms (Carlisle & Trevors, 1988; Krzysko-Lupicka & Sudol, 2008). Studies of glyphosate degrading bacteria have involved selection for, and isolation of pure bacterial strains with enhanced or novel detoxification capabilities for potential uses in biotechnology industry and biodegradation of polluted soils and water. Microorganisms known for their ability to degrade glyphosate in soil and water include *Pseudomonas* sp. strain LBr (Jacob et al., 1988), *Pseudomonas fluorescens* (Zboinska et al., 1992), *Arthrobacter atrocyaneus* (Pipke et al., 1988) and *Flavobacterium* sp. (Balthazor & Hallas, 1986). Soil microbial activity, however, depends on a number of additional parameters, including soil temperature, abundance of air and water, and a number of not yet defined factors, creating rather variable conditions for the decomposition of glyphosate (Stenrød et al., 2005; 2006). Other studies have also shown that soil sorption and degradation of glyphosate exhibit great variation depending on soil composition and properties (de Jonge et al., 2001; Gimsing et al., 2004a, 2004b; Mamy et al.,

2005; Sørensen et al., 2006; Gimsing et al., 2007). Laitinen and co-workers (2006; 2008) reported that phosphorous content in the soil affects the environmental behaviour of glyphosate e.g., its absorbance on soil particles, and its occurrence in surface waters. Weaver and co-workers (2007) claim that its effects on soil microbial communities are short and transient, and that decomposition characteristics of glyphosate do not change significantly in lower soil layers in Mississippi with various tilling methods (Zablotowicz et al., 2009). Outstandingly different result were obtained in an environmental analytical study carried out in Finland, who detected 19% of the applied glyphosate undecomposed and 48% in form of AMPA 20 months after application in Northern European soils of low phosphorous content (Laitinen et al., 2009). This also sheds a light on the high reported glyphosate contamination levels in Scandinavian surface waters (Ludvigsen & Lode, 2001a; 2001b). The phosphorous content of the soil may also play a key role in the low decomposition rate seen through its effect on microbial communities, as soil phosphorous has been shown to be able to stimulate decomposition of glyphosate (Borggaard & Gimsing, 2008). An interesting interaction observed is that persistence of glyphosate significantly increased in soils treated with Cry toxins of *Bacillus thuringiensis* subsp. *kurstaki*, while a similar effect was not seen when soils were treated with purified Cry1Ac toxin (Accinelli et al., 2004; 2006). Therefore, it is worthwhile reconsidering the fate of glyphosate in soils, including sorption, degradation and leachability.

Due to its strong sorption and relatively fast degradation in soil, glyphosate has been claimed to cause very limited risk of leaching to groundwater (Giesy et al., 2000; Busse et al., 2001; Vereecken, 2005; Cox & Surgan, 2006). Yet, other investigations indicates possible leaching and toxicity problems with its use (Veiga et al., 2001, Strange-Hansen, 2004; Kjær, 2005; Landry et al., 2005; Relyea, 2005b; Torstensson et al., 2005; Siimes et al., 2006) and consequent effects on aquatic microbial communities (Pérez et al., 2007; Pesce et al., 2009; Vera et al., 2010; Villeneuve et al., 2011), except cyanobacteria (Powell et al., 1991). Just like soil bacteria, aqueous microorganisms e.g., microalgae may also utilise glyphosate as source of phosphorous (Wong, 2000). An interesting detail is that glyphosate may be formed during water treatment for purification from organic micropollutants. Glyphosate and AMPA were found to be formed during ozonisation of dilute aqueous solution of the complexing agent ethylenediaminetetra(methylenephosphonic acid) (Klinger et al., 1998; Nowack, 2003). The wide use, and hence ubiquity of glyphosate makes great demands on glyphosate safety, i.e. the absence of any harmful environmental effect except on target organisms (the undesirable weeds).

Glyphosate is very stable in higher plants (Putnam, 1976; Zandstra & Nishimoto, 1977; Chase & Appleby, 1979; Gothrup et al., 1976; Wyrill & Burnside, 1976). Through its metabolism, AMPA has been identified as the main metabolite in plants as well e.g., in montmorency cherry (*Prunus cerasus* L.) leaves, field bindweed (*Convolvulus arvensis* L.), henge bindweed (*Convolvulus sepium* L), Canada thistle (*Cirsium arvense* (L) Scop.), tall morning glory (*Ipomea purpurea* (L.) Roth.) and wild buckwheat (*Polygonum convolvulus* L.) (Sandberg et al., 1980; Aizawa, 1982; Aizawa, 1989).

Besides AMPA, its certain derivatives e.g., N-methyl-AMPA or N,N-dimethyl-AMPA have been also found as metabolites, mostly in plants (FAO/WHO, 2006). Decomposition in GT plants is even more complex, as some of these plants have been designed for enhanced degradation of glyphosate. In such plants, further AMPA derivatives e.g., N-acetyl-AMPA,

N-malonyl-AMPA, N-glyceryl-AMPA and various conjugates of AMPA have also been identified (FAO/WHO, 2006).

5.3 Environmental monitoring of glyphosate

Glyphosate shows unique characteristics in soil as compared to other pesticide active ingredients. With predominantly apolar groups pesticides typically bind to the organic matter in soil (Borggaard & Gimsing, 2008). In contrast, glyphosate is of amphoteric (zwitterionic) character, analytical determination of which is to date a great challenge to analytical chemists. As a result of the unusual chemical behaviour of the parent compound (N-phosphonomethylglycine) and its metabolite (AMPA), routine environmental analytical methods do not detect them with sufficient sensitivity. It is also due to the difficult analytical procedure that glyphosate is often not targeted or overlooked in environmental studies, or has been considered of neglectable level. Certain studies, however, report frequent occurrence. In the United States, surface water contamination has been reported due to run-off from agricultural areas (Edwards et al., 1980; Feng et al., 1990) or pesticide drift (Payne et al., 1990; Payne, 1992). Glyphosate has been listed among pesticides of potential concern in surface water contamination in the Mediterranean region of Europe in the mid' nineties (Barceló & Hennion, 1997), and glyphosate and AMPA were found as contaminants in two small tributaries of the river Ruhr in North-Rhine-Westphalia, Germany at up to 590 ng/l concentration (Skark et al., 1998). A monitoring study carried out in Norway found frequent occurrence of glyphosate and its metabolite AMPA in surface water samples. In 54% of the 540 surface water samples collected between 1995 and 1999 glyphosate or AMPA was detected. The maximal concentration was 0.93 µg/l (average 0.13 µg/l) for glyphosate, and 0.2 µg/l (average 0.06 µg/l) for AMPA (Ludvigsen & Lode, 2001a; 2001b). The monitoring study, therefore, indicated broad occurrence of glyphosate and its metabolite at low concentrations. In a study carried out in surface waters of the Midwest in the United States in 2002 glyphosate was detected in 35-40% of the samples (maximal concentration 8.7 µg/l) and AMPA in 53-83% of the samples (maximal concentration 3.6 µg/l) (Battaglin et al., 2005), and both glyphosate and AMPA were detected in vernal snow-flood at concentrations up to 328 and 41 · g/l, respectively, in 2005-2006 in four states of the US (Battaglin et al., 2009). Analysing water samples from 10 wastewater treatment plants in the United States, the U.S. Geological Survey detected AMPA in 67.5% and glyphosate in 17.5% of the samples (Kolpin et al., 2006). The study concluded that urban use of glyphosate contributes to glyphosate and AMPA concentrations in streams in the United States. In a study carried out in Canada in 2004-2005, 21% of the analysed 502 samples contained glyphosate with a maximum concentration of 41 µg/l, and the peak concentration of AMPA was 30 µg/l glyphosate equivalent (Struger et al., 2008). In France, glyphosate and AMPA were detected in 2007 and 2008 due to urban runoff effect (Batta et al., 2009). In fact, Villeneuve et al. (2011) adjudge glyphosate to be one of the herbicides most often found in freshwater ecosystems worldwide, and state that AMPA is the most often detected and glyphosate is the third most frequent pesticide residue in French streams. Elevated glyphosate levels were detected in surface water, soil and sediment samples due to intensive herbicide applications in a GM soybean cultivating area in Argentina (Peruzzo et al., 2008). These studies are warning signs indicating that this herbicide active ingredient of intensive use, that is expected to further expand with the commercial cultivation of GM crops, became an ubiquitous contaminant in surface waters, and therefore, a permanent pollutant factor, which deserves pronounced attention by ecotoxicology.

6. Adverse environmental effects of glyphosate

6.1 Glyphosate and *Fusarium* species

Sanogo and co-workers (2000) observed that crop loss in soy due to infestation by *Fusarium solani* f. sp. *glycines* increased after glyphosate applications. Kremer and co-workers (2005) described a stimulating effect of the root exsudate of GR soy sampled after glyphosate application on the growth of *Fusarium* sp. strains. Treatments caused concentration dependent increase on the mycelium mass of the fungus. Nonetheless, Powel and Swanton (2008) could not confirm these observations in their field study. Kremer and Means (2009) claim that certain fungi utilise glyphosate released from plant roots into the soil as a nutritive, which facilitates their growth. Soil manganese content also affects the above consequence of glyphosate through chelating with the compound and thus, modifying its effects. Considering the fact that numerous plant pathogenic *Fusarium* species produce mycotoxins, an increasing proportion of these species is far not favourable as a side-effect. Johal and Huber (2009) lists numbersome plant pathogens (e.g., *Corynespora cassicola* or *Sclerotinia sclerotiorum* on soy) they claim to grow increasingly after glyphosate treatments, and the list contains several *Fusarium* species (*F. graminearum*, *F. oxysporum*, *F. solani*). They hypothesize that glyphosate causes disturbances in microelement metabolism in plants, and in parallel, deteriorate the defense system of the plants, thereby increasing the virulence of certain plant pathogens. Zobiole and co-workers (2011) confirmed the above effects by their observation that glyphosate treatments facilitate colonisation of *Fusarium* species on the soy roots, but reduces the fluorescent *Pseudomonas* fraction of the rhizosphere, the level of manganese reducing bacteria and of the indoleacetic acid producing rhizobacteria. As a combined result of these effects, root and overall plant biomasses were found to be reduced.

6.2 Toxicity of glyphosate to aquatic ecosystems and amphibians

Substances occurring in surface waters deserve special attention by ecotoxicologists, as they enter a matrix that is the habitat of numerous aqueous organisms and the basis of our drinking water reserves. Drinking water is an irreplaceable essential part of our diet, and is a possible vehicle for chronic exposure (the basis of chronic diseases) in daily contact/consumption.

Glyphosate has been known to cause toxicity to microalgae and other aquatic microorganisms (Goldsborough and Brown 1988; Austin et al., 1991; Anton et al., 1993; Sáenz et al., 1997; DeLorenzo et al., 2001; Ma 2002; Ma et al., 2002; Ma et al., 2003), in fact a green algal toxicity test has been proposed for screening herbicide activity (Ma & Wang, 2002). In contrast, cyanobacteria have been found to show resistance against glyphosate (López-Rodas et al., 2007; Forlani et al., 2008). Tsui and Chu (2003) tested the effect of glyphosate, its most common polyoxyethyleneamine (POEA) type formulating materials, polyethoxylated tallowamines, and the formulated glyphosate preparation (Roundup) on model species from aquatic ecosystems, bacteria (*Vibrio fischeri*), microalgae (*Selenastrum capricornutum*, *Skeletonema costatum*), protozoas (*Tetrahymena pyriformis*, *Euplotes vannus*) and crustaceans (*Ceriodaphnia dubia*, *Acartia tonsa*). The most surprising result of the study was that the assumedly inert detergent formulating agent, POEA was found to be the most toxic component. In light of this it is far not surprising that Cox and Surgan (2006) and Reuben (2010) propounded the question, why tests only on the active ingredients are necessary to be specified in the documentation required by the Environmental Protection Agency of the

Unites States (US EPA), when several of the used formulating components are known to exert biological activity.

Although acute toxicity and genotoxicity of glyphosate have been evidenced to certain fish (Langiano & Martinez, 2008; Cavalcante et al., 2008), glyphosate shows favourable acute toxicity parameters on most vertebrates, and therefore, has been classified as III toxicity category by US EPA. The European discretion is stricter, listing the compound among substances causing irritation (Xi) and severe ocular damage (R41). It has to be noted, however, that that model species of neither amphibians, not reptilians are represented in the toxicological documentations required nowadays. It may not be surprising, therefore, that after atrazine (Hayes et al., 2002; 2010), glyphosate is the second herbicide active ingredient that is questioned due to its detrimental effects on the animal class, considered the most endangered on Earth, amphibians.

Mann and Bidwell (1999) studied the toxicity of glyphosate on tadpoles of four Australian frogs (*Crinia insignifera*, *Heleioporus eyrei*, *Limnodynastes dorsalis* and *Litoria moorei*). The toxicity of Roundup and its 48-hour LC_{50} values were found to be 3-12 mg glyphosate equivalent/l. Tolerance of the adult frogs was substantially greater. A glyphosate-based formulated herbicide preparation (VisionMAX) caused no significant effects on the juvenile adults of the green frogs (*Lithobates clamitans*) when applied at field application doses, only marginal differences in statistics of infection rates and liver somatic indices in relation to exposure estimates (Edge et al., 2011). Chen et al. (2004) observed that the toxicity of glyphosate on the frog species *Rana pipiens* was greatly affected by lacking food resources and the pH of the medium as stress factors. Relyea (2005a) reported tadpole (*Bufo americanus*, *Hyla versicolor*, *Rana sylvatica*, *R. pipiens*, *R. clamitans* and *R. catesbeiana*) mortality related to glyphosate applications. The effect, occurred at 2-16 mg glyphosate equivalent/l concentrations, was linked with the stress caused by the predator of the tadpoles, salamander *Notophthalmus viridescens*. Later Relyea and Jones (2009) included further frog species (*Bufo boreas*, *Pseudacris crucifer*, *Rana cascadea*, *R. sylvatica*) into the study, and found LC_{50} values to be 0.8-2 mg glyphosate equivalent/l. Testing four salamander species (*Amblystoma gracile*, *A. laterale*, *A. maculatum* and *N. viridescens*), the corresponding values ranged between 2.7 and 3.2 mg glyphosate equivalent/l. In this case, glyphosate was formulated with detergent POEA. Further studies also shed light on the fact that another stress factor, population density, playing an important part in the competition of the tadpoles increased the toxic effect of glyphosate (Jones et al., 2010). Lajmanovich and co-workers (2010) detected lowered enzymatic activities (e.g., acetylcholine esterase and glutathion-S-transferase) in a frog species, *Rhinella arenarum* upon glyphosate treatments.

Sparling and co-workers (2006) detected lowered fecundity of the eggs of the semiaquatic turtle, red-eared slider (*Trachemys scripta elegans*) if treated with glyphosate at high doses.

6.3 Teratogenic activity of glyphosate

The teratogenicity of the pesticide preparations containing glyphosate deserves special attention. The very first examples of observed teratogenicity of glyphosate preparations have also been linked to amphibians. Using the so-called FETAX assay, Perkins and co-workers (2000) observed a formulation dependent teratogenic effect of glyphosate on embryos of the frog species *Xenopus laevis*. The concentrations that triggered the effect were relatively high (the highest dose applied in the study was 2.88 mg glyphosate equivalent/l), but not irrealisticly high with respect to field doses of glyphosate, indicating, that high allowed agricultural doses cause glyphosate levels close to the safety margin. Lajmanovich and co-workers (2005) studied the effects of a glyphosate preparation (Glyfos) on the tadpoles of *Scinax nasicus*, and found that a 2-4-day exposure to 3 mg/l glyphosate caused malformation in more than half of the test animals. The treatment was carried out nearly at the LC_{50} level of glyphosate. Dallegrave and co-workers (2003) found fetotoxic effects on rats treated with glyphosate at very high, 1000 mg/l concentration on the 6th-15th day after fertilisation. Nearly half of the newborn rat progeny in the experiments were born with skeletal development disorders.

Testing the effects of glyphosate preparations on the embryos of the sea urchin, *Sphaerechinus granularis*, Marc and co-workers (2004a) observed a collapse of cell cycle control. Inhibition affects DNA synthesis in the G2/M phase of the first cell cycle (Marc et al., 2004b). The authors estimate that glyphosate production workers inhale 500-5000-fold level of the effective concentration in these experiments. A marked toxicity of the formulating agent POEA has also been observed on sea urchins (Marc et al., 2005). The very early DNA damage was claimed to be related to tumour formation by Bellé and co-workers (2007), and the authors consider the sea urchin biotest they developed as a possible experimental model for testing this effect. Jayawardena and co-workers (2010) described nearly 60% developmental disorders on the tadpoles of a Sri Lanka frog (*Polpedates cruciger*) upon treatment with 1 ppm glyphosate.

The teratogenicity of herbicides of glyphosate as active ingredient have been tested lately on amphibian (*X. laevis*) and bird (*Gallus domesticus*) embryos. Applied with direct injection at sublethal doses caused modification of the position and pattern of rhobomeres, the area of the neural crest decreased, the anterior-posterior axis shortened and the occurrence of cephalic markers was inhibited at the embryonic development stage of the nervous system. As a result, frog embryos became of characteristic phenotype: the trunk is shortened, head size is reduced, eyes were improperly or not developed (microphthalmia), and additional cranial deformities occurred in later development. Similar teratogenic effects were seen on embryos of Amniotes e.g., chicken. These developmental disorders may be related to damages of the retinoic acid signal pathway, resulting in the inhibition of the expression of certain essential genes (*shh, slug, otx2*). These genes play crucial roles in the neurulation process of embryogenesis (Paganelli et al., 2010). These findings were later debated by several comments. On behalf of the producers, Saltmiras and co-workers (2011) questioned certain conclusions in the work of Paganelli and co-workers (2010), claiming that the standardised pilot teratogenicity tests, carried out under good laboratory practice (GLP) by the manufacturers, have been evaluated by independent experts of several international organisations. They also considered the dosages used by Paganelli and co-workers exceedingly high, and the mode of application (microinjection) irrealistic in nature. Similar criticism has been voiced by Mulet (2011) and Palma (2011). In his answer, Carrasco (2011) emphasised their opinion that the company representatives ignore scientific facts supporting teratogenicity of atrazine, glyphosate and triadimefon through retinoic acid biosynthesis. He also emphasized that of 180 research reports of Monsanto, 150 are not public, or have never been presented to the scientific community. He also included that they obtained similar phenotypes in their studies with microinjection, than by incubation of the preparations. As a follow-up, Antoniou and co-workers (2011) compiled an extensive review of 359 studies and publications on the teratogenicity and birth defects caused by glyphosate, and heavily criticize the European Union for not banning glyphosate, but rather postponing its re-evaluation until 2015 (European Commission, 2010).

6.4 Genotoxicity of glyphosate

Occupational exposure to pesticides, including glyphosate as active ingredient, may lead to pregnancy problems even through exposure of men (Savitz et al., 1997). Such phenomenon has been first described in epidemiology with Vietnam War veterans exposed to Agent Orange with phenoxyacetic acid type active ingredients contaminated with dibenzodioxins. Although glyphosate has been claimed not to be genotoxic and its formulation Roundup "causing only a week effect" (Rank et al., 1993; Bolognesi et al., 1997), Kale and co-workers (1995) observed mutagenic effects of Roundup in *Drosophila melanogaster* recessive lethal mutation tests. Lioi and co-workers (1998) described increasing sister chromatide exchange in human lymphocytes with increasing glyphosate doses. Walsh and co-workers (2000) detected in murine tumour cells the inhibitory activity of Roundup on the biosynthesis of a protein (StAR) participating in the synthesis of sex steroids. This reduced the operation of the cholesterol – pregnenolon – progesteron transformation pathway to a minimal level. As it often happens in exploring mutagenic effects of chemical substances, additional studies have not found glyphosate mutagenic, and therefore, it is not so listed in the GAP2000 program compiled from US EPA/IARC databases. However, Cox (2004) describes chronic toxicity profile of several substances applied in the formulation of glyphosate.

Studying the activity of dehydrogenase enzymes in the liver, heart and brain of pregnant rats, Daruich and co-workers (2001) concluded that glyphosate causes various disorders both in the parent female and in the progeny. According to results of the study by Benedettia and co-workers (2004), aminotransferase enzyme activity decreased in the liver of rats, impairing lymphocytes, and leading to liver tissue damages. In *in vitro* tests McComb and co-workers (2008) found that glyphosate acts in the mitochondria of the rat liver cells as an oxidative phosphorylation decoupling agent. Mariana and co-workers (2009) observed oxidative stress status decay in the blood, liver and testicles upon injection administration of glyphosate, possibly linked to reproductional toxicity.

Prasad and co-workers (2009) detected cytotoxic effects, as well as chromosomal disorders and micronucleus formation in murine bone-marrow. Poletta and co-workers (2009) described genotoxic effects of Roundup on the erythrocytes in the blood of caimans, correlated with DNA damages.

According to the survey of De Roos and co-workers (2003), the risk of the incidence of non-Hodgkin lymphoma is increased among pesticide users. As the authors found it, this applies to herbicide preparations with glyphosate as active ingredient. Focusing the study solely on glyphosate preparations a year later in the corn belt of the United States, of the majority of malignant diseases, only the incidence of abnormal plasma cell proliferation (*myeloma multiplex, plasmocytoma*) showed a slight rise (De Roos et al., 2004). Myeloma represents approximately 10% of the malignant haematological disorders. Although the cause of the disease is not yet known, its risk factors include autoimmune diseases, certain viruses (*HIV* and *Herpes*), and the frequent use of certain solvents as occupational hazard. On the basis of murine skin carcinogenesis, George and co-workers (2010) reported that glyphosate may act as a skin tumour promoter due to the induction of several special proteins.

6.5 Hormone modulant effects of glyphosate and POEA

Studying chronic exposure of tadpoles of *Rana pipiens*, Howe and co-workers (2004) found that in addition to developmental disorders, gonads in 15-20% of the treated animals developed erroneously, and these animals showed intersexual characteristics. Arbuckle and co-workers (2001) registered increased risk of abortion in agricultural farms after glyphosate applications. In addition, excretion of glyphosate has been determined in the urine of agricultural workers and their family members (Acquavella et al., 2004).

Richard and co-workers (2005) evidenced toxicity of glyphosate on the JEG3 cells in the placenta. Formulated Roundup exerted stronger effect than glyphosate itself. Glyphosate inhibited aromatase enzymes of key importance in estrogen biosynthesis. This effect has also been evidenced in *in vitro* tests by binding to the active site of the purified enzyme. The formulating agent in the preparation enhanced the inhibitory effect in the microsomal fraction. Benachour and co-workers (2007) tested the effect of glyphosate and Roundup Bioforce on various cell lines, and also determined the aromatase inhibiting effect of glyphosate and the synergistic effect of the formulating agent. They suppose that the hormone modulant effect of Roundup may affect human reproduction and fetal development. Testing these human cell lines, Benachour and Séralini (2009) found that glyphosate alone induces apoptosis, and POEA and AMPA applied in combination exert synergistic effects, similarly to the synergy seen for Roundup. The synergy was reported to be further acerbated with activated Cry1Ab toxin related to that produced by insect resistant GM plants, raising concern regarding stacked genetic event GM crops exerting both glyphosate tolerance and Cry1Ab based insect resistance (Mesnage et al., 2011). Moreover, the combined effect caused cell necrosis as well. Effect enhancement is likely to be explained by the detergent activity of POEA facilitating the penetration of glyphosate through cell membranes and subsequent accumulation in the cells. The aromatase inhibitory effect of the formulated preparation was four-fold, as compared to the neat active ingredient. The authors consider it proven, that POEA, previously believed to be inert, is far not inactive biologically. As the authorised MRL of glyphosate in forage is as high as 400 mg/kg, Gasnier and co-workers (2009) studied in various *in vitro* tests, what effects this may cause in a human hepatic cell line. All treatments indicated a concentration-dependent effect in the toxicity tests were found genotoxic in the comet assay for DNA damages, moreover, displayed antiestrogenic and antiandrogenic effects.

6.6 Glyphosate resistance of weeds

Frequent applications of glyphosate and the spread of GT crops outside of Europe escalate the occurrence of glyphosate in the environment, exerting severe selection pressure on the weed species. It has been well known that certain weeds have native resistance against glyphosate e.g., the common lambsquarters (*Chenopodium album*), the velvetleaf (*Abutilon theophrasti*) and the common cocklebur (*Xanthium strumarium*).

The first population of GT *Lolium rigidum* was described in 1996 by Pratley and co-workers in Australia. This was followed in 1997 by GT goosegrass (*Eleusine indica*) in Malaysia (Lee & Ngim, 2000), GT horseweed (*Conyza canadensis*) in the United States (VanGessel, 2001), GT Italian ryegrass (*Lolium multiflorum*) in Chile (Perez & Kogan, 2003). Further known GT weed species include *Echinochloa colona* (2007), *Urochloa panicoides* (2008) and *Chloris truncata*

(2010) in Australia; *Conyza bonariensis* (2003) and ribwort plantain (*Plantago lanceolata,* 2003) in South Africa; ragweed (*Ambrosia artemisifolia,* 2004), *Ambrosia trifida* (2004), *Amaranthus palmeri* (2005), *Amaranthus tuberculatus* (2005), summer cypress (*Bassia scoparia,* 2007) and annual meadow grass (*Poa annua,* 2010) in the United States; *Conyza sumatrensis* (2009) in Spain; Johnsongrass (*Sorghum halepense*) (2005), Italian ryegrass (*Lolium perene,* 2008) in Argentina; *Euphorbia heterophyla* (2006) in Brazil; *Parthenium hysterophorus* (2004) in Colombia and *Digitaria insularis* (2006) in Paraguay (Heap, Epubl). GT Johnsongrass was reported in a continuous soybean field in Arkansas, United States (Riar et al., 2011). Price (2011) claims that agricultural conservation tillage is threatened in the United States by the rapid spread of GT Palmer amaranth (*Amaranthus palmeri* [S.] Wats.) due to wide range cultivation of transgenic, GT cultivars and corresponding broad use of glyphosate. GT amaranths were first identified in Georgia, and later reported in nine states, Alabama, Arkansas, Florida, Georgia, Louisiana, Mississippi, North Carolina, South Carolina, and Tennessee, and a closely related GT amaranth, common waterhemp (*Amaranthus rudis* Sauer) in four states, Illinois, Iowa, Minnesota, and Missouri. Moreover, GT Italian ryegrass populations collected in Oregon, United States appeared to show cross-resistance to another phosphonic acid type herbicide active ingredient, glufosinate (Avila-Garcia & Mallory Smith, 2011).

Powles and co-workers (1998) described a *L. rigidum* population resisting 7-11-fold dosage of glyphosate in Australia. Shrestha and Hemree (2007) found GT subpopulations of 5-8 leaf stage *Conyza canadensis* surviving only 2-4-fold glyphosate doses. According to Powles (2008), it is not coincidental that in countries, where GT crops are on the rise (Argentina and Brazil), the occurrence of GT weeds is more frequent. Moreover, he considers this one of the main obstacles of the spread of GT crops in the agricultural practice. Glyphosate tolerance is an inherited property, therefore, accumulation of weeds in the treated areas is to be expected. Genomics studies of the GT populations revealed that mutation of the gene (*epsps*) encoding the target enzyme responsible for tolerance is not infrequent in nature. (The mutant alleles (*mepsps, 2mepsps*) responsible for tolerance has been found in maize as well, see Table 2.). Reduced or modified uptake or translocation of glyphosate has also been observed, and the metabolic fate of the compound may also become altered in the cell (Shaner, 2009), possibly resulting in GT populations. It is not difficult to predict, that prolonged cultivation of GT crops will necessitate supplemental herbicide administrations with active ingredients other than glyphosate.

7. References

Alibhai, FA & Stallings W.C. (2001). Closing down on glyphosate inhibition – with a new structure for drug discovery. *Proceedings of the National Academy of Sciences of the United States of America,* 98, 2944-2946.

Abu-Irmaileh, B.E. & Jordan, L.S. (1978). Some aspects of glyphosate action in purple nutsedge (*Cyperus rotundus*). *Weed Science,* 26, 700-703.

Accinelli, C.; Screpanti, C.; Vicari, A. & Catizone, P. (2004) Influence of insecticidal toxins from *Bacillus thuringiensis* subsp. *kurstaki* on the degradation of glyphosate and glufosinate-ammonium in soil samples. *Agriculture, Ecosystems and Environment,* 103, 497-507.

Accinelli, C.; Koskinen, W.C. & Sadowsky, M.J. (2006). Influence of Cry1Ac toxin on mineralization and bioavailability of glyphosate in soil. *Journal of Agricultural and Food Chemistry,* 54, 164-169.

Acquavella, J.F.; Alaxander, B.H.; Mandel, J.S.; Gustin, C.; Baker, B.; Chapman, P. & Bleeke, M. (2004). Glyphosate biomonitoring for farmers and their families: results from the farm exposure study. *Environmental Health Perspectives*, 112, 321-326.

Aizawa, H. (Ed.) (1982). *Metabolic Maps of Pesticides*. Vol. 1, p. 140, Academic Press, ISBN 0-12-046480-2, New York.

Aizawa, H. (Ed.) (1989). *Metabolic Maps of Pesticides*. Vol. 2, p. 142, ademic Press, ISBN 0-12-046481-0, New York.

Alferness, P.L. & Iwata, Y. (1994). Determination of glyphosate and (amino methyl)phosphonic acid in soil, plant and animal matrices and water by capillary gas chromatography with mass selective detection. *Journal of Agricultural and Food Chemistry*, 42, 2751-2759.

Altieri, M.A. & Pengue, W. (2006). GM soybean: Latin America's new colonizer. *Seedling*, 2006 (1), 13-17.

Amrhein, N.; Deus, B; Gehrke, P. & Steinrucken, H.C. (1980). The site of the inhibition of the shikimate pathway by glyphosate. II. Interference of glyphosate with chorismate formation *in vivo* and *in vitro*. *Plant Physiology*, 66, 830-834.

Amrhein, N.; Schab, J. & Steinrücken, H.C. (1980). The mode of action of the herbicide glyphosate. *Naturwissenschaften*, 67, 356-357.

Anderson, D.K.; Deuwer, D.L.; & Sikorski, J.A. (1995). Syntheses of new 2-hydroxythiazol-5-yl and 3-hydroxy-1,2,4-triazol-5-ylphosphonic acids as potential cyclic spatial mimics of glyphosate. *Journal of Heterocyclic Chemistry*, 32, 893-898.

Anonymous (2006). Mexican soldiers try to kill herbicide-resistant marijuana. *The Washington Times*, Dec 22, 2006, Available from http://www.washingtontimes.com/news/2006/dec/22/20061222-110251-2050r

Anton, F.A.; Ariz, M. & Alia, M. (1993). Ecotoxic effects of four herbicides (glyphosate, alachlor, chlortoluron and isoproturon) on the algae *Chlorella pyrenoidosa* chick. *Science of the Total Environment*, 134, Supplement 2, 845-851.

Anton, D.; Hedstrom, L.; Fish, S. & Abeles, R. (1983). Mechanism of of enolpyruvyl shikimate-3-phosphate synthase exchange of phosphoenolpyruvate with solvent protons. *Biochemistry*, 22, 5903-5908.

Antoniou, M.; Habib, M. E. E.-D. M.; Howard, C. V.; Jennings, R. C.; Leifert, C.; Nodari, R. O.; Robinson, C. & Fagan, J. (2011)Roundup and birth defects. Is the public being kept in the dark? Earth Open Source, Lancashire, UK, Available from http://www.earthopensource.org/files/pdfs/Roundup-and-birth-defects/RoundupandBirthDefectsv5.pdf

Arbuckle, T.E.; Lin, Z. & Mery, L.S. (2001). An exploratory analysis of the effect of pesticide exposure on the risk of spontaneous abortion in an Ontario farm population. *Environmental Health Perspectives*, 109, 851-857.

Arcuri, H.A.; Canduri, F.; Pereira, J.H.; da Silveira, N.J.F.; Camera, J.C., Jr.; de Oliveira, J.S.; Basso, L.A.; Palma, M.S.; Santos, D.S. & de Azevedo, W.F., Jr. (2004). Molecular models for shikimate pathway enzymes of *Xylella fastidiosa*. *Biochemical and Biophysical Research Communications*, 320, 979-991.

Arhens, W.H. (1994). *Herbicide Handbook*, 7th Edition, pp. 149-152, Weed Science Society of America, ISBN 0-911733-18-3, Champaign, IL, USA.

Arregui, M.C.; Lenardón, A.; Sanchez, D.; Maitre, M.I.; Scotta, R. & Enrique, S. (2004). Monitoring glyphosate residues in transgenic glyphosate-resistant soybean. *Pesticide Management Science*, 60, 163-166.

Ashton, F.M. & Crafts, A.S. (1981). *Mode of Action of Herbicides*. 2nd Edition, pp. 236-253, John Wiley and Sons, ISBN 0-471-04847-X, New York.

Atkinson, D. (1985). Toxicological properties of glyphosate – A summary. In: *The herbicide glyphosate*. Grossbard, E.; Atkinson D. (Eds.), pp. 127-133, Butterworths, London, UK.

Attorney General of the State of New York (1996). False advertising by Monsanto regarding the safety of Roundup herbicide (glyphosate). Assurance of discontinuance pursuant to executive law § 63(15). Attorney General of the State of New York Consumer Frauds and Protection Bureau, Environmental Protection Bureau, New York, USA.

Austin, A.P.; Harris, G.E. & Lucey, W.P. (1991). Impact of an organophosphate herbicide (Glyphosate®) on periphyton communities developed in experimental streams. *Bulletin of Environmental Contamination and Toxicology*, 47, 29-35.

Avila-Garcia, W.V. & Mallory-Smith, C. (2011) Glyphosate-resistant Italian ryegrass (*Lolium perenne*) populations also exhibit resistance to glufosinate. *Weed Science*, 59, 305-309.

Baird, D.D.; Upchurch, R.P.; Homesley, W.B. & Franz, J.E. (1971). Introduction of a new broadspectrum postemergence herbicide class with utility for herbaceous perennial weed control. *Proceedings North Central Weed Control Conference*, 26, 64-68.

Balthazor T.M. & Hallas L.E. (1986). Glyphosate degrading microorganisms from industrial activated sludge. *Applied and Environmental Microbiology*, 51, 432-434.

Barceló, D. & Hennion, M.-C. (1997) Trace determination of pesticides and their degradation products in water. Elsevier, ISBN 0-444-81842-1, Amsterdam.

Battaglin, W.A.; Kolpin, D.W.; Scribner, E.A.; Kuivila, K.M. & Sandtrom, M.W. (2005). Glyphosate, other herbicides, and transformation products in Midwestern streams, 2002. *Journal of the American Water Resource Association*, 41, 323-332.

Battaglin, W.A.; Rice, K.C.; Focazio, M.J.; Salmons, S. & Barry, R.X. (2009). The occurrence of glyphosate, atrazine, and other pesticides in vernal pools and adjacent streams in Washington, DC, Maryland, Iowa, and Wyoming, 2005–2006. *Environmental Monitoring and Assessment*, 155, 281-307.

Bauer, K.H.; Knepper, T.P.; Maes, A.; Schatz, V. & Voihsel, M. (1999). Analysis of polar organic micropollutants in water with ion chromatography-electrospray mass spectrometry. *Journal of Chromatography A*, 837, 1-2

Baur, J.R. (1979). Effect of glyphosate on auxin transport in corn and cotton tissues. *Plant Physiology*, 63, 882-886.

Bellé, R.; Le Bouffant, R.; Morales, J.; Cosson, B.; Cormier, P. & Mulner-Lorillon, O. (2007). L'embryon d'oursin, le point de surveillance de l'ADN endommagé de la division cellulaire et les mécanismes à l'origine de la cancérisation. *Journal de la Société de Biologie*, 201, 317-327.

Benachour, N. & Séralini, G.-E. (2009). Glyphosate formulations induce apoptosis and necrosis in human umbilical, embryonic, and placental cells. *Chemical Research in Toxicology*, 22, 97-105.

Benachour, N.; Sipahutar, H.; Moslemi, S.; Casnier, C.; Travert, C. & Séralini, G.-E. (2007). Time- and dose-dependent effects of Roundup on human embryonic and placental cells. *Archives in Environmental Contamination and Toxicology*, 53, 126-133.

Benedettia, A.L.; de Lourdes Viturib, C.; Gonçalves Trentina, A.; Custódio Dominguesc, M. A. & Alvarez-Silva, M. (2004). The effects of sub-chronic exposure of Wistar rats to the herbicide Glyphosate-Biocarb. *Toxicology Letters*, 153, 227-232.

Bolognesi, C.; Bonatti, S.; Degan, P.; Gallerani, E.; Peluso, M.; Rabboni, R.; Roggieri, P. & Abbondandolo, A. (1997). Genotoxic activity of glyphosate and its technical formulation Roundup. *Journal of Agricultural and Food Chemistry*, 45, 1957-1962.

Borggaard, O.K. & Gimsing, A.L. (2008). Fate of glyphosate in soil and the possibility of leaching to ground and surface waters: a review. *Pesticide Management Science*, 64, 441-456.

Botta, F.; Lavison, G.; Couturier, G.; Alliot, F.; Moreau-Guigon, E.; Fauchon, N.; Guery, B.; Chevreuil, M. & Blanchoud, H. (2009). Transfer of glyphosate and its degradate AMPA to surface waters through urban sewerage systems. *Chemosphere*, 77, 133-139.

Brecke, B.J. & Duke, W.B. (1980). Effect of glyphosate on intact bean plants (*Phaseolus vulgaris* L.) and isolated cells. *Plant Physiology*, 66, 656-659.

Busse, M.D.; Ratcliff, A.W.; Shestak, C.J. & Powers, R.F. (2001). Glyphosate toxicity and effects of long-term vegetation control on soil microbial communities. *Soil Biology and Biochemistry*, 33, 1777-1789.

Campbell, W.F.; Evans, J.O. & Reed, S.C. (1976). Effects of glyphosate on chloroplast ultrastructure of quackgrass mesophyll cells. *Weed Science*, 24, 22-25.

Carlisle, S.M. & Trevors, J.T. (1988). Glyphosate in the environment. *Water, Air & Soil Pollution*, 39, 409-420.

Cavalcante, D.G.S.M.; Martinez, C.B.R. & Sofia, S.H. (2008). Genotoxic effects of Roundup® on the fish *Prochilodus lineatus*. *Mutation Research*, 655, 41-46.

Carrasco, A.E. (2011). Reply to the letter to the editor regarding our article (Paganelli *et al.*, 2010). *Chemical Research in Toxicology*, 24, DOI: 10.1021/tx200072k

Chase, R.L. & Appleby, A.P. (1979). Effects of humidity and moisture stress on glyphosate control of *Cyperus rotundus* L. *Weed Research*, 19, 241-246.

Chen, C.Y.; Hathaway, K.M. & Folt, C.L. (2004). Multiple stress effects of Vision herbicide, pH, and food on zooplankton and larval amphibian species from forest wetland. *Environmental Toxicology and Chemistry*, 23, 823-831.

Christensen, A.M. & Schaefer, J. (1993). Solid-state NMR determination of intra- and intermolecular ^{31}P-^{13}C distances for shikimate 3-phosphate and [1-^{13}C]glyphosate bound to enolpyruvylshikimate-3-phosphate synthase. *Biochemistry*, 32, 2868-2873.

Cikalo, M.G.; Goodall, D.M. & Matthews, W. (1996). Analysis of glyphosate using capillary electrophoresis with indirect detection. *Journal of Chromatography A*, 745, 189-200.

Clegg, S.B.; Stephenson, G.R. & Hall, J.C. (1999). Development of an enzyme-linked immunosorbent assay for the detection of glyphosate. *Journal of Agricultural and Food Chemistry*, 47, 5031-5037.

Cole, D.J.; Dodge, A.D. & Caseley, J.C. (1980). Some biochemical effects of glyphosate on plant meristems. *Journal of Experimental Botany*, 31, 1665-1674.

Cox, C. (2004). Herbicide factsheet: glyphosate. *Journal of Pesticide Reform*, 24, 10-15.

Cox, C. & Surgan, M. (2006). Unidentified inert ingredients in pesticides: implications for human and environmental health. *Environmental Health Perspectives*, 114, 1803-1806.

Cowell, J.E.; Kunstman, J.L.; Nord, P.J.; Steinmetz, J.R. & Wilson, G.R. (1986). Validation of an analytical residue method for analysis of glyphosate and metabolite: an Interlaboratory study. *Journal of Agricultural and Food Chemistry*, 34, 955-960.

Dallegrave, E.; Mantesea, F.D.; Coelho, R.S.; Pereira, J.D.; Dalsenter, P.R. & Langeloh, A. (2003). The teratogenic potential of the herbicide glyphosate-Roundup in Wistar rats. *Toxicology Letters*, 142, 45-52.

Daruich, J.; Zirulnik, F. & Gimenez, M.S. (2001). Effect of the herbicide glyphosate on enzymatic activity in pregnant rats and their fetuses. *Environmental Research*, 85, 226-231.

de Jonge, H.; de Jonge, L.W.; Jacobsen, O.H.; Yamaguchi, T. & Moldrup, P. (2001). Glyphosate sorption in soils of different pH and phosphorus content. *Soil Science*, 166, 230-238.

De Roos, A.J.; Zahm, S.H.; Cantor, K.P.; Weisenburger, D.D.; Holmes, F.F.; Burmeister, L.F. & Blair, A. (2003). Integrative assessment of multiple pesticides as risk factors for non-Hodgkin's lymphoma among men. *Occupational and Environmental Medicine*, 60,   Epubl,   Available   from http://www.ncbi.nlm.nih.gov/pmc/articles/PMC1740618/pdf/v060p00e11.pdf

De Roos, A.J.; Blair, A.; Rusiecki, J.A.; Hoppin, J. A.; Svec, M.; Dosemeci, M.; Sandler, D. P. & Alavanja, M.C. (2004). Cancer incidence among glyphosate-exposed pesticide applicators in the agricultural health study. *Environmental Health Perspectives*, 113: 49-54.

DeLorenzo, M.E.; Scott, G.I. & Ross P.E. (2001). Toxicity of pesticides to aquatic microorganisms: a review. *Environmental Toxicology and Chemistry*. 20, 84-98.

DFG (Deutsche Forschungsgemeinschaft) (1992). Method 405 Glyphosate. In: *Manual of Pesticide Residue Analysis*. Vol. II, pp. 229-304. VCH Publishers Inc., ISBN 3-527-27017-5, New York, USA.

Dill, G.M.; Sammons, R.D.; Feng, P.C.C.; Kohn, F.; Kretzmer, K.; Mehrsheikh, A.; Bleeke, M.; Honegger, J.L.; Farmer, D.; Wright, D. & Haupfear, E.A. (2010). Glyphosate: discovery, development, applications, and properties. Chapter 1. In: *Glyphosate Resistance in Crops and Weeds: History, Development, and Management*, Nandula, V.K. (Ed.), pp. 1-33, Wiley, ISBN 978-0470410318, Hoboken, NJ, USA.

Du, W.; Wallis, N.G. & Payne, D.J. (2000). The kinetic mechanism of 5-enolpyruvylshikimate-3-phosphate synthase from a Gram-positive pathogen *Streptococcus pneumoniae*. *Journal of Enzyme Inhibition and Medicinal Chemistry*, 15, 571-581.

Duke, S.O. & Hoagland, R.E. (1981). Effects of glyphosate on the metabolism of phenolic compounds: VII. Root fed amino-acids and glyphosate toxicity in soybean (*Glycine max*) cultivar hill seedlings. *Weed Science*, 29, 297-302.

Duke, S.O.; Hoagland, R.E. & Elmore, C.D. (1979). Effects of glyphosate on metabolism of phenolic compounds. IV. Phenylalanine ammonia-lyase activity, free amino acids, and soluble hydroxyphenolic compounds in axes of light-grown soybeans. *Physiologia Plantarum*, 46, 307-317.

Duke, S.O. & Powles, S.B. (2008). Glyphosate: a once-in-a-century herbicide. *Pest Management Science*, 64, 319-325.

Edge, C.B.; Gahl, M.K.; Pauli, B.D.; Thompson, D.G. & Houlahan, J.E. (2011) Exposure of juvenile green frogs (*Lithobates clamitans*) in littoral enclosures to a glyphosate-based herbicide. *Ecotoxicology andEnvironmental Safety*, 74, 1363-1369.

Edwards, W.M.; Triplett, G.B. & Kramer, R.M. (1980). A watershed study of glyphosate transport in runoff. *Journal of Environmental Quality*, 9, 661-665.

Egley, G.H. & Williams, R.D. (1978). Glyphosate and paraquat effects on weed seed germination and seedling emergence. *Weed Science*, 26, 249-251.

European Commission 2010. Commission Directive 2010/77/EU of 10 November 2010 amending Council Directive 91/414/EEC as regards the expiry dates for inclusion in Annex I of certain active substances. *Official Journal of the European Union*, L 293, 11.11.2010.

FAO/WHO (2006). Pesticide residues in food – 2005. Evaluations. Part I – Residues. FAO Plant Production and Protection Paper 184/1, pp. 303-500, World Health Organization / Food and Agriculture Organization of the United Nations, ISBN 978-9251054871, Rome, Italy.

Feng, J.C.; Thompson, D.G. & Reynolds, P.E. (1990). Fate of glyphosate in a Canadian forest watershed. 1. Aquatic residues and off target deposit assessment. *Journal of Agricultural and Food Chemistry*, 38, 1110-1118

Fernandez, C.H. & Bayer, D.E. (1977). Penetration, transloca tion, and toxicity of glyphosate in bermudagrass (*Cynodon dactyloni*). *Weed Science*, 25, 396-400.

Folmar, L.C.; Sanders, J.O. & Julin A.M. (1979). Toxicity of the herbicide glyphosate and several of its formulations to fish and aquatic invertebrates. *Archives of Environmental Contamination and Toxicology*, 8, 269-278.

Forlani, G.; Pavan, M.; Gramek, M.; Kafarski, P. & Lipok, J. (2008). Biochemical bases for a widespread tolerance of Cyanobacteria to the phosphonate herbicide Glyphosate. *Plant & Cell Physiology*, 49, 443-456.

Franz, J.E.; Mao, M.K. & Sikorski, J.A. (1997). Glyphosate: A unique global herbicide. ACS Monograph 189. American Chemical Society, ISBN 978-0841234581, Washington, DC, USA.

Fletcher, R.A.; Hildebrand, P. & Akey, W. (1980). Effect of glyphosate on membrane permeability in red beet (*Beta vulgaris*) root tissue. *Weed Science*, 28, 671-673.

Funke, T.; Han, H.; Healy-Fried, M.L.; Fischer, M. & Schönbrunn, E. (2006). Molecular basis for the herbicide resistance of Roundup Ready crops. *Proceedings of the National Academy of Sciences of the United States of America*, 103, 13010-13015.

Gasnier, C.; Dumont, C.; Benachour, N.; Clair, E.; Chagnon, M.-C. & Séralini, G.-E. (2009). Glyphosate-based herbicides are toxic and endocrine distruptors in human cell lines. *Toxicology*, 262, 184-191.

George, J.; Prasad, S.; Mahmood, Z. & Shukla, Y. (2010). Studies on glyphosate-induced carcinogenicity in mouse skin: a proteomic approach. *Journal of Proteomics*, 73, 951-964.

Giesy, J.P.; Dobson, S. & Solomon, K.R. (2000). Ecotoxicological risk assessment for Roundup herbicide. *Reviews of Environmental Contamination and Toxicology*, 167, 35-120.

Gimsing, A.L.; Borggaard, O.K. & Bang, M. (2004a). Influence of soil composition on adsorption of glyphosate and phosphate by contrasting Danish surface soils. *European Journal of Soil Science*, 55, 183–191.

Gimsing, A.L.; Borggaard, O.K.; Jacobsen, O.S.; Aamand, J. & Sørensen, J. (2004b). Chemical and microbial soil characteristics controlling glyphosate mineralisation in Danish surface soils. *Applied Soil Ecology*, 27, 233-242.

Gimsing, A.L. & dos Santos, A.M. (2005). Glyphosate. In: *Biogeochemistry of Chelating Agents*. Chapter 16. ACS Symposium Series, Vol. 910, Nowack B, VanBriesen JM, Eds, pp. 263-277, American Chemical Society, ISBN 978-0841238978, Washington, DC, USA.

Gimsing, A.L.; Szilas, C. & Borggaard, O.K. (2007). Sorption of glyphosate and phosphate by variable-charge tropical soils from Tanzania. *Geoderma*, 138, 127-132.

Goldsborough, L.G. & Brown D.J. (1988). Effect of Glyphosate (Roundup® formulation) on periphytic algal phosotosynthesis. *Bulletin of Environmental Contamination and Toxicology*, 41, 253-260.

González-Martínez, M.A.; Brun, E.M.; Puchades, R.; Maquieira, A.; Ramsey, K. & Rubio, F. (2005). Glyphosate immunosensor application for water and soil analysis. *Analytical Chemistry*, 77, 4219-4227.

Goodwin, L.; Startin, J.R.; Keely, B.J. & Goodall, D.M. (2003). Analysis of glyphosate and glufosinate by capillary electrophoresis-mass spectrometry utilising a sheathless microelectrospray interface. *Journal of Chromatography A*, 1004, 107-119.

Gothrup, O.; O'Sullivan, P.A.; Schraa, R.J. & Vanden W.H. (1976). Uptake, translocation, metabolism and selectivity of glyphosate in Canada thistle and leafy spurge. *Weed Research*, 16, 197-201.

Granby, K.; Johannesen, S. & Vahl, M. (2003). Analysis of glyphosate residues in cereals using liquid chromatography-mass spectrometry (LC-MS/MS). *Food Additives and Contaminants*, 20, 692-698.

Grey, L.; Nguyen, B. & Yang, P. (2001). Liquid chromatography/electrospray ionization/isotopic dilution mass spectrometry analysis of N-(phosphonomethyl) glycine and mass spectrometry analysis of aminomethyl phosphonic acid in environmental water and vegetation matrixes. *Journal of the Association of Official Agricultural Chemists International*, 84, 1770-1780.

Haderlie, L.C.; Slife, F.W. & Butler, H.S. (1978). ^{14}C-glyphosate absorption and translocation in germinating maize (*Zea mays*) and (*Glycine max*) seeds and in soybean plants *Weed Research*, 18, 269-273.

Hammock, B.D.; Székács, A.; Hanzlik, T.; Maeda, S.; Philpott, M.; Bonning, B. & Posse, R. (1989). Use of transition state theory in the design of chemical and molecular agents for insect control. In: *Recent Advances in the Chemistry of Insect Control*, Crombie, L. (Ed.), pp. 256-277; Royal Society of Chemistry, ISBN 9024736684, Cambridge, UK.

Hammock, B.D.; Gee, S.J.; Harrison, R.O.; Jung, F.; Goodrow, M.; Li, Q.-X..; Lucas, A.D.; Székács, A. & Sundaram, K.M.S. (1991). Immunochemical technology in environmental analysis: Addressing critical problems. In: *Immunochemical Methods for Environmental Analysis*, Van Emon, J. and Mumma, R.O. (Eds.), *ACS Symp. Ser.*, Vol 442, pp. 112-139, American Chemical Society, ISBN 978-0851866277, Washington, DC, USA.

Hanke, A.; Singer, H. & Hollender, J. (2008). Ultratrace-level determination of glyphosate aminomethylphosphonic acid and *glufosinate* in natural waters by solid-phase extraction followed by liquid chromatography – tandem mass spectrometry: performance tuning of derivatization, enrichment and detection. *Analytical and Bioanalytical Chemistry*, 391, 2265-2276.

Hayes, T.N.; Collins, A.; Lee, M.; Mendoza, M.; Noriega, N.; Stuart, A.A. & Vonk, A. (2002). Hermaphroditic, demasculinized frogs after exposure to the herbicide *atrazine* at low ecologically relevant doses. *Proceedings of the National Academy of Sciences of the United States of America*, 99, 5476-5480.

Hayes, T.B.; Khoury, V.; Narayan, A.; Nazir, M.; Park, A.; Brown, T.; Adame, K.; Chan, E.; Buchholz, D.; Stueve, T. & Gallipeau, S. (2010). Atrazine induces complete feminization and chemical castration in male African clawed frogs (*Xenopus laevis*). *Proceedings of the National Academy of Sciences of the United States of America*, 107, 4612-4617.

Heap, I. (Epubl.). International survey of herbicide resistant weeds. Herbicide Resistance Action Committee, North American Herbicide Resistance Action Committee, and Weed Science Society of America, Available from http://www.weedscience.org/Summary/UspeciesMOA.asp?lstMOAID=12

Herman, K.M. & Weaver L.M. (1999). The shikimate pathway. *Annual Review of Plant Physiology and Plant Molecular Biology*, 50, 473-503.

Hoagland, R.E. & Duke, S.E. (1982). Biochemical effects of glyphosate. In: *Biochemical Responses Induced by Herbicides*; Moreland, D.E.; St. John, J.B.; Hess, F.D. (Eds.), *ACS Symposium Series* 181; pp. 175-205, American Chemical Society, ISBN 9780841206991, Washington, DC, USA.

Howe, C.M.; Berrill, M.; Pauli, B.D.; Helbing, C.C.; Werry, K. & Veldhoen, N. (2004). Toxicity of glyphosate-based pesticides to four North American frog species. *Environmental Toxicology and Chemistry*, 23, 1928-1938.

Hsu, C.C. & Whang, C.W. (2009). Microscale solid phase extraction of glyphosate and aminomethylphosphonic acid in water and guava fruit extract using alumina-coated iron oxide nanoparticles followed by capillary electrophoresis and electrochemiluminescence detection. *Journal of Chromatography A*, 1216, 8575-8580.

Hudzin, Z.H.; Gralak, D.K.; Drabowicz, J. & Luczak, J. (2002). Novel approach for the simultaneous analysis of glyphosate and its metabolites. *Journal of Chromatography A*, 947, 129-141.

Ibáñez, M.; Pozo, O.J.; Sancho, J.V.; López, F.J. & Hernández, F. (2006). Re-evaluation of glyphosate determination in water by liquid chromatography coupled to electrospray tandem mass spectrometry. *Journal of Chromatography A*, 1134, 51-55.

Jacob, G.S.; Garbow, J.R.; Hallas, L.E.; Kimack, N.M.; Kishore, G.M. & Schaefer, J. (1988). Metabolism of glyphosate in *Pseudomonas* sp. strain LBr. *Applied and Environmental Microbiology*, 54, 2953-2958.

Jayawardene, U.A.; Rajakaruna, R.S.; Navaratne, A.N. & Amerrasinghe, P.H. (2010). Toxicity of agrochemicals to common hourglass tree frog (*Polypedates crugiger*) in acute and chronic exposure. *International Journal of Agriculture and Biology*, 12, 641-648.

Jaworski, E.G. (1972). Mode of action of N-phosphonomethylglycine: inhibition of aromatic amino acid biosynthesis. *Journal of Agricultural and Food Chemistry*, 20, 1195-1198.

Johal, C.S. & Huber, D.M. (2009). Glyphosate effects on diseases of plants. *European Journal of Agronomy*, 31, 144-152.

Jones, D.K.; Hammond, J.I. & Relyea, R.A. (2010). Competitive stress can make the herbicide Roundup more deadly to larval amphibians. *Environmental Toxicology and Chemistry*, 30, 446-454.

Kale, P.G.; Petty Jr. B.T.; Walker, S.; Ford, J.B.; Dehkordi, N.; Tarasia, S.; Tasie, B.O.; Kale, R. & Sohni, Y.R. (1995). Mutagenicity testing of nine herbicides and pesticides currently used in agriculture. *Environmental and Molecular Mutagenesis*, 25, 148-153.

Kantiani, L.; Sanchis, J.A.; Llorca, M.; Rubio, F.; Farré, M. & Barceló, D. (2011). Monitoring of glyphosate residues in environmental groundwater samples by ELISA and LC-MS/MS. In: *Abs. 21st SETAC Europe Meeting, Ecosystem Protection in a Sustainable World.* (Milan, Italy, May 16-19, 2011), p. 157.

Kishore, G.M. & Shah, D.M. (1988). Amino acid biosynthesis inhibitors as herbicides. *Annual Review of Biochemistry*, 57, 627-663.

Kitchen, L.M.; Witt, W.W. & Rieck, C.E. (1981). Inhibition of chlorophyll accumulation by glyphosate. *Weed Science*, 29, 513-516.

Kjær, J.; Olsen, P.; Ullum, M. & Grant, R. (2005). Leaching of glyphosate and amino-methylphosphonic acid from Danish agricultural field sites. *Journal of Environmental Quality*, 34, 608-620.

Klinger, J.; Lang, M.; Sacher, F.; Brauch, H.J.; Maier, D. & Worch, E. (1998). Formation of glyphosate and AMPA during ozonation of waters containing ethylenediaminetetra(methylenephosphonic acid). *Ozone Science and Engineering*, 20, 99-110.

Knuuttila, P. & Knuuttila, H. (1979). Crystal and molecular-structure of N-(phosphonomethyl)-glycine (glyphosate). *Acta Chemica Scandinavica*, 33, 623-626.

Kodama, S.; Ito, Y.; Taga, A.; Nomura, Y.; Yamamoto, A.; Chinaka, S.; Suzuki, K.; Yamashita, T.; Kemmei, T. & Hayakawa K. (2008). A fast and simple analysis of glyphosate in tea beverages by capillary electrophoresis with on-line copper(II)-glyphosate complex formation. *Journal of Health Science*, 54, 602-606.

Kolpin,, D.W.; Thurman, E.M.; Lee, E.A.; Meyer, M.T.; Furlong, E.T. & Glassmeyer, S.T. (2006). Urban contributions of glyphosate and its degradate AMPA to streams in the United States. *Science of the Total Environment*, 354, 191-197.

Kremer, R.J. & Means, N.E. (2009). Glyphosate and glyphosate-resistant crop interactions with rhizosphere microorganisms. *European Journal of Agronomy*, 31, 153-161.

Kremer, R.J.; Means, N.E. & Kim, S. (2005). Glyphosate affects soybean root exudation and rhizosphere microorganisms. *International Journal of Environmental Analytical Chemistry*, 85, 1165-1174.

Krzysko-Lupicka, T. & Sudol, T. (2008). Interactions between glyphosate and autochthonous soil fungi surviving in aqueous solution of glyphosate. *Chemosphere*, 71, 1386-1391.

Laitinen, P.; Siimes, K.; Eronen, L.; Rämö, S.;Welling, L.; Oinonen, S.; Mattsoff, L. & Ruohonen-Lehto, M. (2006). Fate of the herbicide glyphosate, glufosinate-ammonium, phenmedipham, ethofumesate and metamitron in two Finnish arable soils. *Pest Management Science*, 62, 473-491.

Laitinen, P.; Siimes, K.; Rämö, S.; Jauhiainen, L.; Eronen, L.; Oinonen, S. & Hartikainen, H. (2008). Effects of soil phosphorous status on environmental risk assessment of glyphosate and glufosinate-ammonium. *Journal of Environmental Quality*, 37, 830-838.

Laitinen, P.; Rämö, S.; Nikunen, U.; Jauhiainen, L.; Siimes, K. & Turtola, E. (2009). Glyphosate and phosphorous leaching and residues in boreal sandy soil. *Plant and Soil*, 323, 267-283.

Lajmanovich, R.C.; Sandoval, M.T. & Peltzer, P.M. (2005). Induction of mortality and malformation in *Scinax nasicus* tadpoles exposed to glyphosate formulations. *Bulletin of Environmental Contamination and Toxicology*, 70, 612-618.

Lajmanovich, R.C.; Attademo, A.M.; Peltzer, P.M.; Junges, C.M. & Cabana, M.C. (2010). Toxicity of four herbicide formulations with glyphosate on *Rhinella arenarum* (Anura: Bufonidae) tadpoles: B-esterases and glutation-*S*-transferase inhibitors. *Archives in Environmental Contamination and Toxicology*, 60, 681-689.

Landry, D.; Dousset, S.; Fournier, J.-C. & Andreux, F. (2005). Leaching of glyphosate and AMPA under two soil management practices in Burgundy vineyards (Vosne-Romanée, 21-France). *Environmental Pollution*, 138, 191-200.

Langiano, V.C. & Martinez, C.B.R. (2008). Toxicity and effects of a glyphosate-based herbicide on the neotropical fish *Prochilodus lineatus*. *Comparative Biochemistry and Physiology Part C Toxicology & Pharmacology*, 147, 222-231.

Lee, L.J. & Ngim, J. (2000). A first report of glyphosate-resistant goosegrass (*Elusine indica* (L) Gaertn) in Malaysia. *Pest Management Science*, 56, 336-339.

Lee, E.A.; Strahan, A.P. & Thurman, E.M. (2002a). Methods of analysis by the U.S. geological survey organic geochemistry research group-determination of glyphosate, aminomethylphosphonic acid, and *glufosinate* in water using online solid-phase extraction and high-performance liquid chromatography/mass spectrometry.

Open-File Report 01-454. Lawrence, KS, USA, Available from
http://ks.water.usgs.gov/pubs/abstracts/ofr.01-454.abs.html

Lee, E.A.; Zimmerman, L.R.; Bhullar, B.S. & Thurman, E.M. (2002b). Linker-assisted immunoassay and liquid chromatography/mass spectrometry for the analysis of glyphosate. *Analytical Chemistry,* 74, 4937-4943.

Lioi, M.B.; Scarfi, M.R.; Santoro, A.; Barbieri, R.; Zeni1, O.; Salvemini, F.; Di Berardino, D. & Ursini, M.V. (1998). Cytogenetic damage and induction of pro-oxidant state in human lymphocytes exposed *in vitro* to glyphosate, vinclozolin, atrazine, and DPX-E9636. *Environmental and Molecular Mutagenesis,* 32, 39-46.

López-Rodas, V.; Flores-Moya, A.; Maneiro, E.; Perdigones, N.; Marva, F.; Marta, G.E. & Costas, E. (2007). Resistance to glyphosate in the cyanobacterium *Microcystis aeruginosa* as result of pre-selective mutations. *Evolutionary Ecology,* 21, 535-547.

Love, B.J.; Einheuser, M.D. & Nejadhashemi, A.P. (2011). Effects on aquatic and human health due to large scale bioenergy crop expansion. *Science of the Total Environment,* 409, 3215-3229.

Ludvigsen, G.H. & Lode, O. (2001a). "JOVA" – The agricultural environmental pesticides monitoring programme in Norway. In: *Proceeding of the 6th International HCH and Pesticides Forum in Poznan.* 20-22 March. Vijgen, J.; Pruszynski, S. and Stobiecki, S. Eds, pp. 199-206, ISBN 83-913860-7-4,

Ludvigsen, G.H. & Lode, O. (2001b). Results from the agricultural and environmental monitoring program of pesticides in Norway 1995 – 1999. *Fresenius Environmental Bulletin,* 10, 470-474.

Mallat, E. & Barceló, D. (1998). Analysis and degradation study of glyphosate and of aminomethylphosphonic acid in natural waters by means of polymeric and ion-exchange solid-phase extraction columns followed by ion chromatography-post-column derivatization with fluorescence detection. *Journal of Chromatography A,* 823, 129-136.

Ma, J. (2002). Differential sensitivity to 30 herbicides among populations of two green algae *Scenedesmus obliquus* and *Chlorella pyrenoidosa.* *Bulletin of Environmental Contamination and Toxicology,* 68, 275-281.

Ma, J.; Lin, F.; Wang, S. & Xu, L. (2003). Toxicity of 21 herbicides to the green alga *Scenedesmus quadricauda.* *Bulletin of Environmental Contamination and Toxicology* 71, 594-601.

Ma, J.; Xu, L.; Wang, S.; Zheng, R.; Jin, S.; Huang, S. et al. (2002). Toxicity of 40 herbicides to the green alga *Chlorella vulgaris.* *Ecotoxicology and Environmental Safety,* 132, 128-132.

Ma, J. & Wang, S. (2002). A quick, simple, and accurate method of screening herbicide activity using green algae cell suspension cultures. *Weed Science Society of America,* 50, 555-559.

Mamy, L.; Barriuso, E. & Gabrielle, B. (2005). Environmental fate of herbicides trifluralin, metazachlor, metamitron and sulcotrione compared with that of glyphosate, a substitute broad spectrum herbicide for different glyphosate-resistant crops. *Pest Managment Science,* 61, 905-916.

Mann, R.M. & Bidwell, J.R. (1999). The toxicity of glyphosate and several glyphosate formulations to four species of Southwestern Australian frogs. *Archives in Environmental Contamination and Toxicology,* 36, 193-199.

Marc, J.; Mulner-Lorillon, O. & Bellé, R. (2004a). Glyphosate-based pesticides affects cell cycle regulation. *Biology of the Cell,* 96, 245-249.

Marc, J.; Bellé, R.; Morales, J.; Cormier, P. & Mulner-Lorillon, O. (2004b). Formulated glyphosate activities the DNA-response checkpoint of the cell cycle leading to the prevention of G2/M transition. *Toxicological Sciences*, 82, 436-442.

Marc, J.; Le Breton, M.; Cormier, P.; Morales, J.; Bellé, R. & Mulner-Lorillon, O. (2005). A glyphosate-based pesticide impinges on transcription. *Toxicology and Applied Pharmacology*, 203, 1-8.

Mariana, A.; de Alaniz, M. J.T. & Marra, C.A. (2009). The impact of simultaneous intoxication with agrochemicals on the antioxidant defense system in rat. *Pesticide Biochemistry and Physiology*, 94, 93-99.

Marriage, P.B. & Khan, S.U. (1978). Differential varietal tol- erance of peach (*Prunus persica*) seedlings to glyphosate. *Weed Science*, 26, 374-378.

Martins-Júnior, H.A.; Lebre, D.T.; Wang, A.Y.; Pires, M.A.F. & Bustillos, O.V. (2011). Residue analysis of glyphosate and aminomethylphosphonic acid (AMPA) in soybean using liquid chromatography coupled with tandem mass spectrometry. In: Soybean - Biochemistry, Chemistry and Physiology. Ng, T.-B. (Ed.), InTech, ISBN: 978-953-307-219-7, Rijeka, Croatia, Available from
http://www.intechopen.com/articles/show/title/residue-analysis-of-glyphosate-and-aminomethylphosphonic-acid-ampa-in-soybean-using-liquid-chromatog

Marzabadi, M.R.; Font, J.L.; Gruys, K.J.; Pansegrau, P.D. & Sikorski, J.A. (1992). Design & synthesis of a novel EPSP synthase inhibitor based on its ternary complex with shikimate-3-phosphate and glyphosate. *Bioorganic & Medicinal Chemistry Letters*, 2, 1435-1440.

Mesnage, R.; Clair, E.; Gress, S.; Then, C.; Székács, A. & Séralini, G.-E. (2011). Cytotoxicity on human cells of Cry1Ab and Cry1Ac Bt insecticidal toxins alone or with a glyphosate-based herbicide. *Journal of Applied Toxicology*, in press.

McComb, B.C.; Curtis, L.; Chambers, C.L.; Newton, M. & Bentson, K. (2008). Acute toxic hazard evaluations of glyphosate herbicide on terrestrial vertebrates of the Oregon Coast Range. *Environmental Science and Pollution Research*, 15, 266-272.

Mörtl, M.; Maloschik, E.; Juracsek, J. & Székács, A. (2010). Növényvédőszer-maradékok gázkromatográfiás és immunanalitikai meghatározásának eredményei vizekben és talajokban. In. *Komplex monitoring rendszer összeállítása talaj-mikroszennyezők analitikai kimutatására és biológiai értékelésére a fenntartható környezetért. MONTABIO-füzetek IV.* 30-37. old. MTA Növényvédelmi Kutatóintézet, ISBN 978-963-87178-7-0, Budapest, Hungary.

Mulet, J.M. (2011). Letter to the editor regarding the article by Paganelli et al. *Chemical Research in Toxicology*, 24, 609.

Nandula, V.K.; Reddy, K.N.; Rimando, A.M.; Duke, S.O. & Poston, D.H. (2007). Glyphosate-resistant and -susceptible soybean (*Glycine max*) and canola (*Brassica napus*) dose response and metabolism relationships with glyphosate. *Journal of Agricultural and Food Chemistry*, 55, 3540-3545.

Natangelo, M.; Benfenati, E.; De Gregorio, G.; Fanelli, R. & Ciotti, G. (1993). GC-MS analysis of N-phosphonomethylglycine (glyphosate) samples through derivatization with a perfluoroanhydride and trifluoroethanol: Identification of by-products. *Toxicological and Environmental Chemistry*, 38, 225-232.

Nedelkoska, T.V. & Low, G.K.-C. (2004). High-performance liquid chromatographic determination of glyphosate in water and plant material after pre-column derivatisation with 9-fluorenylmethyl chloroformate. *Analytica Chimica Acta*, 511, 145-153.

Nilsson, G. (1977). Effects of glyphosate on the amino acid content in spring wheat plants. *Swedish Journal of Agricultural Research*, 7, 153-157.

Nowack B (2003). Environmental chemistry of phosphonates. *Water Research*, 37, 2533-2546.

Olorunsogo, O.O. (1990). Modification of the transport of protons and Ca^{2+} ions across mitochondrial coupling membrane by N-(phosphonomethyl)glycine. *Toxicology*, 61, 205-209.

Olorunsogo, O.O. & Bababunmi, E.A. (1980). Interference of herbicides with mitochondrial oxidative-phosphorylation - the N-(phosphonomethyl)glycine model. *Toxicology Letters*, 5(Sp 1), 148.

Olorunsogo, O.O.; Bababunmi, E.A. & Bassir, O. (1979). Uncoupling of corn shoot mitochondria by. N-(phosphonomethyl)glycine. *FEBS Letters*, 97, 279-282.

Palma, G. (2011). Letter to the editor regarding the article by Paganelli et al. *Chemical Research in Toxicology*, 24, 775-776.

Paganelli, A.; Gnazzo, V.; Acosta, H.; López, S.L. & Carrasco, A.E. (2010). Glyphosate-based herbicides produce teratogenic effects on vertebrates by impairing retionic acid signaling. *Chemical Research in Toxicology*, 23, 1586-1595.

Parker, R.G.; York, A.C.; Jordan DL (2005) Comparison of glyphosate products in glyphosate-resistant cotton (*Gossypium hirsutum*) and corn (*Zea mays*). *Weed Technology*, 19, 796-802.

Patsias, J.; Papadopoulou A. & Papadopoulou-Mourkidou, E. (2001). Automated trace level determination of glyphosate and aminomethylphosphonic acid in water by on-line anion-exchange solid-phase extraction followed by cation-exchange liquid chromatography and post-column derivatization. *Journal of Chromatography A*, 932, 83-90.

Pauling, L. (1948). Chemical achievement and hope for the future. *American Scietist*, 36, 51-58.

Payne, N.J.; Feng, J.C. & Reynolds, P.E. (1990). Off-target deposits and buffer zones required around water for aerial glyphosate applications. *Pesticide Science*, 30, 183-198.

Payne, N.J. (1992). Off-target glyphosate from aerial silvicultural applications and buffer zones required around sensitive areas. *Pesticide Science*, 34, 1-8.

Penaloza-Vazquez, A.; Mena, G.L.; Herrera-Estrella, L. & Bailey, A.M. (1995). Cloning and sequencing of the genes involved in glyphosate utilization by *Pseudomonas pseudomallei*. *Applied and Environmental Microbiology*, 61, 538-543.

Perez, A. & Kogan, M. (2003): Glyphosate-resistant *Lolium multiflorum* in Chilean orchards. *Weed Research*, 43, 12-19.

Pérez, G.L.; Torremorell, A.; Mugni, H.; Rodríguez, P.; Vera, M.S.; Do Nascimento, M.; Allende, L.; Bustingorry, J.; Escaray, R.; Ferraro, M.; Izaguirre, I.; Pizarro, H.; Bonetto, C.; Morris, D. P. & Zagarese, H. (2007). Effects of the herbicide Roundup on freshwater microbial communities: a mesocosm study. *Ecological Applications*, 17, 2310-2322.

Pérez, G.L.; Vera, M.S. & Miranda, L. (2011). Effects of herbicide glyphosate and glyphosate-based formulations on aquatic ecosystems, herbicides and environment. Kortekamp, A. (Ed.), InTech, ISBN: 978-953-307-476-4, Rijeka, Croatia, Available from http://www.intechopen.com/articles/show/title/effects-of-herbicide-glyphosate-and-glyphosate-based-formulations-on-aquatic-ecosystems

Pesce, S.; Batisson, I.; Bardot, C.; Fajon, C.; Portelli, C.; Montuelle, B. & Bohatier, J. (2009a). Response of spring and summer riverine microbial communities following glyphosate exposure. *Ecotoxicology and Environmental Safety*, 72, 1905-1912.

Perkins, P.J.; Boermans, H.J. & Stephenson, G.R. (2000). Toxicity of glyphosate and *triclopyr* using the frog embryo teratogenesis assay – *Xenopus. Environmental Toxicology and Chemistry*, 19, 940-945.

Peruzzo, P.J.; Porta, A.A. & Ronco, A.E. (2008). Levels of glyphosate in surface waters, sediments and soils associated with direct sowing soybean cultivation in north pampasic region of Argentina. *Environmental Pollution*, 156, 61-66.

Pipke, R.; Schulz, A. & Amrhein, N. (1987). Uptake of glyphosate by an *Arthrobacter* sp. *Applied and Environmental Microbiology*, 53, 974-978.

Poletta, G.L.; Larriera, A.; Kleinsorge, E. & Mudry, M.D. (2009). Genotoxicity of the herbicide formulation Roundup (glyphosate) in broad-snouted caiman (*Caiman latirostris*) evidenced by the comet assay and the micronucleus test. *Mutation Research*, 672, 95-102.

Popp, M.; Hann, S.; Mentler, A.; Fuerhacker, M.; Stingeder, G. & Koellensperger, G. (2008). Determination of glyphosate and AMPA in surface and waste water using high-performance ion chromatography coupled to inductively coupled plasma dynamic reaction cell mass spectrometry (HPIC-ICP-DRC-MS). *Analytical and Bioanalytical Chemistry*, 391, 695-699.

Powell, H.A. ; Kerbby, N.W. & Rowell, P. (1991). Natural tolerance of cyanobacteria to the herbicide glyphosate. *New Phytologist*, 119, 421-426.

Powell, J.R. & Swanton, C.J. (2008). A critique of studies evaluating glyphosate effects on diseases associated with *Fusarium* spp. *Weed Research*, 48, 307-318.

Powles, S.B. (2008). Evolved glyphosate-resistant weeds around the world: lesson to be learnt. *Pest Management Science*, 64, 360-365.

Powles, S.B.; Lorraine-Colwill, D.F.; Dellow, J.J. & Preston, C. (1998). Evolved resistance to glyphosate in rigid ryegrass (*Lolium rigidum*) in Australia. *Weed Science*, 46, 604-607.

Pratley, J.; Baines, P.; Eberbach, R.; Incerti, M. & Broster, J. (1996). Glyphosate resistant annual ryegrass. In: *Proceedings of the 11th AnnualConference of the Grassland Society of New South Wales*. Virgona, J. and Michalk, D. (Eds), p. 126, Wagga Wagga, Australia.

Prasad, S.; Srivastava, S.; Singh, M. & Shukla, Y. (2009). Clastogenic effects of glyphosate in bone marrow cells of swiss albino mice. *Journal of Toxicology*, Epubl, Available from http://www.hindawi.com/journals/jt/2009/308985

Price, A.J.; Balkcom, K.S.; Culpepper, S.A.; Kelton, J.A.; Nichols, R.L. & Schomberg, H. (2011). Glyphosate-resistant Palmer amaranth: A threat to conservation tillage. *Journal of Soil and Water Conservation*, 66, 265-275.

Putnam, A.R. (1976). Fate of glyphosate in deciduous fruit trees. *Weed Science*, 24, 425-430.

Rank, J.; Jensen, A.-G.; Skov, B.; Pedersen, L.H. & Jensen, K. (1993) Genotoxicity testing of the herbicide Roundup and its active ingredient glyphosate isopropylamine using the mouse bone marrow micronucleus test, *Salmonella* mutagenicity test, and *Allium* anaphase-telophase test. *Mutation Research/Genetic Toxicology*, 300, 29-36.

Relyea, R.A. (2005a). The lethal impacts of Roundup and predatory stress on six species of North American tadpoles. *Archives in Environmental Contamination and Toxicology*, 48, 351-357.

Relyea, R.A, (2005b). The lethal impact of Roundup on aquatic and terrestrial amphibians. *Ecological Applications*, 15, 1118-1124.

Relyea, R.A. & Jones, D.K. (2009). The toxicity of Roundup Original Max to 13 species of larval amphibians. *Environmental Toxicology and Chemistry*, 28, 2004-2008.

Reuben, S.H. (2010). Reducing environmental cancer risk: What we can do now. U.S. Department of Health and Human Services, National Institutes of Health, National Cancer Institute, ISBN 9781437934212, Washington, DC, USA, Available from http://deainfo.nci.nih.gov/advisory/pcp/annualReports/pcp08-09rpt/PCP_Report_08-09_508.pdf

Riar, D.S.; Norsworthy, J.K.; Johnson, D.B.; Scott, R.C. & Bagavathiannan, M. (2011) Glyphosate resistance in a Johnsongrass (*Sorghum halepense*) biotype from Arkansas. *Weed Science*, 59, 299-304.

Richard, S.; Moslemi, S.; Sipahutar, H.; Benachour, N. & Séralini, G.-E. (2005). Differential effects of glyphosate and Roundup on human placental cells and aromatase. *Environmental Health Perspectives*, 113, 716-720.

Riechers, D.E.; Wax, L.M.; Liebl, R.A. & Bullock, D.G. (1995). Surfactant effects on glyphosate efficacy. *Weed Technology*, 9, 281-285

Roisch, V. & Lingens, F. (1980). The mechanism of action of the herbicide N-(phosphonomethyl)glycine - its effect on the growth and the enzymes of aromatic amino-acid biosynthesis in *Escherichia coli*. *Hoppe-Seyler's Zeitschrift für Physiologische Chemie*, 361, 1049-1058.

Royer, A.; Beguin, S.; Tabet, J.C.; Hulot, S.; Reding, M.-A. & Communal, P.-Y. (2000). Determination of glyphosate and aminomethylphosphonic acid residues in water by gas chromatography with tandem mass spectrometry after exchange ion resin purification and derivatisation. Application on vegetable matrices. *Analytical Chemistry*, 72, 3826-3832.

Rubio, F.; Veldhuis, L.J.; Clegg, B.S.; Fleeker, J.R. & Hall, J.C. (2003). Comparison of a direct ELISA and an HPLC method for glyphosate determination in water. *Journal of Agricultural and Food Chemistry*, 51, 691-696.

Rueppel, M.L.; Brightwell, B.B.; Schaefer, J. & Marvel, J.T. (1977). Metabolism and degradation of glyphosate in soil and water. *Journal of Agricultural and Food Chemistry*, 25, 517-528.

Rueppel, M.L.; Suba, L.A. & Marvel, J.T. (1976). Derivatization of aminoalkylphosphonic acids for characterization by gas-chromatography mass-spectrometry. *Biomedical Mass Spectrometry*, 3, 28-31.

Sáenz, M.E. ; Di Marzio, W.D. ; Alberdi, J.L. & del Carmen Tortorelli, M. (1997). Effects of technical grade and a commercial formulation of glyphosate on algal population growth. *Bulletin of Environmental Contamination and Toxicology*, 59, 638-644.

Saltmiras, D.; Bus, J.S.; Spanogle, T.; Hauswirth, J.; Tobia, A. & Hill, S. (2011). Letter to the editor regarding the article by Paganelli et al. *Chemical Research in Toxicology*, 24, 607-608.

Samland, A.K.; Amrhein, N. & Macheroux, P. (1999). Lysine 22 in UDP-N-acetylglucosamine enolpyruvyl transferase from *Enterobacter cloacae* is crucial for enzymatic activity and the formation of covalent adducts with the substrate phosphoenolpyruvate and the antibiotic fosfomycin. *Biochemistry*, 38, 13162-13169.

Sammons, R.D.; Gruys, K.J.; Anderson, K.S.; Johnson, K.A. & Sikorski, J.A. (1995). Reevaluating glyphosate as a transition-state inhibitor of EPSP synthase: identification of an EPSP synthase.EPSP.glyphosate ternary complex. *Biochemistry*, 34, 6433-6440.

Sancho, J.V.; Hidalgo, C.; Hernández, F.; López, F.J.; Hogendoorn, E.A. Dijkman, E. (1996) Rapid determination of glyphosate residues and its main metabolite AMPA in soil samples by liquid chromatography. *International Journal of Environmental Analytical Chemistry*,. 62, 53-.

Sandberg, C.L.; Meggitt, W.F. & Penner, D. (1980). Absorption, translocation and metabolism of MC-glyphosate in several weed species. *Weed Research*, 20, 195-200.

Sanogo, S.; Yang, X. B. & Scherm, H. (2000). Effects of herbicides on *Fusarium solani* f. sp. *glycines* and development of sudden death syndrome in glyphosate-tolerant soybean. *Disease Control and Pest Management*, 90, 57-66.

Santillo, D.J.; Leslie, D.M. & Brown, P.W. (1989). Responses of small mammals and habitat to glyphosate applications on clearcuts. *Journal of Wildlife Management*, 53, 164-172.

Savitz, D.A.; Arbuckle, T.; Kaczor, D. & Curtis, K.M. (1997). Male pesticide exposure and pregnancy outcome. *American Journal of Epidemiology*, 146, 1025-1035.

Schloss, J.V. & Aulabaugh, A. (1990). Acetolactate synthase and ketol-acid reductoisomerase: targets for herbicides obtained by screening and *de novo* design. *Zeitschrift für Naturforschung*, 45c, 544-551.

Schönbrunn, E.; Eschenburg, S.; Shuttleworth, V.A.; Schloss, J.V.; Amrheini, N.; Evans, J.N.S. & Kabsch, W. (2001). Interaction of the herbicide glyphosate with its target enzyme 5-enolpyruvylshikimate 3-phosphate synthase in atomic detail. *Proceedings of the National Academy of Sciences of the United States of America*, 98, 1376-1380.

Selvi, A.A.; Sreenivasa, M.A. & Manonmani, H.K. (2011). Enzyme-linked immunoassay for the detection of glyphosate in food samples using avian antibodies. *Food and Agricultural Immunology*, 22, 217-228.

See, H.H.; Hauser, P.C.; Sanagi, M.M. & Ibrahim, W.A. (2010). Dynamic supported liquid membrane tip extraction of glyphosate and aminomethylphosphonic acid followed by capillary electrophoresis with contactless conductivity detection. *Journal of Chromatography A*, 1217, 5832-5838.

Segura, J.; Bingham, S.W. & Foy, C.L. (1978). Phytotoxicity of glyphosate to Italian ryegrass (*Lolium multiflorum*) and red clover (*Trifolium pratense*). *Weed Science*, 26, 32-36.

Shaner, D.L. (2009). Role of translocation as a mechanism of resistance to glyphosate. *Weed Science*, 57, 118-123.

Shrestha, A. & Hemree, K. (2007). Glyphosate-resistant horseweed (*Conyza canadensis* L. Cronq.) biotype found in the South Central Valley. *California Agriculture*, 61, 267-270.

Siimes, K.; Räämö, S.; Welling, L.; Nikunen, U. & Laitinen, P. (2006). Comparison of behaviour of three herbicides in a field experiment under bare soil conditions. *Agricultural Water Management*, 84, 53-64.

Skark, C.; Zullei-Seibert, N.; Schottler, U. & Schlett, C. (1998). The occurrence of glyphosate in surface water. *International Journal of Environmental Analytical Chemistry*, 70, 93-104.

Sørensen, S.R.; Schultz, A.; Jacobsen, O.S. & Aamand, J. (2006). Sorption, desorption and mineralisation of the herbicides glyphosate and MCPA in samples from two Danish soil and subsurface profiles. *Environmental Pollution*, 141, 184-194.

Sparling, D.W.; Matson, C.; Bickham, J. & Doelling-Brown, P. (2006). Toxicity of glyphosate as Glypro and LI700 to red-eared slider (*Trachemys scripta elegans*) embryos and early hatchlings. *Environmental Toxicology and Chemistry*, 25, 2768-2774.

Sprankle, P.; Meggitt, W.F. & Penner, D. (1975). Absorption, action and translocation of glyphosate. *Weed Science*, 23, 235-240.

Stalikas, C.D. & Konidari, C.N. (2001). Analytical methods to determine phosphonic and amino acid group-containing pesticides. *Journal of Chromatography A*, 907, 1-19.

Steinrücken, H.C. & Amrhein, N. (1984). 5-Enolpyruvylshikimate-3-phosphate synthase of *Klebsiella pneumonia*. 2. Inhibition by glyphosate [N-(phosphonomethyl)glycine]. *European Journal of Biochemistry*, 143, 351-357.

Stenrød, M.; Eklo, O.M.; Charnay, M.-P. & Benoit, P. (2005). Effects of freezing and thawning on microbial activity and glyphosate degradation in two Norwegian soils. *Pest Management Science*, 61, 887-898.

Stenrød, M.; Charnay, M.-P.; Benoit, P. & Eklo, O.M. (2006). Spatial variability of glyphosate mineralization and soil microbial characteristics in two Norwegian sandy loam soils as affected by surface topographic features. *Soil Biology and Biochemistry*, 38, 962-971.

Strange-Hansen, R.; Holm, P.E.; Jacobsen, O.S. & Jacobsen, C.S. (2004). Sorption, mineralization and mobility of N-(phosphonomethyl)glycine (glyphosate) in five different types of gravel. *Pest Management Science*, 60, 570-578.

Struger, J.; Thompson, D.; Staznik, B.; Martin, P.; McDaniel, T. & Marvin, C. (2008). Occurrence of glyphosate in surface waters of Southern Ontario. *Bulletin of Environmental Contamination and Toxicology*, 80, 378-384.

Suwannamek, U. & Parker, C. (1975). Control of *Cyperus rotundus* with glyphosate: the influence of ammonium sulphate and other additives.*Weed Research*, 15, 13-19.

Tomlin, C.D.S. (2000). The e-Pesticide Manual, 12th Edition, v.2.0, British Crop Protection Council, ISBN 1-901396 23 1, Brighton, UK.

Torstensson, L.; Börjesson, E. & Stenström, J. (2005). Efficacy and fate of glyphosate on Swedish railway embankments. *Pest Management Science*, 61, 881-886.

Tsui, M.T.K. & Chu, L.M. (2003). Aquatic toxicity of glyphosate-based formulations: comparison between different organisms and the effects of environmental factors. *Chemosphere*, 52, 1189-1197.

Tsunoda, N. (1993) Simultaneous determination of the herbicides glyphosate, glufosinate and bialaphos and their metabolites by capillary gas chromatography-ion-trap mass spectrometry, *Journal of Chromatography*, 637, 167-173.

US FDA (1977). FDA/ACA-77/144A. Pesticide Analytical Manual. Vol. II. U.S. Food and Drug Administration, Rockville, MD, USA.

You, J.; Kaljurand, M. & Koropchak, J.A. (2003). Direct determination of glyphosate in environmental waters using capillary electrophoresis with electrospray condensation nucleation light scattering detection. *International Journal of Environmental Analytical Chemistry*, 83, 797-806.

VanGessel, M.J. (2001). Glyphosate-resistant horseweed from Delaware. *Weed Science*, 49, 703-705.

Veiga, F.; Zapata, J.M.; Marcos, M.L.F. & Alvarez, E. (2001). Dynamics of glyphosate and aminomethylphosphonic acid in forest soil in Galicia, north-west Spain. *Science of the Total Environment*, 271, 135-144.

Vera, M.S.; Lagomarsino, L.; Sylvester, M.; Pérez, G.L.; Rodriguez, P.; Mugni, H.; Sinistro, R.; Ferraro, M.; Bonetto, C.; Zagares, H. & Pizarro, H. (2010). New evidence of Roundup (glyphosate formulation) impact on periphyton community and the water quality of freshwater ecosystems. *Ecotoxicology*, 19, 710-721.

Vereecken, H. (2005). Mobility and leaching of the glyphosate: a review. *Pesticide Management Science*, 61, 1139-1151.

Villeneuve, A.; Larroudé, S. & Humbert, J.F. (2011). Herbicide contamination of freshwater ecosystems: impact on microbial communities. In: *Pesticides – Formulations, Effects,*

Fate, Stoytcheva M. (Ed.), pp. 285-312, InTech, ISBN 978-953-307-532-7, Rijeka, Croatia, Available from
http://www.intechopen.com/articles/show/title/herbicide-contamination-of-freshwater-ecosystems-impact-on-microbial-communities

Vreeken, R.J.; Speksnijder, P.; Bobeldijk-Pastorova, I. & Noij, Th. H.M. (1998). Selective analysis of the herbicides glyphosate and aminomethylphosphonic acid in water by on-line soild-phase extraction-high-performance liquid chromatography-electrospray ionisation mass spectrometry. *Journal of Chromatography A*, 794, 187-199.

Walsh, L.P.; McCormick, C.; Martin, C. & Stocco, D.M. (2000). Roundup inhibits steroidogenesis by distrupting steroidogenic acute regulatory (StAR) protein expression. *Environmental Health Perspectives*, 108, 769-776.

Wan, M.T.; Watts, R.G. & Moul, D.J. (1989). Effects of different dilution water types on the acute toxicity to juvenile Pacific salmonids and rainbow trout of glyphosate and its formulated products. *Bulletin of Environmental Contamination and Toxicology*, 43, 378-385.

Weaver, M.A.; Krutz, L.J.; Zablotowicz, R.M. & Reddy, K.N. (2007). Effects of glyphosate on soil microbial communities and its mineralization in a Mississippi soil. *Pest Management Science*, 63, 388-393.

Wigfield, Y.Y. & Lanquette, M. (1991). Residue analysis of glyphosate and its principal metabolite in certain cereals, oilseeds and pulses by liquid chromatography and post-column fluorescence detection. *Journal of the Association of Official Analytical Chemists*, 74: 842-847.

Winfield, T.W.; Bashe, W.J. & Baker, T.V. (1990) U.S. Environmental Protection Agency Method 547, Determination of Glyphosate in Drinking Water by direct-aqueous-injection b hplc, post-column derivatization, and fluorescence detection. U.S. Environmental Protection Agency, Cincinnatti, OH, USA.

Wolfenden, R. (1969). Transition state analogues for enzyme catalysis. *Nature*, 16, 704-705.

Wong, P.K. (2000). Effects of 2, 4-D, glyphosate and paraquat on growth, photosynthesis and chlorophyll-a synthesis of *Scenedesmus quadricauda* Berb 614. *Chemosphere*, 41, 177-182.

Woodburn, A.T. (2000). Glyphosate: production, pricing and use. *Pest Management Science*, 56, 309-312.

Wyrill, J.B.; III & Burnside, O.C. (1976). Absorption, translocation, and metabolism of 2,4-D and glyphosate in common milkweed and hemp dogbane. *Weed Science*, 24, 557-566.

Yin, G. (2011). Glyphosate: There is no substitute. *Farm Chemicals International*, 2011 (3), Available from
http://www.farmchemicalsinternational.com/magazine/?storyid=2997

Zablotowicz, R.M.; Accinelli, C.; Krutz, L.J. & Reddy, K.N. (2009). Soil depth and tillage effects on glyphosate degradation. *Journal of Agricultural and Food Chemistry*, 57, 4867-4871.

Zandstra, C.H. & Nishimoto, R.K. (1977). Movement and activity of glyphosate in purple nutsedge. *Weed Science*, 25, 268-274.

Zaranyika, M.F. & Nyandoro, M.G. (1993). Degradation of glyphosate in the aquatic environment: an enzymatic kinetic model that takes into account microbial degradation of both free and coloidal (or sediment) particle adsorbed glyphosate. *Journal of Agricultural and Food Chemistry*, 41, 838-842.

Zboinska, E.; Lejczak, B. & Kafarski, P. (1992). Organophosphonate utilization by the wild-type strain of *Pseudomonas fluorescens*. *Applied and Environmental Microbiology*, 58, 2993-2999.

Zhao, P.; Yan, M.; Zhang, C.; Peng, R.; Ma, D. & Yu, J. (2011). Determination of glyphosate in foodstuff by one novel chemiluminescence-molecular imprinting sensor. *Spectrochimica Acta A*, 78, 1482-7486.

Zobiole, L.H. S.; Kremer, R.J.; Oliveira, R.S., Jr. & Constantin, J. (2011). Glyphosate affects micro-organisms in rhizospheres of glyphosate-resistant soybean. *Journal of Applied Microbiology*, 110, 118-127.

3

Prediction of Herbicides Concentration in Streams

Raj Mohan Singh

Department of Civil Engineering, MNNIT Allahabad,
India

1. Introduction

Natural and anthropogenic variables of stream drainage basins such as hydrogeologic parameters (permeability, porosity etc.), amount of agricultural chemicals applied, or percentage of land planted affect agricultural chemical concentration and mass transport in streams. The use of herbicides, pesticides, and other chemicals in agricultural fields increase the concentration of chemicals in streams which severely affects the health of human and environment. The transport of chemical pollutants into river or streams is not straight forward but complex function of applied chemicals and land use patterns in a given river or stream basin. The factors responsible for transport of chemicals may be considered as inputs and chemical concentration measurements in streams as outputs. Each of these inputs and outputs may contain measurement errors. Present work exploited characteristics of fuzzy sets to address uncertainties in inputs by incorporating overlapping membership functions for each of inputs even for limited data availability situations. Soft computing methods such as the fuzzy rule based and ANN (Artificial Neural Networks) is used for characterization of herbicides concentration in streams. The fuzzy c-means (FCM) algorithm is used for the optimization of membership functions of fuzzy rule based models for the estimation of diffuse pollution concentration in streams. The general methodology based on fuzzy, ANN and FCM for estimation of diffuse pollution in streams is presented. The application of the proposed methodology is illustrated with real data to estimate the diffuse pollution concentration in a stream system due to application of a typical herbicide, atrazine, in corn fields with limited data availability. Solution results establish that developed fuzzy rule base model with FCM outperform fuzzy or ANN and capable for the estimation of diffuse pollution concentration values in water matrices with sparse data situations.

Application of pesticides, insecticides and herbicides, cause diffuse pollution, commonly referred to as non-point source pollution in river or streams. Diffuse pollution from agricultural activities is a major cause of concern for the health of human and environment. Diffuse (non-dot, dispersed) pollution generally arises from land-use activities (urban and rural) that are dispersed across a catchment or subcatchment, where as point sources of pollution arise as a process industrial effluent, municipal sewage effluent, deep mine or farm effluent discharge (Novotny 2003, based on CIWEM (D'Arcy et al., 2000)). Potential point sources of pollution is characterised by its location, magnitude and duration of activity; and the sources of pollution is characterized when these parameters are identified

(Mahar and Dattta 2000; Singh and Datta, 2004, and Singh and Datta, 2006a and 2006b). In diffuse sources of pollution or non-point sources of pollution, sources of pollution is moving with polluting media thus making it more difficult and complex problem to solve.

Often diffuse pollution is individually minor but collectively constitutes significant sources at basin scale. Although nonpoint or diffuse sources may contribute many of the same kinds of pollutants, these pollutants are generated in different volumes, combinations, and concentrations (Jha et al., 2005). Thus, diffuse pollution comprises true non-point source pollution together with inputs from a multiplicity of minor point sources. The important characteristics of diffuse pollution are, therefore, not whether anyone can identify the source or sources, but the collective impact of diffuse pollutants and the mechanisms through which they move through the environment. The concept of diffuse pollution is useful because it explains features of pollution in receiving water bodies that differ from the point sources of pollution that are typically well characterized, monitored, and quantified. Some of the characteristics of diffuse pollutants are that the concentrations of some pollutants actually may increase with flow rather than it has diluted, pollution peaks are variable and difficult to predict, and impacts are often slow to develop and become evident years later (e.g. contamination of groundwater). For diffuse pollution, it is the proportion of the land use from which the pollution is derived, is more important.

Agricultural activities such as application of herbicides result in the contamination of surface water with agricultural chemicals. Numerous recent investigations (Goolsby and Battaglin, 1993 and 1995; Schottler et al., 1994; Baker and Richards, 1990) indicate that significant quantities of some herbicides are flushed from cropland to streams each spring and summer during rainfall events following the applications. Peak concentration of several herbicides can exceed 10 µg/l during these events (Coupe et al., 1995; Scribner et al., 1994). Pareira (1990), Crawford (1995, 2001), Capel and Larson (2001), and Smith and Wheater (2004) in their studies on pesticides/herbicides, identified the major factors that control the pollutant transport. Herbicides and pesticides concentrations in surface waters are affected by natural and human factors. For example, concentrations of atrazine, a herbicide widely used on corn fields, tended to be higher in an agricultural basin with permeable, well drained soils, than in an agricultural basin with less permeable, more poorly drained soils (Crawford, 1995). Capel et al. (2001) estimated the annual pollutant transport as percent of use (load as percent of use - LAPU). Larson and Gilliom (2001) developed a regression model for the estimation of pollutants.

Water resources professionals, managers and government authorities involved in surface water management are increasingly pressed to make appropriate decisions on land use and development policies such that these decisions will not adversely affect the health and environment. At the same time, they are constrained by inadequate budgets, limited resources, and incomplete information, which compel them to rely on models to evaluate or to estimate the pollution characteristics in the water bodies, and the implications of their decisions based on those evaluations. In this regard, the role of complex stream quality simulation models e.g. SWAT (Arnold et al. 1983), AGNPS (Young et al., 1989) etc. in evaluating runoff pollution conditions under various agricultural chemicals and land use patterns is also limited. These models incorporate rainfall, catchments, and pollutant characteristics, requiring extensive calibration and verification. However, their results are not without large uncertainties. These uncertainties arise both in the representation of the physical, chemical, and biological processes as well as in the data acquisition and parameters for model algorithms. Consequently, the complexities of these models and their resource-intensive nature are significant obstacles to their application (Charbeneau and Barrett 1998).

There is a need for the development of simpler methods of agricultural stream quality predictions that provide the required information to the analyst and water managers with minimal effort and limited data requirements as compared to complex process models. As an alternative or supplement to complex runoff quality simulation models, fuzzy rule based model with FCM is proposed to estimate pollutant concentration due to applications of agricultural chemical, herbicide, atrazine, in the streams.

The herbicide atrazine (2-chloro-4-[ethylamino]-6-[isopropylamino]-1,3,5-triazine), a chlorinated herbicide, has been one of the most heavily used herbicides in the world. Atrazine is toxic to many living organism. The maximum contaminant level (MCL) of atrazine is restricted to 3 µg/l for drinking water (USEPA, 2001). Because atrazine is water soluble, it has the potential to leach into ground water and run off to surface water. Atrazine is associated with developmental effects (USEPA, 2002), such as birth defects, structural anomalies, and adverse hormone changes. Thus, its accurate estimation in water matrices is imperative.

In this study, a fuzzy rule based model optimized by fuzzy c-Means, is developed to obtain the estimate of atrazine concentrations from agricultural run-off using limited available information. The work discusses the methodology to develop the fuzzy rule base model using annual average use of herbicide atrazine per unit area, extent of herbicide atrazine applied area and herbicide atrazine application season as inputs to fuzzy rule based model and observed herbicide concentration at the basin outlet as the output for the fuzzy model. The data of White River Basin, a part of the Mississippi River system, USA, is used for developing the fuzzy rule base model.

2. Agricultural diffuse pollution concentration simulation in streams

Natural and anthropogenic variables of stream drainage basins such as hydrogeologic parameters (permeability, porosity etc.), amount of agricultural chemicals applied, or percentage of land planted affect agricultural chemical concentration and mass transport in streams. The general form of model that simulates the concentration measurement in a watershed can be represented by (Tesfamichael et al., 2005)

$$C = f(W, H, A) \tag{1}$$

where C is the stream agricultural diffuse pollution observed concentration measurement values; W is a vector of watershed characteristics; and H is a vector of hydrological variables such as precipitation, runoff, etc., and A is a vector of relevant agricultural practices including actual chemical application rate in the field in lb/acre.

For a particular watershed, watershed characteristic, W, may be assumed to be constant. Also, for a particular hydrological unit, H may be assumed to be of similar characteristics. Then, Equation (1), though simplified, may be represented by

$$C = f(A) \tag{2}$$

The **A** may be further represented by

$$A = f(A_C, A_L) \tag{3}$$

where A_C represent the vector of applied agricultural chemical characteristics such as type of agricultural chemical (insecticide, herbicides etc.), application rate, application season etc., and A_L is the land use patterns such as type of crop grown, percentage of cropped area, etc.

Here, agricultural chemical considered is herbicide, atrazine, and crop considered is corn. In this study fuzzy rule based model with FCM simulates the stream system behavior from inputs of agricultural practices and corresponding observed concentration measurement values. In fact the model tries to emulate the mechanism that produced the data set. In this way, the mathematical description of the physical system is learned by the model, and therefore utilized as a tool for stream system simulation. The cluster centers of inputs and outputs obtained using FCM model, in essence, represents a typical characteristics of the system behaviour, and hence utilized in the formation of rule base of the fuzzy model.

3. Methodology

Statistical methodologies have been traditional being utilized for diffuse pollutants predictions in streams. However, transport of herbicides is complex and uncertain phenomena and traditional methods like regression are not able to incorporate uncertainty in model predictions. Present work will discuss methodologies based on recent soft computing techniques like fuzzy, artificial neural network (ANN) and their hybrids. The application of the proposed methodology is illustrated with real data to estimate the diffuse pollution concentration in a stream system due to application of a typical herbicide, atrazine, in corn fields with limited data availability.

3.1 Modeling approach

The models based on fuzzy logic and ANN, also known as intelligent or soft computing models, are potentially capable of fitting a nonlinear function or relationships. Identification of model architecture is decisive factor in the simulation and comparison. The identification of model architecture is crucial in ANN model building process. While the input and output of the ANN model is problem dependent, there is no direct precise way to determine the optimal number of hidden nodes (Nayak et al., 2005).The model architecture is selected through a trial and error procedure (Singh et al.. 2004). The fuzzy model, on the other hand, may be considered as a mapping of input space into output space by partitions in the multidimensional feature space in inputs and outputs. Each partition represents a fuzzy set with a membership function.

3.2 Fuzzy rule based system

Fuzzy logic emerged as a more general form of logic that can handle the concept of partial truth. The pioneering work of Zadeh (1965) on fuzzy logic has been used as foundation for fuzzy modeling methodology that allows easier transition between humans and computers for decision making and a better way to handle imprecise and uncertain information. Human being think verbally, not numerically. As the fuzzy logic systems involves verbal statements and, therefore, the fuzzy logic is more in line with human perception (Zadeh, 2000). Fuzzy logic has an advantage over many statistical methods in that the performance of a fuzzy expert system is not dependent on the volume of historical data available. Since these expert systems produce a result based on logical linguistic rules, extreme data points in a small data set do not unduly influence these models. Because of these characteristics, fuzzy logic may be a more suitable method for diffuse pollution forecasting than the usual regression modeling techniques used by many researchers (e,g. Goolsby and Battaglin (1993); Larson and Gilliom (2001); and Tesfamichael et al. (2005) etc.) for estimation of diffuse pollution concentration in streams or other water bodies.

3.2.1 Fuzzy rule based system architecture

The most common way to represent human knowledge is to form it into natural language expression of the type,

$$\text{IF premise (antecedent), THEN conclusions (consequent)} \qquad (4)$$

The form in expression (4) is commonly referred to as the IF-THEN rule based form (Ross, 1997). It typically expresses an inference such that if a fact (premise, hypothesis, antecedent) is known, then another fact called a conclusion (consequent) can be inferred or derived. Fuzzy logic systems are rule base systems that implements a nonlinear mapping (Dadone and VanLandingham, 2000) between stresses (represented by consequents) and state variables (represented by antecedents). Creating a fuzzy rule based system may be summarized in four basic steps (Ross 1997; Mahabir et al. 2003; Singh and Singh 2005):

a. For each variable, whether an input variable or a result variable, a set of membership functions must be defined. A membership function defines the degree to which the value of a variable belongs to the group and is usually a linguistic term, such as high or low.
b. Statements, or rules, are defined that relate the membership functions of each variable to the result, normally through a series of IF–THEN statements.
c. The rules are mathematically evaluated and the results are combined. Each rule is evaluated through a process called implication, and the results of all of the rules are combined in a process called aggregation.
d. The resulting function is evaluated as a crisp number through a process called defuzzification.

Subjective decisions are frequently required in fuzzy logic modeling, particularly in defining the membership functions for variables. In cases such as in this study, where large data sets are not available to define every potential occurrence scenario for the fuzzification of model, expert opinion is used to create logic in the rule base system.

3.2.2 Membership functions

Membership functions used to describe linguistic knowledge are the enormously subjective and context dependent part of fuzzy logic modeling (Vadiee, 1993). Each variable must have membership functions, usually represented by linguistic terms, defined for the entire range of possible values. The key idea in fuzzy logic, in fact, is the allowance of partial belongings of any object to different subsets of universal set instead of belonging to a single set completely. Partial belonging to a set can be described numerically by a membership function which assumes values between 0 and 1 inclusive. Intuition, inference, rank ordering, angular fuzzy sets, neural networks, genetic algorithms, and inductive reasoning can be, among many, ways to assign membership values or functions to fuzzy variables (Ross, 1997). Fuzzy membership functions may take on many forms, but in practical applications simple linear functions, such as triangular ones are preferable due to their computational efficiency (Khrisnapuram, R,1998). In this study, triangular shapes are utilized to represent the membership functions.

3.3 Fuzzy c-means partitioning

Fuzzy rule based models represent the system behaviour by means of if then fuzzy rules. The basic requirement of fuzzy rule based model is to fuzzify or partition the inputs and outputs representation of a physical system. Assigning the number, shape, overlaps etc. of membership functions is most complex part of the fuzzy rule based model building. In most of the cases the optimality of the membership assigned to different fuzzy variables are not guaranteed. FCM is one of the methods to determine the fuzzy partitions of the available data sets into a predetermined number of groups. The data points are divided into group of points that are close to each other. Each data point belongs to a group or cluster with a membership function. Closeness between data points is defined by a metric distance or data center, and each metric yields a different portioning. This cluster centers are utilized in assigning overlaps of triangular shape membership function in this study.

Fuzzy c-means (FCM) is a method of clustering which allows one piece of data to belong to two or more clusters. The FCM method (developed by Dunn (1973) and improved by Bezdek (1981)) is frequently used in pattern recognition. It is based on minimization of the following objective function:

$$J_m = \sum_{i=1}^{N} \sum_{j=1}^{C_N} u^m{}_{ij} \left\| x_i - c_j \right\|^2, 1 \le m < \infty \tag{5}$$

where m is any real number greater than 1, uij is the degree of membership of xi in the cluster j, xi is the ith of d-dimensional measured data, cj is the d-dimension center of the cluster, and $||*||$ is any norm expressing the similarity between any measured data and the center. The N represents total number of data points, and CN represents the total number of fuzzy centers. Fuzzy partitioning is carried out through an iterative optimization of the objective function shown above, with the update of membership uij and the cluster centers cj by:

$$u_{ij} = \cfrac{1}{\sum\limits_{k=1}^{C_N} \left(\cfrac{\left\| x_i - c_j \right\|}{\left\| x_{i-c_k} \right\|} \right)^{\frac{2}{m-1}}} \tag{6}$$

$$c_j = \cfrac{\sum\limits_{i=1}^{N} u^m{}_{ij} \cdot x_i}{\sum\limits_{i=1}^{N} u^m{}_{ij}} \tag{7}$$

This iteration will stop when maxij{ | uijk+1-uijk | }< ε , where ε is a termination criterion between 0 and 1, where as k are the iteration steps. This study used FCM algorithm (Matlab version 6.5), and ε is equal to 0.1 - 10^{-5} to obtain the pre-specified fuzzy centers.

This study implements FCM algorithm (Matlab version 6.5), m=2, and ε equal to 10^{-5} to obtain the pre-specified fuzzy centers.

3.4 Fuzzy rule based system with FCM for estimation of diffuse pollution concentration in streams

The watershed of the streams plays a vital role in influencing the diffuse pollution concentration in the streams. Basic Steps 1 through Steps 4 as discussed earlier in section Rule Based System are implemented by partitioning the input and output spaces into fuzzy regions with FCM, generation of fuzzy rules from available data pairs, assigning a degree to each rule, construction of a combined fuzzy rule base, and mapping from the input space to the output space using the rule base and a defuzzification (Wang and Mendel, 1992).

The vector AC and AL as represented by equation (2) are characterized for the specified watershed of the streams. As explained earlier, AC represents the vector of applied agricultural chemical characteristics such as type of agricultural chemical (insecticide, herbicides etc.), application rate, application season etc. The AL is the land use patterns such as type of crop grown, percentage of cropped area, etc. and C is the stream agricultural diffuse pollution observed concentration measurement values. Patterns were generated using a known set of input-output data pairs. The input data pairs AC and AL values and corresponding output values of C for a particular year constitutes a pattern. While AC and AL are constant for a particular year, the C is temporally and spatially varying at each of the monitoring station sites.

Fuzzy rules are building-blocks of fuzzy rule base systems. Partitioning the fuzzy variables into linguistic variables is necessary step towards designing the rule base system. Fuzzy partitions for the input and output variables are defined or generated according to the type of data as discussed in the membership section (Singh, 2008). In this work, FCM model is utilized to supply optimum number data centers to partition the input and output fuzzy variables.

It is absolutely possible to obtain the redundant and inconsistent rules from the data patterns having same antecedent parts. As mentioned, each rule is assigned a degree or weight by multiplying the membership functions of inputs and outputs for that rule. In the standard approach the rule having largest degree is adopted (Wang and Mendel, 1992). As an improvement, the degree of each rule is multiplied by a redundancy index to obtain the effective degree for that rule. The redundancy index may be defined as:

$$\text{Redundancy Index (R.I.)} = \frac{r_i}{T_r} \tag{8}$$

where, ri represents the redundant rule with same i antecedents; and Tr represents the sum of all the redundant rules. Final fuzzy rule base includes the rules having the highest effective degree.

The fuzzy inference mechanism uses the fuzzified inputs and rules stored in the rule base for processing the incoming inputs data and produces an output. The fuzzy rules are processed by fuzzy sets operations as discussed in rule based section as basic steps for fuzzy rule base system. The fuzzy rule based design is accepted to be satisfactorily completed when its performance during training and testing satisfies the stopping criteria based on some statistical parameters.

3.5 ANN based methodology for estimation of diffuse pollution concentration in streams

The ANN learns to solve a problem by developing a memory capable of associating a large number of example input patterns, with a resulting set of outputs or effects. ANN is discussed in ASCE Task Committee (2000), etc. An overview of artificial neural networks and neural computing, including details of basics and origins of ANN, biological neuron model etc. can be found in Hassoun (1999), Schalkoff (1997), and Zurada (1997). The details of ANN model building process and selection of best performing ANN model for a given problem is available in (Singh et al., 2004).

As illustrated in the fuzzy model building for estimation of diffuse pollution concentrations in streams, the AC and AL values for a particular year in a watershed are inputs, and corresponding C values in the stream is out put for the ANN model. The values of AC, AL and C for a particular year constitute a data pattern. A standard back propagation algorithm (Rumelhart et al., 1986) with single hidden layer is employed to capture the dynamic and complex relationship between the inputs and outputs utilizing the available patterns. The ANN architecture that perform better than other evaluated architectures based on certain performance evaluation criteria, both in training and testing, was selected as the final architecture.

3.6 Performance evaluation criteria

The performance of the developed models are evaluated based on some performance indices in both training and testing set. Varieties of performance evaluation criteria are available (e.g. Nash and Sutcliffe 1970; WMO 1975; ASCE Task Committee on Definition of Criteria for Evaluation of Watershed Models1993 etc.) which could be used for evaluation and inter comparison of different models. Following performance indices are selected in this study based on relevance to the evaluation process. There can be other criteria for evaluation of performance.

3.6.1 Correlation coefficient (R)

The correlation coefficient measures the statistical correlation between the predicted and actual values. It is computed as:

$$R = \frac{\sum_{i=1}^{n}(Xai - \overline{X}ai)(Xpi - \overline{X}pi)}{\sqrt{\sum_{i=1}^{n}(Xai - \overline{X}ai)^2 \sum_{i=1}^{n}(Xpi - \overline{X}pi)^2}} \qquad (9)$$

where Xai and Xpi are measured and computed values of diffuse pollution concentration values in streams; $\overline{X}ai$ and $\overline{X}pi$ are average values of Xai and Xpi values respectively; i represents index number and n is the total number of concentration observations.

The correlation coefficient measures the statistical correlation between the predicted and actual values. A higher value of R means a better model, with a 1 meaning perfect statistical correlation and a 0 meaning there is no correlation at all.

3.6.2 Root sean square error (RMSE)

Mean-squared error is the most commonly used measure of success of numeric prediction, and root mean-squared error is the square root of mean-squared-error, take to give it the same dimensions as the predicted values themselves. This method exaggerates the prediction error - the difference between prediction value and actual value of a test case. The root mean squared error (RMSE) is computed as:

$$RMSE = \sqrt{\frac{1}{n}(\sum_{i=1}^{n}(Xai - Xpi)^2)} \tag{10}$$

For a perfect fit, Xai = Xpi and RMSE = 0. So, the RMSE index ranges from 0 to infinity, with 0 corresponding to the ideal.

3.6.3 Standard error of estimates (SEE)

The standard error of estimate (SEE) is an estimate of the mean deviation of the regression from observed data. It is defined as (Allen, 1986):

$$SEE = \sqrt{\frac{\sum_{i=1}^{n}(Xai - Xpi)}{(n-2)}} \tag{11}$$

3.6.4 Model efficiency (Nash–Sutcliffe coefficient)

The model efficiency (MENash), an evaluation criterion proposed by Nash and Sutcliffe (1970), is employed to evaluate the performance of each of the developed model. It is defined as:

$$ME_{Nash} = 1.0 - \frac{\sum_{i=1}^{n}(Xa_i - X_{pi})^2}{\sum_{i=1}^{n}(X_{ai} - \overline{X}ai)^2} \tag{12}$$

A value of 90% and above indicates very satisfactory performance, a value in the range of 80–90% indicates fairly good performance, and a value below 80% indicates an unsatisfactory fit.

4. Data synthesis and architecture identification of models

In this work, the diffuse pollution concentration in stream is considered due to herbicide atrazine application in corn fields of the watershed. Concentration measurements data were obtained from the National Water Quality Assessment (NAWQA) program of the U S Geological Survey (USGS) (http://water.usgs.gov/nawqa/naqamap.html) for the period 1992 to 2002. The stream considered is White River, and monitoring site for the atrazine concentration measurement, is Hazeltone (Crawford, C.G, 1995), the outlet site of the watershed of White River Basin in Indiana State. At Hazeltone site, Latitude is 38°29'23", and Longitude is 87°33'00" and Drainage area 11,305.00 square miles. The White River basin is a part of the Mississippi River system where the application of atrazine accounts for 24 percent of all agricultural herbicides. The major agricultural chemical characteristics, AC, which contribute to the atrazine concentration at the watershed outlet are identified as its application rate (lb/Acre) and application time. The major land use patterns, AL, is the extent of cropped area (percentage of cultivated area (Pareira, 1990; Crawfard, 2001; and Capel and Larson, 2001).

Time series of data (average monthly values) from 1992-2001 are utilized for model building and validation. The major agricultural chemical characteristics, AC, which contribute to the atrazine concentration at the watershed outlet are identified as its application rate (lb/acre) and application time. The major land use pattern, AL, is the extent of cropped area (percentage of cultivated area (Crawford, 2001, 1995).These data are utilized for identification of fuzzy and ANN based models architectures by applications of the methodologies discussed in previous sections. The performance evaluations criteria are utilized to judge the predictive capability of the best performing fuzzy and ANN models. The procedure of developing fuzzy logic rule based model is implemented using the data of atrazine application rate as first input, atrazine application season as second input, and the percentage area applied with atrazine as third input. The atrazine concentration measurement values observed at the monitoring site is the output for the fuzzy rule based model. The weighted average of herbicide application rates and percentage of area applied of the corn and soybean cropped area are given in Table 1. The seven years data (1992-1998) are utilized for training and the three years data (1999-2001) (Table 1) are utilized for testing models.

Year	Weighted Percentage Area	Application Rate (lb/Acre)
1992	79	1.35
1993	91	1.31
1994	87	1.35
1995	87	1.31
1996	91	1.31
1997	84	1.33
1998	89	1.36
1999	91	1.26
2000	80	1.41
2001	94	1.35

Table 1. Agricultural Herbicide Atrazine Application Rate and Percentage Area Applied for the Corn Crop.

4.1 Evaluation of fuzzy c-means centers

The FCM model represented by equation (5) is used to partition the input data into fuzzy partitions. The FCM algorithm is implemented using MATLAB version 6.5 for ε equal to 10^{-5} to obtain the pre-specified fuzzy centers. The 3, 4, and 5 fuzzy centers for the inputs application rate and weighted percentage area obtained using the FCM model is shown in Table 2. Instead of iterating for the optimal number of fuzzy centers, a prior knowledge about the fuzzy partitioning for the fuzzy rule based models were utilized in implementing fuzzy c-means algorithm.

Fuzzy Partition centers by FCM Model		
Fuzzy Partitions	Input Application Rate (lb/Acre)	Application Rate (lb/Acre)
3-Fuzzy Centers	1.26	80.38
	1.31	86.68
	1.37	90.75
4-Fuzzy Centers	1.26	79.50
	1.31	84.02
	1.33	87.21
	1.36	90.88
5-Fuzzy Centers	1.26	80.00
	1.31	86.67
	1.33	87.00
	1.35	89.17
	1.41	91.0

Table 2. Different Fuzzy Partition Centers Using FCM Model

4.2 Training and testing the fuzzy rule based model with FCM

The seven years data (1992-1998) are utilized for training and the three years data (1999-2001) are utilized for testing the fuzzy rule based model with FCM. The model is assumed to be performing satisfactory when model efficiency coefficient (MENash) as given by equation (12) is greater than 90 percent, and other performance indices are also improved. Although arbitrary, it may be used as stopping criteria to limit the processing of large number of rules with increase in linguistic fuzzy variables for the inputs.

Performance of fuzzification of inputs application rate and weighted percentage area were studied by assigning 3, 5, and 7 fuzzy variables without using FCM (Singh, 2008). Though performance of fuzzifiction with 7 variables worked better than fuzzification with 3 and 5 variables; fuzzification by 5 fuzzy variables are comparable to fuzzification with 7 variables as shown in Table 3. Fuzzy rule based models with 3, 5 and 7 fuzzy variables are represented by Fuzzy_3M, Fuzzy_5M, and Fuzzy_7M models respectively in the Table 3. As 3 partitions are not adequate, four fuzzy partitions were specified for the use of fuzzy rule based system with FCM model. The four centers as shown in Table 2, obtained using FCM are partitioned into four linguistic fuzzy variables as low, medium, high, and very high. A sample schematic representation of membership function is shown for the input atrazine application rate in Figure 1.

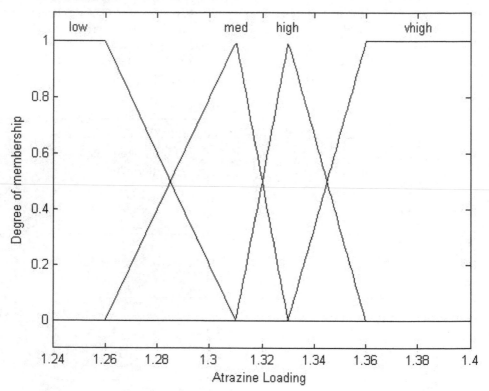

Fig. 1. A sample representation of linguistic variables membership function for first input.

The input application season is assigned 12 fuzzy variables, S1-S12 corresponding to each month of a year. The output concentration measurement values of atrazine is represented by 25 fuzzy centers by FCM model and represented by fuzzy variables, C1-C25, so that all the ranges of atrazine concentration measurement values in the data set for the period 1992-2001, is adequately represented. All the fuzzy variables in inputs and outputs are represented by triangular shape, except at the domain edges, where they are semi trapezoidal. This representation has been selected based on literature due to their computational efficiency (Khrisnapuram R 1998; Guillaume and Charnomordic, 2004). A sample representation of the membership functions is shown in Figure 1 for the first input. Of course, other divisions of the inputs and output domain regions and other shapes of membership functions are possible. The total number of rules in case of 4 linguistic variables for inputs application rate and weighted percentage area, and 12 fuzzy variables for seasons are 192. The total number of rules was much high i.e. 588 when 7 fuzzy variables were used for inputs application rate and weighted percentage area. The model building process is completed by creating combined fuzzy rule base using inputs-output pair values of training set data. Finally, the defuzzification converts fuzzy output produced by the fuzzy rule base model as crisp output corresponding to any new inputs.

5. Concentration measurement estimation results

The performance of the FCM based fuzzy rule based model is evaluated based on performance indices as described in performance evaluation criteria. These include root mean square error (RMSE), correlation coefficient (R) between the actual and estimated monthly average concentration measurement values of atrazine herbicides, standard error of estimate (SEE) and MENash. The performance evaluation results of the fuzzy rule based model with four fuzzy variables obtained using FCM, represented as Fuzzy_4_FCM, is also compared with that of the fuzzy rule based models with 3, 5, 7 linguistic variables for both of the input 1 and input 3. The performance of the Fuzzy_4_FCM model is also compared with solution results of an artificial neural network (ANN) based model using back propagation algorithm (Rumelhart et al. 1986) as represented by ANN_M in Table 3.

Models	Training Error (1992-198)				Testing Error (1999-2001)			
	RMSE	R	SSE	ME$_{Nash}$	RMSE	R	SSE	ME$_{Nash}$
Fuzzy_3M	1.318	0.891	1.377	0.550	0.703	0.886	0.771	0.623
Fuzzy_5M	0.836	0.969	0.837	0.894	0.455	0.952	0.498	0.855
Fuzzy_7M	0.706	0.970	0.775	0.915	0.342	0.975	0.375	0.914
ANN_M	1.153	0.918	1.264	0.752	0.906	0.759	0.993	0.446
Fuzzy_4M_FCM	0.492	0.998	0.539	0.967	0.725	0.968	0.416	0.901

Table 3. Comparison of training and testing errors for different models.

It can be noted from the Table 3 that the error statistics are better for Fuzzy_4M_FCM model than those of Fuzzy_3M, Fuzzy_5M and ANN_M model in both the training and testing in prediction in atrazine concentration measurement values. Its performance is even better than Fuzzy_7M model in training. Model efficiency (MENash) in training is 94.3 percent whereas it is 91.5 percent for Fuzzy_7M model. Similarly, RMSE, R, and SSE values are also comparable. In testing, results are also comparable though error statistics for Fuzzy_7M model is slightly better than Fuzzy_4_FCM. Thus, the FCM optimized fuzzy membership functions partitions in Fuzzy_4_FCM model are performing comparable to almost double the fuzzy partitions without FCM in Fuzzy_7M model. Figure 2 shows better RMSE value by Fuzzy_4_FCM model in comparison to other models.

It can also be noted from Table 3 that performances of fuzzy rule based model is better than those obtained using an ANN model with 2 inputs (atrazine application rate and weighted percentage area), 12 outputs (average monthly concentration measurements), and 11 hidden nodes (selected on the basis of experimentation) represented by ANN_M model. The poor performance by ANN_M model may be due to inadequate training patterns for experimentation, as the total number of free parameters become more than the number of training patterns even for 1 hidden node in hidden layer.

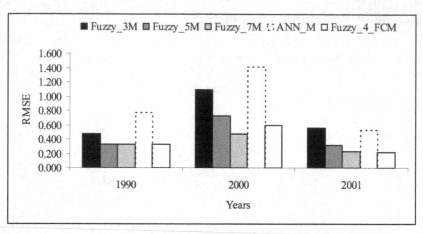

Fig. 2. Performance comparison of models.

Scatter plots of average monthly observed and predicted atrazine concentration measurement in the stream for model Fuzzy_4_FCM are plotted for the testing period 1999, 2000, and 2001. Comparison of actual and model estimated values are also presented for average monthly variations of atrazine concentration in the stream during the testing period, 1999-2001. Figure 3 represents scatter plot, and Figure 4 represents comparison of actual and Fuzzy_4_Model estimated values for the period 1999. Scatter plots between the observed and Fuzzy_4_FCM predicted average atrazine concentration measurement values in stream followed a 1:1 line except for a few cases of high magnitudes. The high values of coefficient of determination, R^2 (0.933), indicate that there is a good match between the observed and model predicted atrazine concentration. Figure 4 shows a comparison of observed and, Fuzzy_4_FCM model predicted average monthly atrazine concentration measurement values in the stream. The observed and Fuzzy_4_FCM predicted values match well except for the occurrence of peak value.

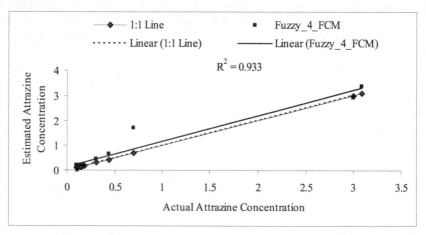

Fig. 3. Scatter plot of observed and Fuzzy_4_FCM Model predicted average monthly atrazine concentration for the testing period year 1999.

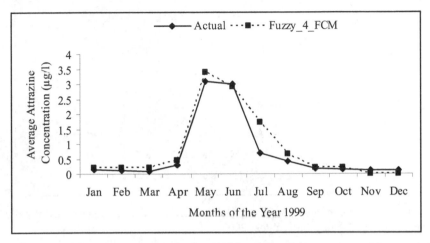

Fig. 4. Comparison of observed and Fuzzy_4_FCM predicted average monthly atrazine concentration for the testing period year 1999.

Figure 5 represents scatter plot of observed and Fuzzy_4_FCM predicted values, and Figure 6 represents comparison of observed and Fuzzy_4_FCM predicted atrazine concentration values for the period 2000. Scatter plots between the observed and Fuzzy_4_FCM predicted average atrazine concentration measurement values in stream followed a 1:1 line with R^2 value of 0.95. In this case though initial and final months values matches well, intermediate months values including peak value does not mach well as shown in Figure 6.

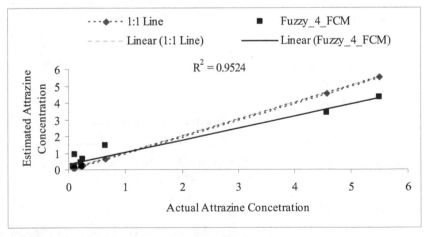

Fig. 5. Scatter plot of observed and Fuzzy_4_FCM Model predicted average monthly atrazine concentration for the testing period year 2000.

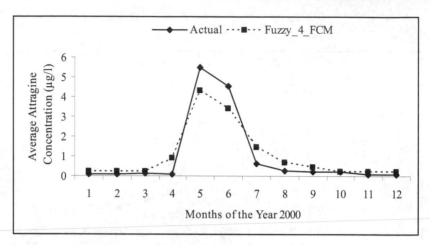

Fig. 6. Comparison of observed and Fuzzy_4_FCM predicted average monthly atrazine concentration for the testing period year 2000.

Figure 7 represents scatter plot of observed and Fuzzy_4_FCM predicted values, and Figure 8 represents comparison of observed and Fuzzy_4_FCM model predicted atrazine concentration values for the period 2001. Scatter plots between the observed and Fuzzy_4_FCM predicted average atrazine concentration measurement values in stream followed a 1:1 line with high value R^2 (0.93).

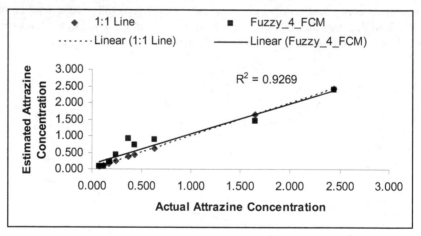

Fig. 7. Scatter plot of observed and Fuzzy_4_FCM Model predicted average monthly atrazine concentration for the testing period year 2001.

6. Discussion of results

The performance evaluation results presented in this study establish the potential applicability of the developed methodology in estimation of monthly atrazine concentration measurement values using fuzzy rule based models with FCM. However, the comparative

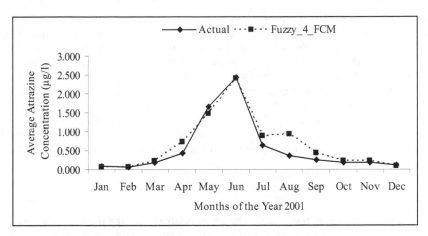

Fig. 8. Comparison of observed and Fuzzy_4_FCM model predicted average monthly atrazine concentration for the testing period year 2001.

performance of the methodology in different evaluation periods, under or over prediction of peak values, fuzzy rule based model control parameters (shape, total number of fuzzy centers, overlaps etc. of membership functions; fuzzy set operations i.e, defuzzification methods etc.) needs to be investigated further.

The performance of fuzzy rule based model with FCM is better than those without FCM model with even more number of fuzzy partitions. This is inferred by comparison of performances of Fuzzy_4_FCM model with Fuzzy_3M, Fuzzy_5M, and Fuzzy_7M models. In all the evaluation results obtained by Fuzzy_4_FCM model for the period 1999-2001, the R^2 values from scatter plots, and MENash values obtained from observed and model predicted values are high (around 0.9). This implies good match between the observed and model predicted values. The fuzzy rule with FCM model also performed better than the ANN based model. It establishes that the developed fuzzy rule based model with FCM is potentially suitable for estimation of concentration measurement values with limited data availability. The performances of the developed models are better in comparison to performance of regression models developed for the Mississippi River Systems (Battaglin and Goolsby, 1997). Their study show that multiple linear regression models estimate the concentration of selected agricultural chemicals with maximum R-squared value is 0.514, and in the case of atrazine, R-squared value is 0.312. In this study, almost all the developed models have R-squared value greater than 0.55. However, this comparison is limited as the White River basin considered in this study is only a part of (one of 10 basins) of Mississippi River Systems considered by them (Battaglin and Goolsby, 1997).

The estimation results obtained using fuzzy rule based models are encouraging but not conclusive. In almost all the evaluations, though initial months and final months concentration measurement values matches well, the intermediate values including the peak values are either over predicted or under predicted except for the year 2001 where peak predicted value matched well with the observed value. As the intermediate months, from April to July observes most of the changes in atrazine observed concentration measurement values, the same dynamics are exactly not reflected in model predictions. Thus, though the FCM model works better than ANN model in case of limited data availability, its performance is also affected due to limited data sets. In the present study, the inputs were assigned with triangular shape. Further improvement in the performance of the methodology may be possible with more extensive evaluations of membership functions shape, number of data centers for membership functions for each variables, and overlap between two membership functions. Present methodology utilized centroid method for defuzzification. Performance of other defuzzification method also need to be investigated. The error in prediction of peak values shows the limitation of the methodology. However, these results show potential applicability of the proposed methodology. The main advantage of the developed methodology is incorporate some prior knowledge into the model frame work, and its ability to perform in case of limited availability of data than other methods such as ANN.

7. Conclusions

The present study describes the framework for evaluating average monthly concentration of agricultural non point source pollution due to herbicide atrazine in streams by fuzzy rule based model with FCM utilizing limited amount of data. The values of statistical performance evaluation criteria indicate the model is able to simulate the behaviour of diffuse pollution sources from agricultural fields like attrazin in streams. The fuzzy rule based model with FCM performs comparatively better than the fuzzy rule based model without FCM and even with more fuzzy partitions. The proposed methodology also performs better than the ANN model when applied to the same problem. However, the model predicts with lesser accuracy for the intermediate months concentration measurement values including peak values. An extensive evaluation of the effect of more number of FCM based fuzzy centers and shapes of membership functions may fully establish the applicability of the methodology.

However, the proposed fuzzy rule based approach with FCM uses least amount of information in terms of number of inputs required, incorporate prior knowledge about fuzzy partitions, and also uses linguistic variables which make it relatively easy to interpret the rules. Prior knowledge about the physical system in the form of rule base can also be directly incorporated in the suggested approach. This preliminary study shows that the developed fuzzy rule based approach with FCM is potential suited to estimation of diffuse pollution concentration like atrazine in streams.

8. References

ASCE Task Committee on Definition of Criteria for Evaluation of Watershed Models (1993) Criteria for evaluation of watershed models. *J. Irrig. Drain. Eng.* ASCE 119(3): 429–442.

Arnold, J.G., Allen, P.M., Bernhardt, G. (1983) A comprehensive surfacegroundwater flow model. *Journal of Hydrology* 142: 47–69.

Baker D.B., Richards R.P. (1990) Transport of soluble pesticides through drainage networks in large agricultural river basins, In: Kurtz D.A. (Ed.), Long Range Transport of Pesticides. Lewis Publishers, Inc., Chelsea, MI.

Battaglin, W. A. and Goolsby, D. A., 1997. Statistical modeling of agricultural chemical occurrence in Midwestern Rivers. *Journal of Hydrology*, 196, 1-25.

Bezdek J. C. (1981) Pattern Recognition with Fuzzy Objective Function Algoritms. Plenum Press, New York.

Capel Paul D., Larson Steven J. (2001) Effect of Scale on the Behavior of Atrazine in Surface Waters. *Envron. Sci. Technol.* 35(4): 648-657, 2001.

Charbeneau R., Barrett M. (1998) Evaluation of methods for estimating storm water pollutant loads, *Water Environment Research:* 70(7), 1295-1302.

Crawfard C.G. (2001) Factors Affecting Pesticide Occurrence And Transport In A Large Mid-Western River Basin. *Journal of American Water Res. Asstn.* 37(1): 1-15, 2001.

Crawford, C.G. (1995) Occurrence of pesticides in the White River, Indiana, 1991-95: U.S. *Geological Survey Fact Sheet No. 233-95.*

Coupe R. H., Goolsby D. A., Iverson J. L., Zaugg S.D., Markovchick D. J. (1995) Pesticide, Nutrient, Streamflow and Physical Property Data for the Mississippi River, and Major Tributaries, April 1991- September, 1992. *U.S. Geol. Surv. Open-File* Rep. 93-657.

Dadone P, VanLandingham H.F. (2000) On the differentiability of fuzzy logic systems. *Proceedings of IEEE Conference on Systems, Man and Cybernetics*, Nashville, TN, 2703-2708.

D'Arcy B.J., Ellis J.B., Ferrier R.C., Jenkins A., Dils R. (eds.) (2000) Diffuse pollution impacts, the environmental and economic impacts of diffuse pollution in the UK, *Chartered Institution of Environmental Management(CIWEM), Lavenhum Press.*

Dunn J. C. (1973) A Fuzzy Relative of the ISODATA process and its use in detecting compact well-separated clusters. *Journal of Cybernetics* 3: 32-57

Goolsby D. A., Battaglin, W. A. (1993) Occurrence, distribution, and transport of agricultural chemicals in surface waters of the midwestern United States, In: Goolsby D. A., Boyer L.L, Mallard G.E. (Eds.), Selected papers on agricultural chemicals in Water Resources Midcontinental United States. *U S Geological Survey Open*-File Rep. 93-418: 1-24

Goolsby D. A., Battaglin, W. A. (1995) Occurrence, and distribution of pesticides in rivers of the midwestern United States, In: Leng, M.L., Leovey, E.M.K., Zubkoff, P.L.(Eds.), *Agrochemical Environmental Fate: State of the Art.* CRC Press Inc., Boca Raton, FL: 159-173.

Guillaume S, Charnomordic B. (2004) Generating an interpretable family of fuzzy partitions from data. *IEEE transactions on Fuzzy Systems,* 12 (3): 324-335.

Jha, R., Ojha, C.S.P, Bhatia, K.K.S., 2005. Estimating nutrient outflow from agricultural watersheds to the river Kali in India. *Journal of Environmental Engineering*, ASCE, 131(12), 1706–1715.

Khrisnapuram R. (1998) Membership function elicitation and learning. In: Ruspini E.H., Bonissone P.P., Pedrryez W. (eds.). *Handbook of Fuzzy Computation*, Institute of Physics Publishing, Dirac House, Temple Bath, Bristol.

Larson Steven J. and Gilliom Robert J. (2001) Regression Models For Estimating Herbicides Concentrations In U.S. Streams From Watershed Characteristics, *Journal of American Water Res. Asstn.* 3(5): 1349-1367.

Mahabir C, Hicks, F.E., Fayek A.R. (2003) Application of fuzzy logic to forecast seasonal runoff. *Hydrol. Process.* 17: 3749-3762.

Mahar, P.S., and Datta, B. (2000) Identification of pollution sources in transient groundwater system. *Water Resource Management* 14(6): 209 - 227.

Nash J. E., and Sutcliffe, J. V. (1970) River flow forecasting through conceptual models. Part 1-A: Discussion principles. *J. Hydrol.* 10: 282–290

Novotny V., Water Quality (2nd edition) (2003), *Diffuse Pollution & Watershed Management.* John Wiley and Sons, New York. ISBN 0-471-39633-8.

Pareira E. Wilfred and Rostad Colleen E. (1990) Occurrence, Distribution and Transport of herbicides and their Degradation Products in the Lower Mississippi River and Its Tributaries, *Envron. Sci. Technol.* 24 (9): 1400-1406.

Ross T.J. (1997) *Fuzzy Logic with Engineering Applications*. McGraw-Hill, Book Co., Singapore.

Rumelhart, D.E., Hinton, G.E., Williams, R.J. (1986) Learning internal representation by error propagation. *Parallel Distributed Processin. 1:* 318-362, MIT Press, Cambridge, Mass.

Schottler S. P., Eisenreich S. J., Capel P. D. (1994) Atrazine, Alachlor, and Cyanazine in a Large Agricultural River System. *Environ. Sci. Technol.,* 28, p 1079-1089.

Scribner, E. A., Goolsby D. A., Thurman E. M., Meyer M. T., Pomes M. L. (1994) Concentrations of Selected Herbicides, Two Triazine Metabolites, and Nutrients in Storm Runoff From Nine Stream Basins in the Midwestern United States, 1990-92. *U.S. Geol. Surv. Open-File Rep.* 94-396.

Singh, R. M. (2008). Fuzzy Rule Based Estimation of Agricultural Diffuse Pollution Concentration in Streams. *Journal of Environmental Science and Engineering* (ISSN036-827 X), 50 (2), 147-152.

Singh, R. M., Datta, B. (2006a) Identification of unknown groundwater pollution sources using genetic algorithm based linked simulation optimization approach. Journal of Hydrologic Engineering ASCE 11(2): 101-109.

Singh R. M., and Datta, B. (2006b) Artificial neural network modeling for identification of unknown pollution sources in groundwater with partially missing concentration observation data. *Water Resources Management (Springer),* 21 (3), 557-572.

Singh, R.M., Singh, R. (2005). Agricultural diffuse pollution estimation using fuzzy rule based model. In: *Proceedings of the International Symposium on Recent Advances in Water Resources Development and Management (RAWRDM-2005),* IIT Roorkee, India, 1010-1018.

Smith, R.M. S., Wheater H.S. (2004) Multiple objective evaluation of a simple phosphorus transfer model *Hydrol. Process.* 18(9): 1703-1720.

Tesfamichael A.A., Caplan A.J., Kaluarachchi J.J. (2005) Risk-cost-benefit analysis of atrazine in drinking water from agricultural activities and policy implications. *Water Resources Research,* 41, W05015, 1-13.

U.S. Environmental Protection Agency (USEPA). 2001. Drinking water exposure assessment for atrazine and various chloro-triazine and hydroxyl-triazine degradates, report, Environ. Fate and Effects Div., Washington, D.C.

U.S. Environmental Protection Agency (USEPA). 2002. Cost of illness handbook, report, Off. of Pollut. Prevention and Toxics, Washington, D.C.

Vadiee, N. (1993). Fuzzy rule-based expert systems –I and II. In M. Jamshidi, N. Vadiee and T. Ross (eds.), *Fuzzy logic and control: software and hardware applications,* Prentice Hall, Englewood Cliffs, N.J., Chapters 4 and 5.

Wang Li-Xin, and Mendel J.M. (1992) Generating Fuzzy rules by Learning from Examples. *IEEE transactions on Systems, Man and Cybernetics* 22 (6): 1414-1427.

World Meteorological Organization (WMO). (1975) Intercomparison of conceptual models used in operational hydrological forecasting. Tech. Rep. No. 429, WMO, Geneva, Switzerland.

Young, R.A., Onstad, C.A., Boesch, D.D., Anderson, W.P. (1989) AGNPS—a nonpoint-source pollution model for evaluating agricultural watersheds. *Journal of Soil and Water Conservation* 44 (2): 168–173.

Zadeh, L.A. (2000). Towards a Perception-Based Theory of Probabilistic Reasoning- Abstract of a plenary address at the conference of North American Fuzzy Information Processing Society, 2000.

Zadeh, L. A. (1965) Fuzzy sets. *Information and Control* 8: 338-353.

4

A Critical View of the Photoinitiated Degradation of Herbicides

Šárka Klementová
Faculty of Science University of South Bohemia
Czech Republic

1. Introduction

The application of herbicides to agricultural soil is a well established and effective practice to control weed growth. Another areas of herbicide application are roads and railways where herbicides are used to mantain the quality of the track and a safe working environment for railway personnel (Torstenson, 2001). Some of total herbicides are used in urban areas, or as algicides in paints and coatings (Lindner et al.,2000). Among the wide range of herbicides available, phenyl-urea and triazine derivatives represent a prominent group, the variety and use of which having increased markedly during the past decades. Many of the compounds in both families are biorecalcitrant, i.e. their microbiological degradation is slow or totally ineffective, they therefore persist in the environment for many weeks or even months after application.

The partial water solubility of triazines and phenylurea herbicides results in their leaching or washing into surface and ground waters from the place of application.

For many important classes of pesticides including phenylurea and triazine herbicides, photoinitiated transformation may be the only relevant elimination process in surface waters. In waste-waters, advanced photochemical oxidation processes (EPA Handbook, 1998) using oxidative agents/UV combination have been under study.

2. Photoinitiated reactions

Each reaction started by an absorption of radiation may be classified as a photochemical or photoinitiated reaction. According to the mechanism of the photoinitiated reaction, photolytic, photosensitized and photocatalytic reactions can be distinguished.

A photolytic reaction is usually understood as a reaction in which the quantum of radiation absorbed has enough energy to cause the breaking of a covalent bond in the substrate compound. Usually highly energetic UV radiation (less than 250 nm) is necessary for this purpose. These reactions cannot proceed on the Earth´s surface since solar radiation reaching the Earth´s surface contains wavelengths greater than 290 nm.

A photosensitized reaction needs a sensitizer molecule. This is a molecule that is able to absorb radiation and to transfer the absorbed excitation energy onto another molecule. The energy can be transferred either onto an organic molecule, substrate (e.g. herbicide molecule), or onto an oxygen molecule as shown in Eqs. 1 - 5.

$$^1Sens + h\nu \rightarrow {}^1Sens^* \tag{1}$$

$$^1Sens^* + {}^1Substrate \rightarrow {}^1Substrate^* + {}^1Sens \rightarrow Product + {}^1Sens \tag{2}$$

$$^1Sens^* \rightarrow through\ ISC \rightarrow {}^3Sens^* \tag{3}$$

$$^3Sens^* + {}^3O_2 \rightarrow {}^1O_2 \tag{4}$$

$$^1O_2 + {}^1Substrate \rightarrow Oxidized\ product \tag{5}$$

Eq.1 represents excitation of the sensitizer from the ground state (which is always a singlet state, i.e. all electrons in the molecule are paired) to the first excited singlet state. Eq. 2 represents energy transfer onto the substrate and its subsequent reaction into a product. Eq. 3 represents the conversion of the sensitizer from the first excited singlet state (all electrons are paired in the molecule in a singlet state) into the first triplet state (where two electrons are unpaired) through so called intersystem crossing (ISC). The sensitizer in the triplet state is able to react with molecular oxygen dissolved in the reaction mixture (Eq.4) because the ground state of molecular oxygen with its two unpaired electrons is a triplet state. If this ISC process did not occur, the reaction would not proceed since a reaction between a singlet and a triplet state molecule is spin-forbidden. The reaction results in the formation of a powerful oxidative species, singlet oxygen that oxidizes organic substrate molecules (Eq. 5).

Photocatalysis may occur as a homogeneous process or as a heterogeneous process. In homogeneous photocatalytic reactions light produces a catalytically active form of a catalyst. E. g. ferric ions may be reduced photochemically in the presence of an electron donor to ferrous ions that exhibit much higher catalytic activity. The subsequent catalytic reaction of a substrate is a ´dark´ reaction, i.e. not photochemical, since the reaction does not need light. Heterogeneous photocatalysis includes photochemical reactions on semiconductors. It proceeds via the formation of an electron-hole pairs under irradiation. These holes and electrons react with the solvent (water) and dissolved oxygen to produce an oxidative species, mainly OH radicals (Eqs. 6 – 11).

$$h^+ + H_2O \rightarrow HO^\bullet + H^+ \tag{6}$$

$$h^+ + OH^- \rightarrow HO^\bullet \tag{7}$$

$$O_2 + e^- \rightarrow O_2^{\bullet-} \tag{8}$$

$$O_2^{\bullet-} + H^+ \rightarrow HO_2^\bullet \tag{9}$$

$$2HO_2^\bullet \rightarrow H_2O_2 + O_2 \tag{10}$$

$$H_2O_2 + O_2^{\bullet-} \rightarrow HO^\bullet + O_2 + OH^- \tag{11}$$

3. Characterisation of s-triazine and phenylurea herbicides

S-triazine herbicides contain an aromatic ring with three N heteroatoms. The formula of a triazine herbicide, atrazine, is shown in Fig. 1., the formula of a phenylurea herbicide, chlorotoluron, in Fig. 2.

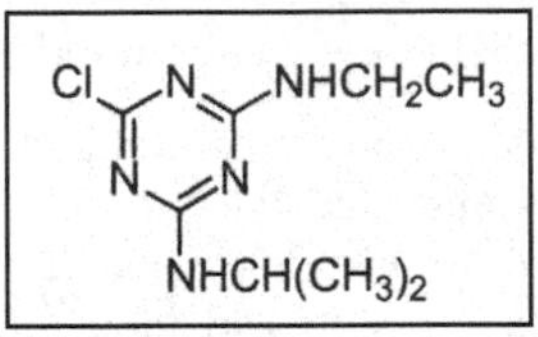

Fig. 1. The structural formula of a triazine herbicide, atrazine.

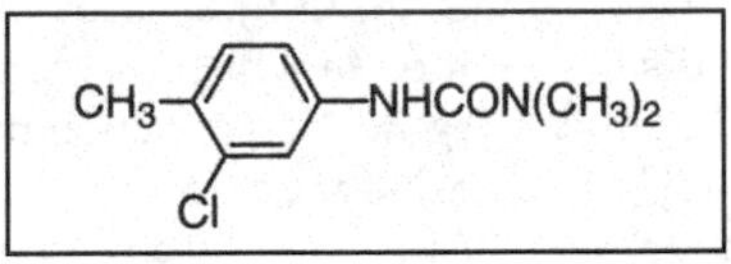

Fig. 2. The structural formula of a phenylurea herbicide, chlorotoluron .

The triazine herbicides were introduced in the 1950s (Gysin & Knüsli, 1957, Gast et al., 1956, both in Tomlin, 2003), phenylurea pesticides a decade later (L'Hermite et al., 1969, in Tomlin 2003).

The solubilities of these herbicides in water are in milligrams or at most tens of milligrams per liter as shown for three trazine and one phenylurea herbicide in Table 1. Table 1 also summarizes the DT_{50} values for the selected herbicides. DT_{50} signifies 50% dissipation time, i.e. the amount of time required for 50% of the initial pesticide concentration to dissipate. Unlike half-life dissipation time does not assume a specific degradation model.

Herbicide	solubility (mg/l)	DT_{50} (days)
Atrazine	33 (22°C)	field: 16 – 77, median 41 natural waters: 10 –105 groundwaters: 105 - >200
Propazine	5.0 (20°C	soil: 80 - 100
Simazine	6.2 (20°C)	soil: 27 - 102
chlorotoluron	74 (25°C)	soil: 30- 40 water: >200

Table 1. Solubilities and DT_{50} values of selected triazine and phenylurea herbicides as given in Tomlin (2003).

All these herbicides are photosynthetic electron transport inhibitors at the photosystem II receptor site. They are all also systemic herbicides. Systemic herbicides (in comparison with contact herbicides) are translocated through the plant, either from foliar application down to the roots or from soil application up to the leaves. They are capable of controlling perennial plants and may be slower in action but ultimitaly more effective than contact herbicides.

4. Biodegradation of selected triazine and phenylurea herbicides

4.1 Biodegradation of triazines

In spite of the fact that triazine and phenylurea herbicides persist in the natural environment for a long time and do not undergo biodegradation easily there are some higher plants and microorganisms capable of metabolizing these compounds.

In tolerant plants triazines as well as phenylurea herbicides are readily metabolized. Plant metabolites include the hydroxy- and dealkylated derivatives of parental compounds. Atrazine (6-chloro-N^2-ethyl-N^4-isopropyl-1,3,5-triazine-2,4-diamine) is metabolized in tolerant plants to hydroxyatrazine and amino acid conjugates, with further decompositon of hydoxyatrazine by degradation of the side-chains. The resulting amino acids on the ring are hydrolyzed and mineralized (i.e. degraded to CO_2). In sensitive plants, unaltered atrazine accumulates, leading to chlorosis (a condition in which leaves produce insufficiant amounts of chlorophylls) and death. The similar degradation or action pathways apply for propazine (6-chloro-N^2,N^4-di-isopropyl-1,3,5-triazine-2,4-diamine) and simazine (6-chloro-N^2,N^4-diethyl-1,3,5-triazine-2,4-diamine). With chlorotoluron (3-(3-chloro-p-tolyl)1,1-dimethylurea), metabolites found in winter wheat include 3-chloro-p-toluidine,3- (3-chloro-4-methylphenyl)-1-methylurea and 1-(3-chloro-4-methylphenyl)urea (Tomlin, 2003).

Behki and Khan studied agricultural soils to which atrazine was applied for a long time. They isolated three bacteria strains (*Pseudomonas* family) capable of utilizing atrazine as the sole source of carbon (Behki & Khan, 1986). Those bacteria use the side-chain carbon, thus N-dealkylation resulting in desisopropylatrazine and desethylatrazine was observed. Two bacterial strains were able to cause the splitting of chlorine from atrazine as well as from the dealkylated metabolites. The same authors proved the capacity to degrade atrazine, propazine, and simazine in the bacteria of *Rhodococcus* species (Behki & Khan, 1994), the degradation rates being however lower than in *Pseudomonas* bacteria.

Not only bacteria but also other organisms such as soil fungal communities have been found to be able to attack and degrade triazines (Kodama et al., 2001).

A *Pseudomonas* bacterial strain was used to degrade atrazine by Wenk (Wenk et al., 1998). The rate of atrazine disappearance was shown to depend on the water content of the soil and on the number of inoculated bacteria; the time necessary for atrazine removal differed ranging from 1 to 25 days. A partial mineralisation of atrazine into CO_2 was also observed.

Such results are in agreement with the findings of Crawford and his coworkers (Crawford et al., 2000), who concluded that the biodegradation rate is affected by the properties of soils and sediments, by agricultural cultivation practices and by the history of triazine application onto the particular soil.

Two genes responsible for s-triazine degradation have been found in four bacterial phyla (Jason Krutz et al., 2010).

4.2 Biodegradation of phenylurea chlorotoluron

Biotransformation of phenylurea herbicides by soil microorganisms (bacterial and fungi) has been reported by several authors (Badawi et al., 2009; Khadrani et al., 1999; Sørensen et al., 2003; Tixier et al., 2002). Bacteria degrade phenylurea herbicides by successive N-dealkylation to substituted aniline products. Fungal pathways result in successive dealkylated metabolites as well as aniline derivatives, but Badawi (Badawi et al., 2009) reported the detection of a new major metabolite which (according to thin layer chromatography and nuclear magnetic resonance spectrometry) is a non-aromatic diol.

Biodegradation by some bacterial and fungal strains leads to the formation of very toxic substituted anilines which have even higher levels of LD_{50} - the dose required to kill half the members of a tested population after a specified test duration time (Tixier et al., 2000a; Tixier et al, 2009). The same applies to products of photochemical degradation (Tixier et al., 2000b).

5. Photochemical degradation of triazine and phenylurea herbicides

5.1 Possible photoinitiated pathways for herbicide degradation

An organic substrate may undergo the following photoinitiated reactions under natural sunlight or artificial source irradiation:

- direct sunlight photodegradation;
- homogeneous photocatalytic degradation in the presence of dissolved metal ions;
- heterogeneous photocatalytic degradation on particulate metal compounds in natural waters;
- heterogeneous photocatalytic degradation on semiconductors;
- photosensitized reaction - reaction in the presence of sensitizers;
- photolytic degradation by short-wavelength irradiation.

For a pollutant the processes given above are schematically visualized in Fig. 3.

5.2 Direct sunlight photodegradation

Direct sunlight photodegradation can proceed with substrates that are able to absorb the solar action spectrum. Solar radiation reaching the Earth´s surface has wavelengths ranging from about 300 nm upwards. Triazine and phenylurea compounds, which absorb at range well below 300 nm (absorption maxima at 220 – 235 nm) cannot therefore undergo direct sunlight photodegradation.

5.3 Homogeneous photocatalytic degradation in the presence of dissolved metal ions

Homogeneous photocatalytic reactions of triazine herbicides in the presence of dissolved metal ions were studied for ferric, copper, and manganese ions (Klementova & Hamsova, 2000). Cupric and manganese (II) ions exhibited only small activities, and only in high concentrations. Table 2 shows the results for atrazine degradation in aqueous solutions under irradiation at a range of wavelengths from 300 to 350 nm. When no metal ions are added, no reaction occurs.

In the case of atrazine the addition of Cu (II) or Mn(II) ions results in conversion below 15 % or less. Ferric ions in comparable concentration cause the conversion of practically all the atrazine in 90 minutes of irradiation. The degradation of atrazine was shown to be strongly dependent on the ferric ion concentration (Fig. 4). Simazine and propazine did not show such a strong dependence on the added ferric ions.

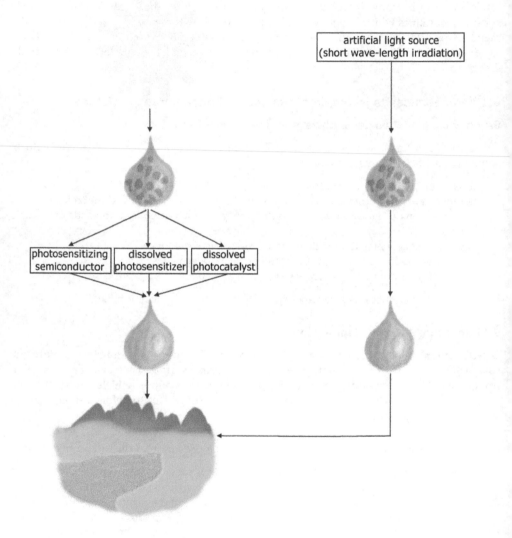

Fig. 3. Scheme of possible degradation pathways of a pollutant non-absorbing solar radiation.

In order to prove the photocatalytic mechanism of the degradation in the triazine solutions, formation of Fe^{2+} ions was measured in the reaction system. The results are set out in Fig. 5.

time of irrad. (minutes)	atrazine consumption (% of initial concentration)						
	no added metal ions	Cu(II) 3.3*10⁻⁴ mol/l	Cu(II) 1.0*10⁻³ mol/l	Mn(II) 1.6*10⁻⁴ mol/l	Mn(II) 1.0*10⁻³ mol/l	Fe(III) 1.0*10⁻⁴ mol/l	Fe(III) 3.3*10⁻⁴ mol/l
0	0	0	0	0	0	0	0
30	0	6	8	1	7	30	97
60	0	8	12	4	8	64	98
90	0	14	15	6	9	98	99

Table 2. Degradation of atrazine in photoinitiated reaction in air saturated aqueous solution in the presence of metal ions. Initial concentration of atrazine $5.0*10^{-5}$ mol/l. Irradiation: Rayonet photochemical reactor RPR 100, lamps 3000Å, emission up to 290 nm filtered by optical glass. (From Klementová & Hamsová, 2000.)

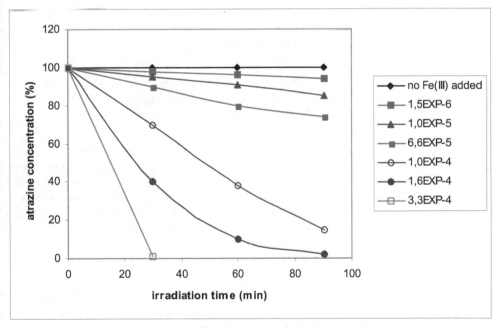

Fig. 4. Effect of ferric ions concentration on atrazine photochemical degradation (conditions of irradiation see Tab.2). Initial concentration of atrazine $5.0*10^{-5}$ mol/l. (From Klementová & Hamsová, 2000).

The photoreduction of ferric to ferrous ions occurs quickly under the irradiation of all three triazines, atrazine, propazine and simazine, though the reaction mixtures were saturated by the air. In the steady state, about 23% of added ferric ions are present in the reduced form in the reaction mixture of atrazine, about 70% in the reaction mixture of propazine, and nearly 90% in the reaction mixture of simazine.

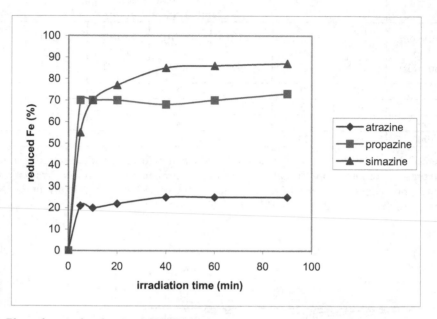

Fig. 5. Photochemical reduction of Fe(III) in the reaction systems with atrazine, propazine and simazine, resp., in the air saturated reaction mixtures. Concentration of substrates $5.0*10^{-5}$ mol/l, concentration of initial Fe^{3+} ions $1.0*10^{-4}$ mol/l. Conditions of irradiation – see Table 2. (From Klementová & Hamsová, 2000).

Homogeneous photocatalytic reactions in the presence of ferric ions may provide a possible pathway for the photochemical degradation of atrazine in water bodies; the problem being that the iron content in natural surface waters is about $1*10^{-5}$ mol/l, a relatively ineffective concentration for atrazine degradation. Other triazine derivatives, propazine and simazine, seem not to be affected by homogeneous photocatalytic degradation in the presence of the ions that are most abundant in natural waters (iron and manganese).

5.4 Heterogeneous photocatalytic degradation

There are no data on the heterogeneous photocatalytic degradation of herbicides with particulate matter in natural waters. Ample studies deal on the other hand with heterogeneous photochemical degradation in relation to semiconductors especially in the context of decontamination option for drinking water and in waste-water treatment.

Semiconductor photocatalysis uses solid catalytic systems where five discrete stages associated with conventional heterogeneous catalysis can be distinguished:

a. transfer of liquid or gaseous phase reactant to the catalytic surface by the diffusion;
b. adsorption of the reactant on the catalyst surface;
c. reaction of the adsorbed molecules;
d. desorption of products;
e. removal of products from the interface region by the diffusion.

The photocatalytic reaction occurs in the stage where the reactants are absorbed on the catalyst surface, the activation of the reaction being photonic activation. The semiconductor is activated by irradiation from a light source of appropriate wavelength depending on the band gap energy of the semiconductor. The activation generates a pair of charge carriers, a hole, h^+, and an electron, e^-; the charge carriers generated photochemically can react with molecules on the surface of the semiconductor (Eqs. 6 – 11 and Fig. 6).

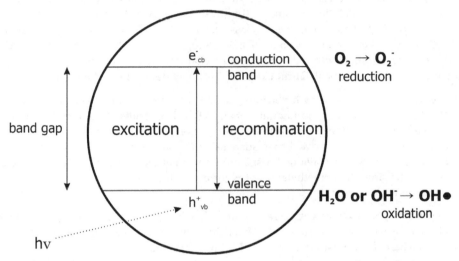

Fig. 6. Scheme of oxidative species production on semiconductors under irradiation.

Various metal oxides, e.g. TiO_2 (Hashimoto et al, 2005; Héquet et al., 2001; Konstantinou et al., 2001a; Linsebigler et al., 1995; Pelizzetti et al, 1990; Penuela& Barceló, 2000)   ZnO (Byrappa et al., 2006), CeO_2 (Yongging Zha et al., 2007), ZrO_2 (Bota et al., 1999), WO_3 (Guo et al., 2007) and many other composites of semiconductors or doped semiconductors have been used as catalysts in semiconductor photocatalytic reactions (e.g. Dunliang et al. , 2009).

TiO_2 – the most widely used semiconductor in contaminant photocatalysis – occurs in three distinct polymorphs: anatase, rutile and brookite. Of these three forms only anatase is functional as a photocatalyst. Anatase is a typical n-type semiconductor with a band gap of about 3.2 eV. Photons with a wavelength shorter then 385 nm have enough energy to excite electrons from the valence band to the conduction band of this material. Since the 1970s, anatase has been a popular choice as semiconductor photocatalyst in research efforts because it is non-toxic and mechanically stable, has high photo-activity and low cost, and exhibits a reasonable overlap with the ultra-violet portion of the solar spectrum which makes it attractive for solar applications. Up to now a multitude of compounds have been investigated as target pollutants in photocatalytic oxidation studies on TiO_2. The studies have been performed at bench scale using small reactors operating as batch or flow reactor systems. Besides pollutant degradation successful tests for the treatment of bacteria, viruses, fungi, and tumor cells have been reported. Construction materials coated with TiO_2 exhibit self-cleaning properties (Devilliers, 2006).

Triazine herbicides photocatalytic degradation on TiO_2 has been studied by several authors, e.g. Héquet et al., 2001; Konstantinou et al., 2001;Pelizzetti et al., 1990; Penueala & Barceló,

2000), in some cases with the addition of oxidative species such as hydrogen peroxide or photo-Fenton system, $H_2O_2/Fe(III)$, providing hydroxyl radicals. Atrazine was found to be degraded to desethylatrazine and desisopropylatrazine, i. e. the same compounds that are metabolites of biodegradation. These metabolites are not easily further degraded in the photocatalytic process on TiO_2.

In our group (Klementová, 2011), we compared the degradation of atrazine in the homogeneous photocatalytic reaction in the presence of Fe (III) and the photocatalytic degradation on TiO_2 (batch experiment, glass coated with TiO_2, irradiation by Philips TLD 15 W 08 lamps). The reaction constant of the heterogeneous photocatalytic reaction (0.018 min^{-1}) was comparable with the reaction constant in reaction mixtures with higher concentrations of ferric ions (0.021 min^{-1} for Fe(III) concentration 1.4*10^{-4} mol/l).

The degradation of phenylurea herbicides on TiO_2 has been studied e.g. by Amorisco et al.(2006), Haque et al. (2006) and Lhomme et al. (2005). The results of such studies show the importance of operational conditions (adsorption capacity, initial concentrations chlorotoluron, TiO_2 forms – coated or in suspension (Lhomme et al., 2005). The pathway of chlorotoluron degradation contained a substitution of chloride ion by the hydroxyl group on the aromatic ring, the demethylation of N group on the side chain, and in some cases a breaking down of the aromatic ring was observed.

Heterogeneous photocatalysis may represent a feasible pathway for the degradation of herbicides in waste-water treatment or even drinking water treatment, especially under conditions where the aromatic ring structure is broken down.

5.5 Photosensitized reactions

Photosensitized reactions may proceed in natural waters in the presence of natural sensitizers such as humic substances. Humic substances originate from the decay of plant and animal biomass and humification reactions in the decaying material. The molecules of humic substances are of variable structure and size (molecular weight ranging from several hundreds to several hundreds of thousands). Humic substances are classified into three operational classes:

- humic acids, which are non-soluble under low pH values,
- fulvic acids, which are soluble at all pH values,
- humins, which is the insoluble fraction.

Humic acids and fulvic acids have an acidic character due to their substential content of carboxylic and phenolic functional groups (Schnitzer & Khan, 1972); Dojlido & Best, 1993). The basic structural features of humic and fulvic acids are shown in Fig. 7.

Humic and fulvic acids have featureless absorption spectra with increasing absorption from the short-wavelengths of visible light through the ultraviolet radiation range.

Photosensitizing properties resulting in the production of singlet oxygen molecules ($^1O_2$), superoxide anions (O_2^-), hydroxyl radicals (HO$^•$), peroxyradicals (ROO$^•$), and hydrated electrons (e_{aq}^-) have been well established (Cooper et al., 1989; Hoigné et al., 1989; Mill T., 1989; Simmons & Zepp, 1986).

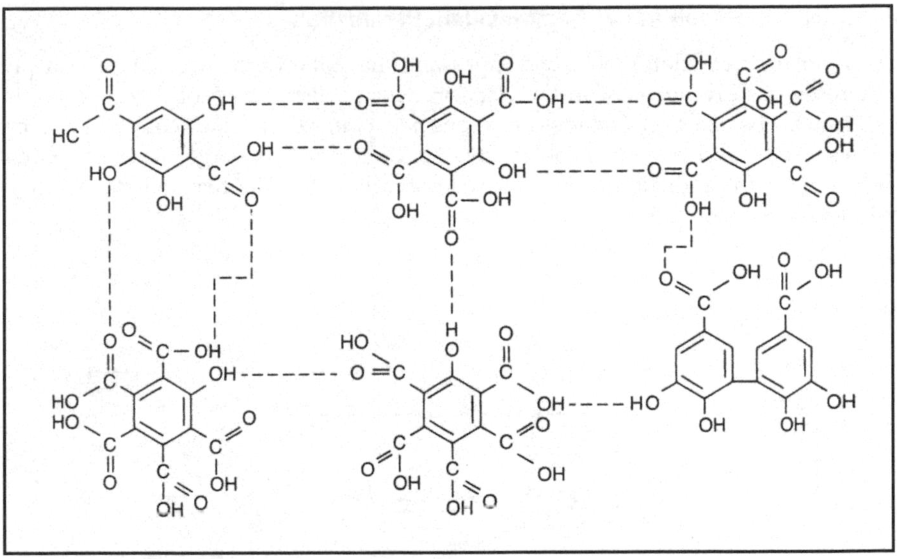

Fig. 7. Structure of fulvic acids. (From Dojlido & Best, 1993).

The photosensitized degradation of triazine and phenylurea herbicide in the presence of humic substances has been studied by several authors, e.g. Amine-Khodja et al. (2006); Comber (1999); Gerecke et al. (2001); Klementova & Piskova (2005); Konstantinou et al. (2001b), Minero et al. (1992) and Schmitt et al. (1995). The results suggest that there is no unambiguous answer about the influence of humic substances. Some authors report better degradation of the substrates, other report decrease in reaction rates in the presence of humic substances. The explanation probably lies in the combination of absorption characteristics of humic samples, their concentrations and the light sources used in the studies. In concentrated humic waters, inner filtration (i. e. the absorption of a significant part of the radiation energy by the photosensitively inactive parts of humic molecules) may play an important role and cause a decrease in the reaction rate of degradation. The heterogeneous chemical character of humic fractions may also be responsible for the variable photosensitizing activities of individual humic samples.

Two groups of artificial sensitizers which provide defined oxidative species were studied in our group for triazine and triazine metabolite degradation: phthalocyanines, i.e. photosensitizers providing singlet oxygen, and anthraquinonesulfonate causing formation of superoxide anions (Klementová & Hamsová, 2000). To our surprise phthalocyanines (aluminium-chloro-phthalocynanine-disulfonate and zinc-phthalocyanine-trisulfonate) showed no observable effect. Anthraquinonesulfonate presence in the aqueous solutions of triazine herbicides (atrazine, propazine, simazine) and the two of atrazine metabolites (desethylatrazine and desisopropylatrazine) resulted in a relatively swift degradation (Fig. 8). Anthraquinonesulfonate was repeatedly added to the reaction mixtures since its molecules are degraded by UV light. This result suggests that triazine herbicides are readily degradable by superoxide species. Nevertheless, the aromatic ring is not broken down so the decomposition is incomplete as it is in other sensitized and catalyzed reactions.

5.6 Photolytic degradation by short-wavelength radiation

Direct photolytic degradation is a decomposition that follows the absorption of a photon (and therefore a rearrangement in the electron density distribution of the molecule in the excited state). The reaction includes only one reactant, i.e. the molecule that undergoes photolysis. The products of a photolytic splitting may undergo another photolytic decompositon if the radiation is of a suitable wavelength. The reaction follows the first order kinetics scheme (Eq. 12).

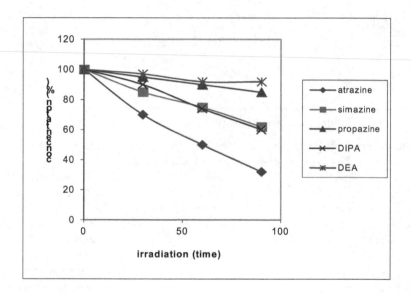

Fig. 8. Photosensitized degradation of triazine herbicides atrazine, simazine and propazine, and atrazine metabolites desethylatrazine (DEA) and desisopropylatrazine (DIPA) with anthraquinone sulfonate as the sensitizer. Initial concentration of individual substrates: $5.0*10^{-5}$ mol/l. Concentration of anthraquinonesulphonate after addition: $1*10^{-4}$ mol/l, addiotions each 30 minutes. Irradiation: Rayonet photochemical reactor RPR 100, lamps 3000Å, emission up to 290 nm filtered by optical glass. (From Klementová & Hamsová, 2000).

$$A \rightarrow B \tag{12}$$

To achieve a photolytic decomposition highly energetic radiation is necessary. Usually a low pressure mercury lamp (emitting most radiation energy at the 254 nm wavelength) is used in these experiments. It is therefore obvious that such processes cannot contribute to herbicide degradation on the Earth's surface, but have their potential in waste-water and drinking water treatment.

Photolytic degradation of triazine and phenylurea herbicides has been studied by several authors. Frimmel & Hessler (1994) irradiated atrazine, desethyatrazine and simazine by low pressure mercury lamp. The rate constants of individual reaction were identical ($1.9*10^{-4}$ s^{-1}). Palm & Zetzsch (1996) carried out kinetic experiments with atrazine, propazine and simazine irradiated by xenon lamp in quartz vessels. Their kinetic evalutation gave the rate constants similar to those calculated by Frimmel & Hessler (1994); slightly higher rate constants and differing for the individual substrates studied were gained by Klementová & Píšková (2005) who irradiated atrazine, simazine, propazine, desethylatrazine and desisopropylatrezine by RPR 3000Å lamps (wavelength range 250 – 350 nm) – see Table 3.

triazine	atrazine	propazine	simazine	DEA	DIPA
rate constant (s^{-1})	$4.64*10^{-4}$	$4.35*10^{-4}$	$5.45*10^{-4}$	$5.86*10^{-4}$	$6.33*10^{-4}$

Table 3. First-order kinetics rate constant for photolytic UV degradation (lamps RPR 3000Å) of triazine and triazine derivatives. DEA – desethylatrazine; DIPA – desisopropylatrazine.

Phenylurea herbicides UV photolysis has been studied e.g. by Benitez et al. (2006) for chlorotoluron, diuron, isoproturon, and by Klementová & Zemanová (2008) for chlorotoluron. Benitez et al. (2006) reported a dependence of the reaction rate on the pH value of the solution; the results published by Klementova & Zemanová (2008) did not support the reported pH dependence, the degradation was pH independent in the range of pH values from 2 to 11.

Measuring the content of dissolved organic carbon (DOC) by DOC analyzer revealed that photolysis in solutions saturated with air results in the partial mineralization of organic substrates, i.e. decomposition of the organic carbon into CO_2. About 20 % of organic carbon was mineralized in 90 minutes of irradiation.

Photolytic degradation by short-wavelength radiation therefore apparently represents a powerful tool for herbicides degradation in waste-water and drinking water treatment, since it leads to total decomposition of organic matter.

5.7 Photochemical degradation of triazine and phenylurea herbicides – common features

In all cases where photochemical degradation was observed in our experiments, the initial step of the degradation of the triazine and phenylurea herbicides and triazine herbicide metabolites was dechlorination and hydroxyderivative formation. Chlorine was found in the solution as chloride ions, Cl$^-$, that were detected in the reaction mixtures by ion chromatography. Hydroxyderivatives were detected by high performance liquid chromatography with a mass spectrometer as an analyzer. Fig. 9 shows one example of herbicide (chlorotolurone) degradation, and chloride ions and hydroxyderivative formation. In this case, as well as in the case of other triazine substrates, the plots of the substrate decomposition and the chloride formation are perfectly symmetrical. Hydroxyderivatives are intermediates that decompose further with a reaction rate constant nearly equal to that of the original substrate decomposition.

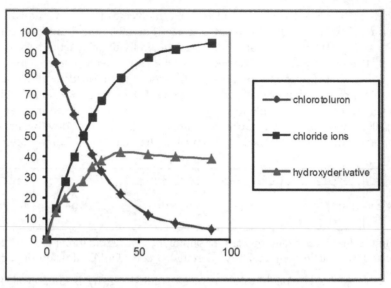

Fig. 9. Chloride ion release and hydroxyderivative formation in chlorotoluron photodecomposition.

6. Conclusions

In order to summarize the findings presented in this chapter on the photoinitiated degradation of triazine and phenylurea herbicides we can conclude:

- Direct sunlight photodegradation cannot proceed in natural surface waters since the substrates absorption maxima do not correspond to the solar action spectrum.
- In most cases natural (humic) sensitizers do not seem to have significant effects on degradation of the substrates. If the concentration of humic sensitizers is low, only a small amount of reactive oxidative species is formed and the degradation is ineffective. If the concentration of humic sensitizers is high, they absorb a lot of radiation themselves, thus radiation is reduced due to inner filtration and cannot reach molecules under the thin surface layer.
- Some artificial sensitizers cause herbicide degradation, but their application in wastewater and drinking water treatment cannot be expected; such sensitizers are expensive for other than small scale laboratory experiments and they themselves together with their degradation products would contaminate the water to which they were applied.
- Homogenous photocatalytic degradation seems to be able to contribute to photodegradation of the substrates in the natural water environment, the typical iron concentrations in natural waters are however not sufficient to bring about a significant conversion of the substrates.
- Heterogeneous photocatalysis with immobilised semiconductors and photolysis remain the only potentially helpful methods for the removal of the recalcitrant herbicides from waste-waters and perhaps even from contaminated drinking waters. The obstacles connected with the use of these two approaches on a larger scale arise from the three-dimensional nature of water purification: in assuring the delivery of sufficient amounts

of light energy to enable purification of higher columns of solutions. With heterogenous photocatalysis the three-dimensionality has one more aspect: the photocatalytic reaction on a semiconductor is a surface process, thus the reactant must be captured by the photocatalyst surface.

- With all processess demanding artificial irradiation the cost of lamps and energy must be taken into consideration.

Nevertheless, environmental pollution including water and soil pollution with herbicides is an increasingly grave problem, and with herbicides resistent to biodegradation and persisting for a long time in the environment the possibilities of photochemical degradation will not cease to attract attention. The possibilities for further development are open especially in the area of heterogeneous photocatalysis. An important key to success will be the utilisation of nano-sized photocatalyst powders dispersed on substrates with extremely large surface areas. Another approach is the modification of TiO_2 to make it sensitive to visible light. So far the researchers investigating in this field are struggling with the issue of low reproducibility and chemical stability, nonetheless heterogeneous photocatalysis represents a promising prospect for 21 century.

7. Acknowledgement

I would like to thank my son David Klement for his help with formulas and schemes drawing.

8. References

Amine-Khodja A., Trubetskaya O. Trubetskoj O., Cavani L., Ciavatta C., Guyot G. & Richard C. (2006). Humic-like substances extracted from composts can promote the photodegradation of Irgarol 1051 in solar light. *Chemosphere*, Vol. 62, No. 6, pp. 1021 – 1027.

Amorisco A., Losito I., Carbonara T., Palmisano F. & Zamboni P.G. (2006). Photocatalytic degradation of phenyl-urea herbicides chlorotoluron and chloroxuron: Characterisation of the by-products by liquid chromatography coupled to electrospray ionization tandem mass spectrometry. *Rapid Comm. Mass Spectrom.*, Vol. 20, pp. 1569 – 1576.

Badawi N., Rønhede S., Olsson S., Kragelund B. B., Johnsen A.H., Jacobsen O.S. & Aamand J. (2009). Metalolites of phenylurea herbicides chlorotoluron, diuron, isoproturon and linuron produced by the soil fungus Mortierella sp. *Environ. Poll.*, Vol. 157, No. 10, pp. 2806 – 2812.

Behki R.M. & Khan S.U. (1986). Degradation of Atrazine by Pseudomonas: N-dealkylation and dehalogenation of atrazine and its metabolites. *J. Agric. Food Chem.*, Vol. 34, pp. 746 – 749.

Behki R.M. & Khan S.U. (1994). Degradation of atrazine, propazine and simazine by Rhodococcus Strain B-30. *J. Agric. Food. Chem.*, Vol. 42, pp. 1237 – 1241.

Botta S.G., Navío J.A., Hidalgo M.C., Restrepo G.M & Litter M.J. (1999). Photocatalytic properties of ZrO_2 and Fe/ZrO_2 semiconductors prepared by a sol-gel technique. *J. Photochem. Photobiol. A: CHem.*, Vol. 129, No 1 –2, pp. 89 – 99.

Byrappa K., Subramani A.K., Ananda S., Lokanatha Rai K.M., Dinesh R. & Yoshimura M. (2006). Photocatalytic degradation of rhodamine B dye using hydrothermally synthesized ZnO. *Bull. Mater. Sci.*, Vol. 29, No. 5, pp. 433 – 438.

Comber S.D.W. (1999). Abiotic persistence of atrazine and simazine in water. *Pestic. Sci.*, Vol. 55, pp. 696 – 702.

Cooper W.J., Zika R.G., Petasne R.G. & Fischer A.M. (1989). Sunlight-induced photochemistry in Humic substances in natural waters: Major reactive species. *Aquatic humic substances: Influence on fate and treatment of pollutants*. Eds.: Suffet J.H., Mac Carthy P. ACS Symposium Series 219, American Chemical Society, Washington D.C.

Crawford J.J., Traina S.J. & Tuovinen O.H. (2000). Bacterial degradation of atrazine in redox potential gradients in fixed-film Sand columns. *Soil Sci. Soc. Am. J.*, Vol.64, pp. 624 – 634.

Devilliers D. (2006). Semiconductor Photocatalysis: Still an Active Research Area Despite Barriers to Commercialization. *Energia , CAER – University of Kentucky, Center for Applied Energy Research*, Vol. 17., No. 3, pp. 1 – 3.

Dojlido J. & Best G.A. (1993). *Chemistry of water and water pollution*. Ellis Horwood Limited, ISBN 0-13-878919-3, Chichester, England.

Dunliang Jian, Pu-Xian Gao, Wenjie Cai, Bamidele S. Allimi, Pamir Alpai S., Yong Ding, Zhong Lin Wang & Brooks C. (2009). Synthesis, characterization, and photocatalytic properties of ZnO/ (La, Sr) CoO$_3$ composite nanorod arrays. *J. Mater. Chem.*, Vol. 19, pp. 970 – 975.

EPA Handbook (1998). *Advanced Photochemical Oxidation Processes*. EPA/625/R-98-004, Washington, DC.

Frimmel F.H. & Hessler D.P. (1994). Photochemical degradation of triazine and anilide pesticides in natural waters. *Aquatic and surface photochemistry*. Eds. Helz G.R., Zepp R.G., Crosby D.G. CRC Press, Inc., Boca Raton, Florida.

Gerecke A.C., Canonica S., Müller S. R., Schärer M. & Schwarzenbach R.P. (2001). Quantification of dissolved natural organic matter (DOM) mediated phototransformation of phenylurea herbicides in lakes. *Environ. Sci. Technol.*, Vol. 35, No. 19, pp. 3915 – 3923.

Guo Y., Quan X., Lu N. Zhao H. & Chen S. (2007). High photocatalytic capability of self-assembled nanoporous WO$_3$ with preferential orientation of (002) planes. *Environ. Sci. Technol.*, Vol. 41, No. 12, pp. 4422 – 4427.

Haque M.M., Muneer M. & Bahnemann D.W. (2006). Semiconductor-mediated photocatalysed degradation of a herbicide derivative, chlorotoluron, in aqueous suspensions. *Environ. Sci. Technol.*, Vol. 40, pp. 4765 – 4770.

Hashimoto K., Irie H. & Fujishima A. (2005). TiO$_2$ photocatalysis: A historical overview and future prospects. *Jap. J. Appl. Physics*, Vol. 44, No. 12, pp. 8269 – 8285.

Héquet V., Gonzalez C. & Le Cloirec P. (2001). Photochemical processes for atrazine degradation: Methodological approach. *Water Res.*, Vol. 35, No. 18, pp. 4253 – 4260.

Hoigné J., Faust B.C., Haag W.R., Scully F.E., Jr. & Zepp R.G. (1989). Aquatic humic substances as sources and sinks of photochemically produced transient reactants. *Aquatic humic substances: Influence on fate and treatment of pollutants*. Eds.: Suffet J.H., Mac Carthy P. ACS Symposium Series 219, American Chemical Society, Washington D.C.

Jason Krutz L., Shaner D.L.,Weaver M.A., Webb R.M., Zablotowicz R.M., Reddy K.N., Huang Y. & Thomson S.J. (2010). Agronomic and environmental implication of enhanced s-triazine degradation. *Pest. Manag. Sci.*, Vol. 66, No. 5, pp. 461 – 481.

Khadrani A., Seigle-Murandi F., Steiman R. & Vroumsia T. (1999). Degradation of three phenylurea herbicides (chlorotoluron, isoproturon and diuron) by micromycetes isolated from soil. *Chemosphere*, Vol. 38, pp. 3041 – 3050.

Klementová (2011). Photocatalytic degradation of triazine and phenylurea herbicides on TiO_2. *In preparation*.

Klementova S. & Hamsova K. (2000). Catalysis and sensitization in photochemical degradation of triazines. *Res. J. Chem. Environ.*, Vol. 4, pp. 7 – 12.

Klementova S. & Piskova V. (2005). UV photodegradation of triazine pesticides and their metabolites. *Res. J. Chem. Environ.*, Vol. 9, pp. 20 – 23.

Klementová S. & Zemanová M. (2008). UV Photochemical degradation of a phenyl-urea herbicide chlorotoluron. *Res. J. Chem. Environ.*, Vol. 12, pp. 5 - 11.

Kodama T., Ding L., Yoshida M. & Yajima M. (2001). Biodegradation of an s-triazine herbicide, simazine. *J. Molecul. Catalysis B: Enzymatic*, Vol. 11, 1073 – 1078.

Konstantinou I.K., Sakellarides T.M., Sakkas V.A & Albanis. T.A. (2001a). Photocatalytic degradation of selected s-triazine herbicides and organophosphorus insecticides over aqueous TiO_2 suspensions. *Environ. Sci. Technol*, Vol. 35, pp. 398 – 405.

Konstantinou I.K., Zarkadis A.K. & Albanis T.A. (2001b). Photodegradation of selected herbicides in various natural waters and soils under environmental conditions. *J. Environ. Qual.*, Vol. 30, pp. 121 – 130.

Lhomme L., Brosillon S., Wolbert D. & Dussaud J. (2005). Photocatalytic degradation of a phenylurea, chlorotoluron, in water using an industrial titanium dioxide coated media. *Appl. Catal. B: Environ.*, Vol. 61, pp. 227 – 235.

Lindner W., Rohermel J., Taschenbrecker E. & Wohner G. (2000). Algicide combination. *Patent No 6117817 (US)*.

Linsebigler A.L., Guangquan Lu & Yates J.T., Jr. (1995). Photocatalysis on TiO_2 Surfaces: Principles, mechanismsms, and selected results. *Chem. Rev.*, Vol. 95, pp. 735 –758.

Mill T. (1989). Structure – activity relationship for photooxidation processes in the environment. *Environ. Toxicology & Chemistry*, Vol. 8, No. 1. pp. 31 – 45.

Minero C., Pramauro E., Pelizzetti E., Dolci M. & Marchesini A. (1992). Photosensitized transformation of atrazine under simulated sunlight in aqueous humic acid solution. *Chemosphere*, Vol. 24, No. 11, pp. 1597 – 1606.

Palm W.U. & Zetzsch C. (1996). Investigation of the photochemistry and quantum yields of triazines using polychromatic irradiation and UV-spectroscopy as analytical tool. *Intern. J. Environ. Anal. Chem.*, Vol. 65, pp. 313 – 329.

Pelizzetti E., Maurino V., Minero C., Carlin V., Praumaro E., Zebinatti O. & Tosato M.L. (1990). Photocatalytic Degradation of Atrazine and other s- triazine herbicides. *Environ. Sci. Technol.*, Vol. 24, pp. 1559 – 1565.

Penuela G. A. & Barceló D. (2000). Comparative photodegradation study of atrazine and desethylatrazine in water samples containing titanium dioxide/hydrogen peroxide and ferric chloride/hydrogen peroxide. *J. AOAC Int.*, Vol. 83, No. 1. pp. 53 – 60.

Schmitt P., Freitag D., Sanlaville Y., Lintelmann J. & Kettrup A. (1995). Capillary electrophoretic study of atrazine photolysis. *J. Chromatogr. A*, Vol. 709, pp. 215 – 225.

Schnitzer M. & Khan S.U. (1972). *Humic substances in the environment.* Marcel Dekker, Inc., ISBN 0-8247-1614-0, New York.

Simmons M.S. & Zepp R.G. (1986). Influence of humic substances on photolysis of nitroaromatic compounds in aqueous systems. *Water Res.*, Vol. 20, No. 7, pp. 899 – 904.

Sørensen S.R., Bending G.D., Jacobsen C.S., Walker A. & Aamand J. (2003). Microbioal degradation of isoproturon and related phenylurea herbicides in and below agricultural fields. *VEMS Micobiol. Ecology*, Vol. 45, pp. 1 – 11.

Tixier C., Bogaerts P., Sancelme M., Bonnemoy F., Twagilimana L. & Cuer A. (2000a). Fungal biodegradation of a phenylurea herbicide, diuron: Structure and toxicity of metabolites. *Pest. Manag. Sci.*, Vol. 56, pp. 455 – 462.

Tixier C., Meunier L., Bonnemoy F. & Boule P. (2000b). Phototransformation of three herbicides: chlorotoluron, isoproturon, and chlorotoluron: Influence of irradiation on toxicity. *Int. J. Photoenergy*, Vol. 2, pp. 1 – 8.

Tixier C., Sancelme M., A ï t-A ï ssa S., Widehem P., Bonnemoy F. Cuer A., Trufaut N., & Veschambre H. (2002). Biotransformation of phenylurea herbicides by a soil bacterial strain, *Arthrocacter*, sp. N2: Structure, ecotoxicity and fate of diuron metabolite with soil fungi. *Chemosphere*, Vol. 46, pp. 519 – 526.

Tixier C., Sancelme M., Bonnemoy F., Cuer A. & Veschambre H. (2009). Degradation of a phenylurea herbicide, diuron: Synthesis, ecotoxicity, and biotransformation. *Environ. Toxicol.Chem.*, Vol. 20, No. 7, pp. 1381 – 1389.

Tomlin C.D.S. (2003). *The Pesticide Manual.* BCPS (British Crop Protection Council), ISBN 1 901396 13 4, Hampshire, UK.

Torstenson, L. (2001). Use of Herbicides on Railway Tracks in Sweden. *Pest. Outlook.*, Vol.12, pp. 16 – 21.

Wenk M., Baumgartner T., Dobovsek J., Fuchs T., Kucsera J., Zopfi J. & Stucki G. (1998). Rapid atrazine mineralisation in soil slurry and moist soil by inoculation of an atrazine-degrading Pseudomonas sp. strain. *Appl. Microbiol. Biotechnol,* Vol. 49, pp. 624 – 630.

Yongging Zha, Shaoyang Zhang & Hui Pang (2007). Preparation, characterization and photocatalytic activity of CeO_2 nanocrystalline using ammonium bicarbonate as precipitant. *Material Letters*, Vol. 61, No. 8 – 9, pp. 1863 – 1866.

5

Oxidative Stress as a Possible Mechanism of Toxicity of the Herbicide 2,4-Dichlorophenoxyacetic Acid (2,4-D)

Bettina Bongiovanni[1], Cintia Konjuh[1],
Arístides Pochettino[1] and Alejandro Ferri[2]
[1]*Laboratorio de Toxicología Experimental, Departamento de
Ciencias de los Alimentos y Medio Ambiente;*
[2]*Departamento de Química Analítica, Facultad de Ciencias Bioquímicas y Farmacéuticas,
Universidad Nacional de Rosario, Rosario,
Argentina*

1. Introduction

Chlorophenoxy herbicides are widely used in agriculture and forestry, for the control of broad-leaved weeds in pastures, cereal crops, as well as along public rights of way. Structurally, these herbicides consist of a simple aliphatic carboxylic acid moiety attached to a chlorine-substituted aromatic ring via an ether linkage. One of the most commonly used herbicides of this type is 2,4-dichlorophenoxyacetic acid (2,4-D) (Fig. 1). In congruence with the similitude between its molecular structure and that of the plant hormone indole-acetic acid, 2,4-D acts as a plant growth regulator that can interfere with normal hormonal action and plant growth (Munro et al., 1992).

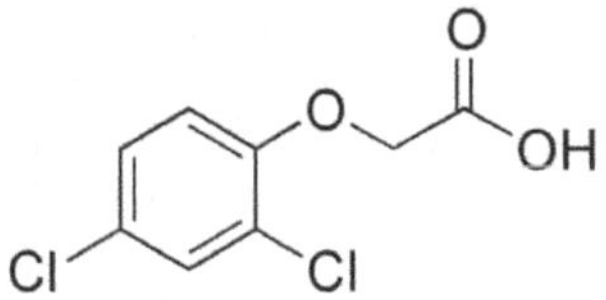

Fig. 1. Structure of the 2,4-Dichlorophenoxyacetic Acid.

2.4-D was synthesized for the first time in 1941 and commercially marketed in the United States (U.S.) in 1944 (IARC, 1986) and worldwide since 1950 (Munro et al. 1992). The widespread use of 2,4-D as a domestic herbicide and as a component of Orange Agent encouraged the study of its toxicity.

Human exposure to chlorophenoxy herbicides may occur through inhalation, skin contact or ingestion. The predominant route for occupational exposure to 2,4-D has been the absorption of spills or aerosol droplets through the skin.

Several studies have shown that doses of 50, 70 or 100 mg/kg body weight (bw)/day of 2,4-D produce a wide range of toxic effects on the embryo and on the reproductive and neural systems in animal (mostly rat) and human models (Rosso et al., 2000; Barnekow et al., 2001; Charles et al., 2001). Doses of 50 mg/kg bw/day of 2,4-D have been reported to increase ventral prostate weight in rats. Treatment of human prostate cancer cell cultures with 10 nM 2,4-D enhanced the androgenic activity of dihydroxytestosterone (DHT) on cell proliferation and transactivation (Kim et al., 2005). In cultured chinese-hamster ovary cells, 2.0 to 10.0 µg/ml 2,4-D were reported to produce DNA damage and sister chromatid exchange (Gonzalez et al., 2005). Importantly, although the 2,4-D toxicity in low doses is controversial, the U.S. Environmental Protection Agency (U.S. EPA, 2006) established a LD50 of 639 mg/kg based on rat studies.

There could be particular situations in which the susceptibility of a population exposed to environmental pollutants can be dangerously enhanced. This may be the case for many rural populations subjected to some specific nutritional deficiencies, as often observed in developing countries. Such situation may be worthy of attention during the development stage, especially concerning the endocrine and nervous systems.

It has been recently found that 2,4-D administered to lactating rats can pass to suckling pups, an can also inhibit the suckling-induced hormone release in the mother. Thus, gestational and lactational periods –including the neonatal and prepubertal stages– seem to be particularly favorable for the induction of 2,4-D effects in rodents (Stürtz et al., 2000; 2006).

2. Adverse effects on developing nervous system

In human studies, prenatal exposure to 2,4-D was associated with mental retardation of the children (Casey, 1984). Comparable animal experiments in chicken and rats showed that prenatal exposure altered some behavioral patterns of the offspring (Sanders & Rogers, 1981; Sjoden & Soderberg, 1972).

In the rat, one critical period for normal maturation during growth seems to be that corresponding to the perinatal development of the brain—"the brain growth spurt"— spanning the first 3 or 4 weeks of life (Diaz & Samson, 1980). Therefore, exposure of rats to pesticides during the first weeks of life would have adverse effects on growth and behavior, as well as on the locomotor activity, as affected by anatomical changes. Noteworthy, the age at exposure is an important factor (Kolb & Wishaw, 1989).

This selective susceptibility of the developing nervous system may be due to several toxicokinetic factors and a partial lack of a blood–brain barrier (BBB) in the fetus. In humans, the BBB is not fully developed until the middle of the first year of life (Rodier, 1995).

Gupta et al. (1999) have shown that different classes of pesticides are able to change the permeability characteristics of the BBB in rats when administered during some susceptible periods of the BBB development, and that this effect may persist after exposure for variable periods. An altered BBB may render the nervous system more vulnerable to other toxics that would not be able to pass the BBB otherwise.

Therefore, although the developing nervous system has some capacity to adapt to or compensate for early perturbations, many chemical agents have been shown more toxic on the developing than on the adult nervous system (Tilson, 1998).

In the last two decades many different alterations have been reported in neonatal rats exposed to 2,4-D through breast milk, at a dose producing no overt signs of toxicity in dams. Alterations in astroglial cytoarchitecture and neuronal function (Brusco et al., 1997) as well as neuro-behavioral changes were observed in pups and adult rats after an early exposition to the herbicide (Bortolozzi et al., 1999, 2001). Other reported effects in neonate rats were a deficit in myelin lipid deposition (Konjuh et al., 2008) and changes in the ganglioside pattern in some brain regions (Rosso et al., 2000).

2.1 Metals and monoamines levels

Studies in well-fed or undernourished rat offsprings showed that the mechanisms for the induction of the above effects would include some changes in brain monoaminergic system (Ferri et al., 2000) and in iron (Fe), copper (Cu) and zinc (Zn) brain levels (Ferri et al., 2003).

Importantly, the combination of neonatal undernourishment plus mothers' exposure at 2,4-D low dose (70 mg/kg bw) induced a higher modification of the measured parameters than those induced by undernourishment or 2,4-D exposure alone. The data showed a different pup's brain areas susceptibility to the 2,4-D effects and an increased vulnerability to the herbicide, including an increased mortality at a higher dose (100 mg/kg bw), a feature which was not observed in well-nourished animals.

In addition, the results suggest that malnutrition or exposure to 2,4-D exert their effects independently (Tables 1 & 2) (Ferri et al., 2003) and the fact that the alterations observed are very different according to the area involved, reinforces the idea of a selective susceptibility for each brain region.

2.2 Oxidative stress

Different studies suggest some functional relationships between the oxidative status of the Central Nervous System (CNS) and the protecting level of catecholamines (Kumiko et al., 2001) and metals, like Fe and Cu, the major generators of reactive oxygen species –ROS- in Alzheimer's disease (Huang et al., 1999), related with a decreased glutathione (GSH) content (Dringer, 2000) and also involved in Fenton's and Haber Weiss' redox reactions . (Halliwell & Gutteridge, 1998; Milton, 2004). Other data have shown that 2,4-D affects the redox chain, thus altering cell energetic metabolism and redox balance (Palmeira et al., 1994; Sulik et al., 1998; Bukowska et al., 2003; Duchnowicz et al., 2002).

In rat pups, exposure to 2,4-D through breast milk induced a number of changes in different brain areas, such as disparate changes in the activity of some protective enzymes, an increase in reactive oxygen species (ROS) levels, and a depletion of reduced glutathione (GSH) content (Tables 3, 4 & 5, respectively) (Ferri et al., 2007).

Therefore, as long as a high oxygen consumption by the CNS increases its sensitivity to oxidative stress (Emerit et al., 2004), the observed changes in the levels of metal ions and neurotransmitters, particularly catecholamines, as well as the oxidative status imbalance, would point out oxidative stress as one possible mechanism of adverse 2,4-D effects on the CNS.

AREA	Treatment	NE	DA	DOPAC	HVA	TRP	5-HT	5-HIAA
PFc	DMSO	0.93 ± 0.04	3.20 ± 0.44	0.97 ± 0.09	0.28 ± 0.03	20.71 ± 0.61	1.07 ± 0.13	1.08 ± 0.07
	2,4-D	1.10 ± 0.06* (↑20%)	2.01 ± 0.30* (↓37%)	0.90 ± 0.17	0.25 ± 0.03	11.26 ± 0.51** (↓46%)	1.48 ± 0.09* (↑38%)	1.03± 0.05
Str	DMSO	4.24 ± 0.42	20.79 ± 1.61	9.37 ± 0.48	3.19 ± 0.12	27.46 ± 1.61	2.98 ± 0.31	2.84 ± 0.29
	2,4-D	2.36 ± 0.44** (↓44%)	17.03 ± 3.31	6.96 ± 0.69** (↓26%)	2.05 ± 0.32** (↓36%)	28.06 ± 1.76	1.85 ± 0.25* (↓38%)	2.81 ± 0.42
Hipp	DMSO	0.91 ± 0.12	0.70 ± 0.09	0.44 ± 0.07	0.30 ± 0.04	5.55 ± 0.35	0.74 ± 0.09	1.63 ± 0.11
	2,4-D	1.67 ± 0.24* (↑83%)	0.92 ± 0.09* (↑31%%)	0.58 ± 0.06* (↑31%)	0.50 ± 0.05* (↑66%)	3.68 ± 0.20** (↓34%)	1.10 ± 0.13* (↑49%)	1.75 ± 0.10
Hyp	DMSO	9.05 ± 1.19	1.58 ± 0.35	1.46 ± 0.21	1.08 ± 0.20	3.81 ± 0.26	1.52 ± 0.13	3.09 ± 0.59
	2,4-D	13.57 ± 1.44* (↑50%)	1.80 ± 0.33	1.03 ± 0.17	1.14 ± 0.18	2.96 ± 0.35	2.08 ± 0.31	2.68 ± 0.26
MB	DMSO	3.16 ± 0.57	1.54 ± 0.31	0.65 ± 0.14	0.36 ± 0.06	31.88 ± 1.21	2.79 ± 0.21	3.54 ± 0.40
	2,4-D	3.96 ± 0.17	1.96 ± 0.17	0.78 ± 0.12	0.24 ± 0.02	23.34 ± 0.97** (↓27%)	4.14 ± 0.19** (↑48%)	4.78 ± 0.37* (↑35%)
Cereb	DMSO	1.58 ± 0.12	0.17 ± 0.06	0.31 ± 0.01	0.09 ± 0.01	7.39 ± 0.55	0.46 ± 0.03	0.48 ± 0.03
	2,4-D	2.44 ± 0.08** (↑54%)	0.21 ± 0.04	0.29 ± 0.01	0.13 ± 0.01	4.86 ± 0.21** (↓34%)	0.43 ± 0.04	0.43 ± 0.02

Monoamine content is expressed as pMol/mg of tissue. Values indicate means ± SEM. Values between brackets are % of increase (↑) or decrease (↓), respectively, with respect to each DMSO control value.*p < 0.05; **p < 0.01; n= 6/group; 100 mg 2,4-D/kg cw of mother. PFc (Pre frontal cortex), Str (Striatum), Hipp (Hippocampus), Hyp (Hypothalamus), MB (Midbarin), Cereb (Cerebellum), NE (Norepinephrine), DA (Dopamine), DOPAC (3,4-Dihydroxyphenylacetic acid), HVA (Homovanillic Acid), TRP (Tryptophan), 5-HT (Serotonin) and 5-5-HIAA (Hydroxyindoleacetic acid); other abbreviations as indicated in the text.

Table 1. Monoamine levels in different brain areas of 25-day-old, 2,4-D-exposed pups.

Metal	Treatment	PFc	Str	Cereb	Hipp	MB	Hyp
Fe	DMSO	19.58 ± 2.36	17.07 ± 0.90	16.43 ± 1.27	14.23 ± 0.73	18.79 ± 2.03	19.48 ± 2.25
	2,4-D 70 mg/kg	24.48 ± 1.83* (↑25.05 %)	16.65 ± 2.65	17.53 ± 0.86	12.86 ± 1.20	15.54 ± 0.55	20.95 ± 1.76
	2,4-D 100 mg/kg	29.48 ± 2.19* (↑50.56 %)	15.75 ± 2.65	20.26 ± 0.68* (↑23.31%)	13.98 ± 1.80	14.95 ± 0.35	21.31 ± 2.68
Zn	DMSO	12.51 ± 1.20	34.10 ± 2.40	24.40 ± 1.90	25.40 ± 2.10	32.65 ± 1.10	29.80 ± 3.40
	2,4-D 70 mg/kg	14.64 ± 2.20	28.63 ± 2.40* (↓16.04 %)	21.50 ± 0.85	27.89 ± 1.60* (↑9.80%)	29.55 ± 1.74	27.11 ± 2.56
	2,4-D 100 mg/kg	17.93 ± 2.00* (↑43.32%)	13.10 ± 2.00** (↓61.58%)	24.40 ± 0.70	34.30 ± 3.50* (↑35.04%)	24.35 ± 1.54** (↓25.42%)	26.37 ± 3.20
Cu	DMSO	1.85 ± 0.08	2.25 ± 0.11	1.88 ± 0.07	1.84 ± 0.02	2.21 ± 0.19	1.95 ± 0.08
	2,4-D 70 mg/kg	1.97 ± 0.18	2.17 ± 0.13	2.00 ± 0.10	2.01 ± 0.20	2.16 ± 0.21	1.91 ± 0.16
	2,4-D 100 mg/kg	2.31 ± 0.21* (↑ 24.86%)	2.38 ± 0.21	2.20 ± 0.08** (↑17.02%)	2.23 ± 0.17* (↑21.19%)	2.14 ± 0.20	1.91 ± 0.15

Metal contents are expressed as micrograms per gram of wet tissue. Values indicate means ± SEM. Values between brackets are % of increase (↑) or decrease (↓), respectively, with respect to each DMSO control value. *p < 0.05 with reference to DMSO control values. **p <0 .01 with reference to DMSO control values; n= 6/group. 100 mg 2,4-D/kg cw of mother. PFc (Pre frontal cortex, Str (Striatum), Cereb (Cerebellum), Hipp (Hippocampus), MB (Midbarin), Hyp (Hypothalamus), and other abbreviations as in the text.

Table 2. Effects of 2,4-D on iron, zinc and copper levels in different brain areas of well-nourished pups.

Enzime	Treatment	Brain	PFc	Str	Cereb	Hipp	MB	Hyp
Cu,Zn-SOD	DMSO	2330 ± 119	1950 ± 200	2460 ± 150	2730 ± 330	2330 ± 200	1950 ± 120	2320 ± 160
	2,4-D	2400 ± 93	2450 ± 310* (↑ 25.6%)	2540 ± 220	2610 ± 240	2980 ± 320* (↑ 27.9%)	2140 ± 90	2400 ± 120
Mn-SOD	DMSO	250 ± 110	250 ± 80	310 ± 120	320 ± 130	270 ± 90	240 ± 120	370 ± 140
	2,4-D	284 ± 135	150 ± 70	280 ± 100	280 ± 90	390 ± 60	290 ± 110	350 ± 110
CAT	DMSO	2556 ± 150	2950 ± 250	2740 ± 200	2300 ± 130	2580 ± 160	2360 ± 200	2740 ± 150
	2,4-D	1978 ± 133* (↓ 22.5%)	2200 ± 200* (↓ 25.4%)	2250 ± 150* (↓ 17.9%)	2530 ± 170	2690 ± 210	1850 ± 120* (↓ 21.6%)	2810 ± 210
Se-GPx	DMSO	31.52 ± 1.24	30.00 ± 1.43	28.19 ± 2.10	25.62 ± 9.10	29.05 ± 1.14	31.93 ± 1.06	28.89 ± 1.85
	2,4-D	26.76 ± 1.14* (↓ 15.10%)	24.19 ± 1.90* (↓ 19.4%)	22.29 ± 2.86* (↓ 20.9%)	29.52 ± 2.10* (↑ 15.2%)	32.29 ± 1.00* (↑ 11.1%)	27.57 ± 1.03* (↓ 13.6%)	27.97 ± 1.56
noSe-GPx	DMSO	17.71 ± 0.69	20.80 ± 1.08	18.55 ± 0.62	18.18 ± 0.98	20.94 ± 0.69	22.65 ± 0.77	15.81 ± 0.69
	2,4-D	20.32 ± 0.94* (↑ 14.7%)	18.99 ± 0.99	15.65 ± 0.46* (↓ 15.6%)	17.95 ± 1.05	19.90 ± 0.87	19.35 ± 0.82* (↓ 14.6%)	17.01 ± 0.71

Enzyme activities are expresed as miliUnits per miligram of protein. Values indicate means ± SEM. Values between brackets are % of increase (↑) or decrease (↓), respectively, with respect to each DMSO control value. *$p < 0.05$, n= 6/group. 100 mg 2,4-D/kg cw of mother. PFc (Pre frontal cortex, Str (Striatum), Hipp (Hippocampus), Hyp (Hypothalamus), MB (Midbarin), Cereb (Cerebellum), Cu,Zn-SOD (Copper,Zinc superoxide dismutase), Mn-SOS (Manganese superoxide dismutase), CAT (catalase), Se-GPx (selenium-glutathione peroxidase), noSe-GPx (non selenium-glutathione peroxidase),and other abbreviations as in the text

Table 3. Protective Enzymes Activities in brain areas of 25-old-day pups lactationally exposed to 2,4-D.

	Brain	PFc	Str	Cereb	Hipp	MB	Hyp
DMSO	45.1±2.5	17.8±0.7	22.6±0.8	24.0±0.9	18.0±1.0	20.3± 1.0	22.6±1.2
2,4-D	38.0±1.4* (↓ 15.7%)	20.6±0.5* (↑ 15.7%)	25.1±0.7* (↑ 11.1%)	23.7±1.1	18.1±1.1	23.8±0.7* (↑ 17.2%)	21.1±1.3

ROS levels are expressed as IF per mg of protein. Values indicate means ± SEM. Values between brackets are % of increase (↑) or decrease (↓), respectively, with respect to each DMSO control value.; *$p < 0.05$, n= 6/group. 100 mg 2,4-D/kg cw of mother. PFc (Pre frontal cortex, Str (Striatum), Hipp (Hippocampus), Hyp (Hypothalamus), MB (Midbarin), Cereb (Cerebellum), other abbreviations as in the text.

Table 4. ROS levels in brain areas of 25-old-day pups lactationally exposed to 2,4-D.

	Brain	PFc	Str	Cereb	Hipp	MB	Hyp
DMSO	1.22±0.40	1.23±0.29	1.31±0.24	0.79±0,19	0.88±0.19	1.06±0.12	1.07±0.41
2,4-D	1.25±0.38	1.29±0.20	0.82±0.18* (↓ 37.4%)	0.80±0.22	0.94±0.24	0.70±0.15* (↓ 34.0%)	1.08±0.30

GSH levels are expressed as microgram per miligram of protein. Values indicate means ± SEM. Values between brackets are % of increase (↑) or decrease (↓), respectively, with respect to each DMSO control value. *$p < 0.05$, n= 6/group. 100 mg 2,4-D/kg bw of mother. PFc (Pre frontal cortex, Str (Striatum), Hipp (Hippocampus), Hyp (Hypothalamus), MB (Midbarin), Cereb (Cerebellum), other abbreviations as in the text.

Table 5. GSH levels in brain areas of 25-old-day pups lactationally exposed to 2,4-D.

3. Prostate, ovary and breast

The endocrine system of many vertebrate embryos seems to be particularly susceptible to a variety of substances or either natural or anthropogenic origin, including pesticides (Crews et al., 2000). However, there are few studies on developmental toxicology that focus on the 2,4-D's effects on hormone-sensitive organs such as the prostate, ovary and breast.

Free radicals are associated with oxidative stress and are also thought to play some significant roles in reproduction. Induction of oxidative stress by many environmental contaminants—such as pesticides—has also been pointed out during the last decade as a possible mechanism of some toxic effects on the reproductive system (Bagchi et al., 1992; Abdollahi et al., 2004). It is already known that reproductive cells and tissues will remain stable only when antioxidant and oxidant status are in balance (Lee et al., 2010). ROS levels are a double-edged sword, as long as they not only serve as key signal molecules in physiological processes, but also have a role in pathological processes involving the female reproductive tract (Agarwal et al., 2005).

On the other hand, there are diverse environmental chemical contaminants which can be potentially harmful to the mammary gland in association with estrogens. Oxidative catabolism of both estrogen and those compounds, a mechanism mediated by the same enzymes, generates reactive free radicals that can cause oxidative damage. Xenobiotic chemicals may exert their pathological effects through generation of reactive free radicals (Mukherjee et al., 2006).

There is growing evidence that free radicals can exert a wide spectrum of deleterious effects on the reproductive system and asocciated glands (Saradha et al., 2008). Thus, Pochettino et al. (2010) investigated the effect of 2,4-D on oxidative stress and antioxidative system and on some hormone-sensitive organs such as ventral prostate, ovaries and breasts, exposed to the herbicide during the pre- and the postnatal period, as described next (Pochettino et al., 2010).

3.1 Prostate

In rat ventral prostate, 2,4-D caused oxidative stress during the whole development, through a significant increase in lipid peroxides, hydroxyl radical levels and protein oxidation. Morevover, the antioxidant enzyme activity was increased at any age, as shown for Glutathione S-transferase (GST), catalase (CAT) and selenium-glutathione peroxidase (Se-GPx), with the exception of Se-GPx administered at the 90th postnatal day (PND 90). Nevertheless, at PND 90 a reduced activity of Glutathione Reductase (GR) was detected (Table 6).

GST is relevant to detoxification of endogenous compounds and xenobiotic substances such as environmental pollutants, drugs, and natural toxins (Pietsch et al., 2001; Padros et al., 2003; Cazenave et al., 2006). Several studies have demonstrated that enhanced GST activity by ROS in the testis could represent an adaptative response to oxidative stress, probably targeted to achieve a detoxification of peroxide-containing metabolites (Kaur et al., 2006).

As far as the testis is intimately related to the prostate, this interpretation looks coherent with the observed ROS-induced increase in GST activity in the prostate.

		PND 45	PND 60	PND 90
Hydroxyl radical	Control	3.25±0.34	2.97±0.39	1.09±0.13
	2,4-D	8.75±0.61* (↑169%)	6.53±0.09* (↑119%)	2.03±0.18* (↑85%)
Carbonyl groups	Control	3.54±0.12	10.66±1.07	7.02±0.88
	2,4-D	4.84±0.11* (↑37%)	15.01±1.32* (↑47%)	12.42±1.11* (↑77%)
Total Thiols	Control	491±12	642±86	341±24
	2,4-D	520±14	748±14	333±8
MDA	Control	29.09±0.32	27.74±3.74	38.27± 2.14
	2,4-D	41.91±3.05* (↑44%)	42.69±3.13* (↑54%)	47.48 ± 2.54* (↑24%)
GST	Control	8.93±0.67	10.53±2.53	13.37±2.09
	2,4-D	19.07±2,45* (↑113%)	15.95±1.04* (↑45)	18.14±0.26* (↑36)
CAT	Control	10.93±1,20	5.44±0.21	5.99±0.21
	2,4-D	14.74±1.26* (↑35%)	11.44±0.34* (↑110%)	7.77±0.39* (↑28%)
Se-GPx	Control	312±18	518±57	562±32
	2,4-D	436±33* (↑36)	880±41* (↑70%)	530±13
GR	Control	10.05±0.86	33.01±2.52	31.09±4.36
	2,4-D	10.47±1.78	37.75±3.48	10.57±0.05* (↓72%)

Hydroxy radical are expresed as 2,3 dihydroxybenzoic acid/salicilc acid rario; carbonyl groups and total thiols are expresed as micromol per miligram of protein; MDA is expresed as nanomol per microgram of protein. GST, CAT and GR activities are expressed as Units per miligram of protein; and Se-GPx is expresed as miliUnits per miligram of protein. Each value is the mean ± SEM. Values between brackets are % of increase (↑) or decrease (↓); *p < 0.05, n= 6/group. 70 mg 2,4-D/kg cw of mother. Abbreviations as in the text.

Table 6. Oxidative parameters in ventral prostate.

Therefore, the 2,4-D-induced increase in all ROS level, lipid peroxidation and protein oxidation may have caused some critical oxidative stress in ventral prostate. Nevertheless, the increased activity of some antioxidant enzymes in the prostate could have not been strong enough as to counteract the oxidative stress produced by the herbicide at different stages of rat development. Moreover, it is not a general rule that increase in oxidative species stimulates antioxidant activity (Celik & Tuluce, 2007).

3.2 Ovary

The complex ovarian structure varies widely during differentiation. Free radicals play important regulating roles during the ovarian follicular cycle, possibly through inhibition of steroid production (Behrman et al., 2001). There is also a delicate balance between ROS and antioxidant enzymes in the ovarian tissues (Agarwal et al., 2005). Non-physiological effects of free radicals include premature ovarian follicular atresia via cell apoptosis. Many pesticides— e.g. the xenoestrogen pesticide methoxychlor — can induce oxidative stress and apoptosis in the ovary (Gupta et al., 2006). Moreover, clinical studies have reported increased levels of reactive oxygen species associated to a decreased female fertility (Agarwal et al., 2006).

		PND 45	PND 60	PND 90
Hydroxyl radical	Control	3.65±0.26	1.89 ± 0.22	1.09 ± 0.13
	2,4-D	8.75±0.89	1.98 ± 0.13	4.35 ± 0.53* (↑93%)
Carbonyl groups	Control	14.77±2.75	5.74 ± 0.13	6.22 ± 0.94
	2,4-D	23.71±0.47* (↑60%)	8.80±0.72* (↑55%)	5.64±0.73
Total Thiols	Control	1462±162	672±24	519±38
	2,4-D	1360±176	676±39	537±22
MDA	Control	87.71±14.02	34.12±2.24	34.68±1.31
	2,4-D	192.5±17.8* (↑119%)	42.49±1.35* (↑24%)	39.39±0.89* (↑14%)
GST	Control	38.51±0.41	10.59±0.81	10.99±0.18
	2,4-D	25.15±1.37 (↓34.6%)	7.97±0.54* (↓24.7)	9.91±0.57
CAT	Control	42.89±3.14	27.86±1.08	16.08± 0.42
	2,4-D	43.41±0.67	15.34±0.43* (↓44.9%)	16.38±0.71
Se-GPx	Control	691±97	411±48	514±29
	2,4-D	1622±117* (↑135)	549±24* (↑33%)	593±23* (↑15%)
GR	Control	14.48±3.44	17.15±1.67	28.64±2.31
	2,4-D	12.67±2.61	16.88±1.45	19.62±1.75* (↓31%)

The parameters are expresed as in Table 7. Each value is the mean ± SEM. Values between brackets are % of increase (↑) or decrease (↓); *$p < 0.05$, n= 6/group. 70 mg 2,4-D/kg cw of mother. Abbreviations as in the text.

Table 7. Oxidative parameters in ovary.

On analyzing the 2,4-D toxic effects on the ovary, Pochettino et al. (2010) found an increase in lipid peroxide (LPO) evidenced by augmented levels of malondialdehyde (MDA) and decrease antioxidant enzyme activity. These effects differed with age, while an increase in Se-GPx activity was exceptionally observed at all ages (Table 7). These effects could reflect the natural diversity of rat ovarian cell types at different ages. Another explanation would be the well-known, protecting effect of estrogens against apoptosis and oxidative stress in a variety of tissues and cells (Spyridopoulos et al., 1997; Tomkinson et al., 1997; Garcia-Segura et al., 1998; Pelzer et al., 2000). Estrogens increase all ovarian weight, follicular growth, and the mitotic index of granulose cells, and also control granulosa cell apoptosis (Richards, et al., 1980; Bendell & Dorrington, 1991) and have exerted varied antioxidant effects (Chatterjee & Chatterjee 2009). Further studies are needed to analyze the time-course of the effects observed.

3.3 Breast

Pocchetino et al. (2010) observed that 2,4-D increased MDA levels at all ages (Table 8). It is known that MDA reflects the extent of oxidant status and is considered a good marker of oxidative stress (Wen et al., 2006). Both, singlet oxygen and hydroyl radicals have a high potential to initiate free-radical chain reactions in lipid peroxidation (Celik & Tuluce, 2007). As the hydroyl radical level was unchanged in that study, 2,4-D could have stimulated LPO by increasing singlet oxygen levels. In addition, 2,4-D inhibited the activity of anti-oxidative enzymes such as CAT, Se-GPx, GR and GST (Table 9).

		PND 45	PND 60	PND 90
Hydroxyl radical	Control	2.61±0.11	3.04±0.11	4.31±0.45
	2,4-D	2.65±0.44	3.49±0.52	4.49 ± 0.11
Carbonyl groups	Control	19.25±0.82	28.57±3.86	59.38±10.69
	2,4-D	21.34±5.47	23.31±5.49	57.37±14.89
Total Thiols	Control	942±5	1072±77	3551±757
	2,4-D	951±25	667±46* ($\downarrow$62%)	1560±226* ($\downarrow$56%)
MDA	Control	52.62±1.57	71.07±4.68	158.41±2.59
	2,4-D	70.65±7.48* ($\uparrow$34%)	139.2±17.94* ($\uparrow$96%)	217.8±18.95* ($\uparrow$37%)
GST	Control	17.18±0.59	19.41±1.51	72.81±7.41
	2,4-D	10.41±1.91* ($\downarrow$40%)	13.54±0.92*($\downarrow$30%)	32.35±5.98* ($\downarrow$55%)
CAT	Control	59.38±3.03	137.62±10.73	358.21±36.31
	2,4-D	62.55±1.57	81.27 ± 2.55* ($\downarrow$41%)	122.11±17.42* ($\downarrow$66%)
Se-GPx	Control	538±44	1430±31	5596±1015
	2,4-D	198±19* ($\downarrow$63%)	695±15* ($\downarrow$51%)	2257±474* ($\downarrow$60%)
GR	Control	15.94±0.91	90.75±5.51	228.81±14.31
	2,4-D	7.39±1.54* ($\downarrow$53.6%)	60.02±9.05* ($\downarrow$34%)	76.02±10.95* ($\downarrow$67%)

The parameters are expresed as in Table 7. Each value is the mean ± SEM. Values between brackets are % of increase ($\uparrow$) or decrease ($\downarrow$); *p < 0.05, n= 6/group. 70 mg 2,4-D/kg cw of mother. Abbreviations as in the text.

Table 8. Oxidative parameters in breast

Therefore, the decreased activity of anti-oxidative enzymes may decrease the protection against oxidants (Amstad et al., 1991).

In that regard, Dimitrova et al. (1994) suggested that the superoxide radicals, either by themselves or after transformation to H_2O_2, stimulate cysteine oxidation and inhibit the activity of the enzymes. Furthermore, Regoli & Principato (1995) demonstrated that the flux of superoxide radicals inhibits CAT activity. Consequently, the decreased CAT activity might have reflected a flux of superoxide radicals promoted by 2,4-D. Moreover, GR also plays an important role in cellular antioxidant protection, catalyzing the reduction of glutathione disulfide (GSSG) to GSH (Kim et al., 2010).

Thus, the decrease in thiol groups could reflect GSH depletion in the breast. Therefore, 2,4-D produced oxidative imbalance, mainly during puberty and adulthood, probably because the gland is more sensitive to xenobiotics at these stages of development.

4. In vitro studies

It has been observed that 2.4-D concentrations of 1 to 2 mM impaired neurite outgrowth, disrupted the cytoskeleton, and disorganized the Golgi apparatus in cultured cerebellar granule cells (CGC) (Rosso et al., 2000). Futhermore, Kaioumuva et al. (2001b) have demonstrated that the dimethylamonium salt of 2,4-D (DMA 2,4-D) at 0.1 to 5 mM induces apoptosis in a dose- and time-dependent pattern in peripheral blood lymphocytes of healthy individuals and in Jurkat cells. Whereas, Tuschl & Schwab (2003) showed that 4 to 16 mM 2,4-D induces cytotoxic effects and apoptosis in HepG2 cells.

In rat CGC, either 1 or 2 mM 2,4-D induced similar increases of cellular death. The herbicide decreased significantly mean neuronal survival (46.4%) after 48 h, while no affect was observed after 24 h of treatment (Bongiovanni et al., 2007, 2011) (Fig. 2).

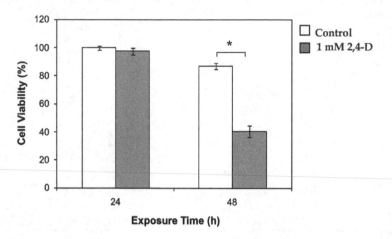

Fig. 2. Effect of 2,4-D on rat cerebellar granule cell viability. Cell cultures were incubated for 24 or 48 h in presence or ausence of 1 mM 2,4-D. Values are means ± SEM; * indicates p< 0.001 vs. control group; n= 10/group.

Bongiovanni et al. (2007, 2010) studied oxidative stress as a possible mechanism of toxicity aiming to elucidate the mechanism of death induction by 2,4-D. Oxidative stress parameters were altered: ROS level and Se-GPx activity increased whereas CAT activity decreased at both treatment times (24 and 48 h). GSH content was reduced only after 48 h of 2,4-D treatment. However, neither Mn-SOD nor Cu,Zn-SOD activities nor reactive nitrogen species (RNS) levels were affected (Tables 9 & 10). Interestingly, although the oxidative parameters evaluated were modified at the two time-limits studied, the cell viability only decreased at 48 h of treatment. This finding could be explained by a time dependency of this latter alteration.

Parameters	24 h		48 h	
	Control	1 mM 2,4-D	Control	1 mM 2,4-D
ROS	1.03 ± 0.25	2.30 ± 0,22* (↑ 123%)	2.28 ± 0.35	4.13 ± 0.32* (↑ 81%)
RNS	7.45 ± 1.13	8.23 ± 1.85	4.82 ± 0.27	6.05 ± 0.47
GSH	2.408 ± 0.09	2.225 ± 0.09	1.508 ± 0.061	1.125 ± 0.031* (↓ 25%)

Parameters are expresed as micrograms per miligram of protein. Values between brackets are % of increase (↑) or decrease (↓); *p < 0.001, n= 10/group. Abbreviations are indicated in the text.

Table 9. ROS, RNA and GSH levels (means ± SEM) in rat cerebellar granule cell in culture for 24 or 48 h in presence or ausence of 1 mM 2,4-D.

Enzimes	24 h		48 h	
	Control	1 mM 2,4-D	Control	1 mM 2,4-D
CAT	30.97 ± 1.26	15.80 ± 1.23* (↓ 49%)	15.82 ± 1.59	6.52 ± 0.83* (↓ 59%)
(Zn,Cu) SOD	10.43 ± 1.23	10.56 ± 1.45	8,49 ± 1,20	7.05 ± 1.65
(Mn) SOD	4.45 ± 0,88	5.97 ± 1.90	2,86 ± 1,90	1.98 ± 1.00
Se-GPx	9.71 ± 1.20	39.75 ± 2.90* (↑ 309%)	12.73 ± 1.75	33.73 ± 4.31* (↑165%)

Parameters are expresed as Units per miligram of protein. Values between brackets are % of increase (↑) or decrease (↓); *$p < 0.001$, n= 10/group. Abbreviations are indicated in the text.

Table 10. CAT, SODs and GPx activities (means ± SEM) in rat cerebellar granule cells in culture for 24 or 48 h in presence or ausence of 1 mM 2,4-D.

On using a PC-12 cell model, other authors have been previously shown that a depletion of mitochondrial and cytoplasmatic GSH results in increased ROS levels, disruption of the mitochondial transmembrane potential, rapid loss of mitochondial function, decrease in the ATP concentration, and eventually a higher cell death rate (Nieminen et al., 1995; Wüllner et al., 1999).

Therefore, the alteration in oxidative parameters suggest that the possible mechanisms of chlorophenoxy herbicide toxicity could involve dose-dependent cell membrane damage, uncoupling of oxidative phosphorylation, acetylcoenzyme disruption (Bradberry et al., 2000), and an indirect disruption of mitochondrial transmembrane potential which may lead to caspase inactivation (Kaioumova et al., 2001a). Mitochondrial structural modifications and increased permeability of the pores were also reported in association with a ROS increase (Belizário et al., 2007). In contrast, other studies suggest that 2,4-D cytotoxic effects are exerted by apoptosis induction via a direct effect on mitochondria (Tuschl & Schwab, 2003).

In this regard, Bongiovanni et al. (2011), in agreement with De Moliner et al. (2002), demonstrated that 2,4-D induces apoptosis and necrosis in CGC. While De Moliner et al. (2002) showed that 2,4-D-induced apoptosis is associated with and increase in caspase-3 activity preceded by cytochrome-c release from mitochondria, the quantification of ultrastructural changes showed that 1 mM 2,4-D stimulated neuronal death. As much as 49% of necrotic cells and 20% of apoptotic cells were observed, while only 31% of CGC presented normal growth with respect control group (p<0.001; Fig. 3 compared with Fig. 4) (Bongiovanni et al., 2011).

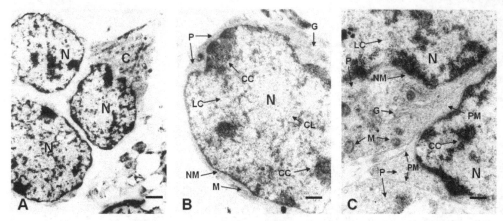

Fig. 3. Electron photomicrographies showing cerebellar granular neurons cultured in a control medium (NaCl 0.9%) for 48 h. a–b. Cell morphology is preserved (nucleus with laxe chromatin, dense chromatin patch close to the nucleus envelope, scarce cytoplasm, and the presence of neurites). Bars correspond to 1 μm in (a) and 160 nm in (b); c. Cells show preserved ultrastructural characteristics (Golgi apparatus, polyribosomes and mitochondrial characteristics of normal granular cerebellar cells). Bars correspond to 320 nm in (c). C cytoplasm, CC dense chromatin, G Golgi apparatus, LC laxe chromatin, M mitochondria, N nucleus, NM nuclear membrane, P polyribosome, PM plasmatic membrane.

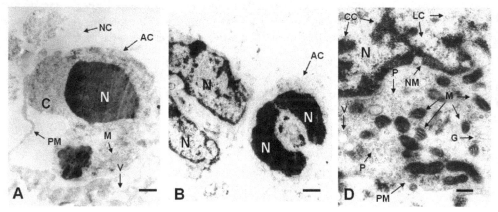

Fig. 4. Electron photomicrographies showing the ultrastructural cytoplasmatic characteristics of cerebellar granular cells after 2,4-D addition to the medium for 48 h. a–b. An apoptotic cell (nuclear fragmentation and very dense chromatinic accumulus), a necrotic cell (cytoplasm very scarce, no nucleus), and cells with scarce cytoplasm and small nucleus are shown, allowing comparison with the control group (Cf Figs. 3a, b). Bars correspond to 1 μm. c. A cell with cytoplasmatic protutions, vacuoles, disorganization of the cytoplasmatic reticulum, distended cisterns of the Golgi apparatus, and mitochondrial swelling. Bars correspond to 400 nm. AC apoptotic cell, NC necrotic cell, V vacuole, and other abreviations in Fig. 3.

In these studies, melatonin and amphetamine were used as phamacological tools aiming to improve the analysis of oxidative stress as a mechanism of toxicity, by assessing whether these compounds could be effective in preventing the toxic effect of 2,4-D in the redox balance of CGC *in vitro* (Bongiovanni et al., 2007, 2011).

A remarkable body of evidence indicates that melatonin exerts antioxidative protection in cell culture and *in vivo* systems (Pandi-Perumal et al., 2006). Regarding to 2,4-D toxicity, the oxidative stress induced by 1 mM 2,4-D was counteracted by the concomitant addition of 0.1 or 0.5 mM melatonin in CGC cultures (Bongiovanni et al., 2007).

On the other hand, amphetamine has constistently been reported to accelerate the recovery of several functions in animals and humans with brain injury (Goldstein, 2000; Martinsson & Eksborg, 2004). Amphetamine was also shown to stimulate both the dendritic growth in the ventral tegmental area (Mueller et al., 2006) and the neurotrophic and neuroplastic responses after brain damage (Moroz et al., 2004; Adkins & Jones 2005). However, few data are available regarding any possible protective effect of amphetamine. In this regard, Bongiovanni et al., (2011) demonstrated that 1 or 10 µM amphetamine reverted the 2,4-D-induced apoptosis and oxidative stress in CGC. Nevertheless, amphetamine alone induced no significant changes with respect to the control culture. Noteworthy, at 1 µM AMPH plus 2,4-D, 39% of the cells were normal; 53% were necrotic, and 8% showed apoptosis. At 10 µM AMPH plus 2,4-D, 57% of the cells were normal, 43% were necrotic, and no apoptotic cells were observed by electron microscopy (Fig. 4 compared with Fig. 5).

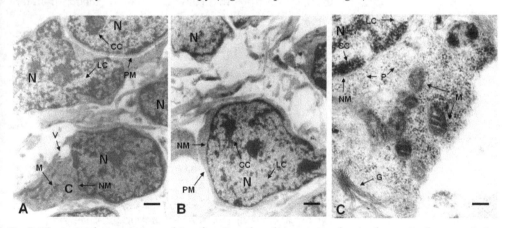

Fig.5. Electron photomicrographies showing the ultrastructural cytoplasmatic characteristics of cerebellar granular cells after 2,4-D and 10 µM AMPH addition to the medium for 48 h. a–b. Cells present more conserved morphology (nucleus and cytoplasm) than those treated with 2,4-D alone (Cf Figs. 4a, b). Bars correspond to 1 µm. c. The cell shows mitochondria and Golgi cisterns more preserved than those of the cells treated with 2,4-D alone (Cf Fig. 4c). Bars correspond to 600 nm. AC apoptotic cell, NC necrotic cell, V vacuole and other abreviations in Fig. 3.

The collected evidence would indicate a protective effect of melatonin and amphetamine against 2,4-D-induced cell death, possibly due to an inhibition of the oxidative mechanisms, as judged by the close relationship between ROS and apotosis induction (Carmody &

Cooter, 2001). While apoptosis and necrosis present some early features that may be common to both, mithocondrial disorders could be irreversibly compromised in necrotic, but not in apoptotic neurons (Nicotera & Leist, 1997). This could explain why amphetamine decrease apoptosis but not necrosis in 2,4-D-treated cells.

In summary, 2,4-D would induce necrosis and apoptosis, the latter being possibly mediated by an oxidative imbalance.

5. Concluding remarks

A great body of evidence suggests that exposure to 2,4-D or to its ester or salt formulations is associated with a wide range of adverse effects in human and different animal species (Berkley & Magee, 1963; Bortolozzi et al, 2001, 2003; Ferri et al., 2003, 2007; Konjuh et al., 2008; Stürtz et al., 2010).

Oxidative stress may affect the cells as a result of imbalance between the (physiological) production of potentially toxic ROS and some (physiological) scavenging activities (Park et al., 1999). Xenobiotics that interact with one or several complexes of the mitochondrial electron transport system, impairing the normal electron flow, may enhance ROS generation, leading to an imbalance between prooxidant species and cellular antioxidants (Jurado et al., 2011).

This review has analyzed the oxidative stress as a possible mechanism of toxicity by the herbicide 2,4-D. The collected evidence confirms that 2,4-D is an environmental pollutant that induces oxidative stress and could determine important deleterious changes in the development of the neural and reproductive systems in the studied models (Ferri et al., 2007; Bongiovanni et al., 2007, 2011; Pocchettino et al., 2010).

While the reported results showed that 2,4-D induces both necrosis and apoptosis, the evidence suggests that apoptosis would be mediated by or associated to an oxidative imbalance (Bongiovanni et al., 2011). Then, the oxidative stress would produce cytochrome-c release from mitochondria and a consequent activation of caspase-3 in the affected cells (De Molliner et al., 2002). However, as mitochondria contribute to both apoptosis and necrosis, intracellular ATP and GSH could determine cell death by one or both of these mechanisms (Leist et al., 1997; Yutaka et al., 1997; Qian et al., 1999; Nieminen, 2003; Bongaerts, 2008). Therefore, the 2,4-D cytotoxic actions may involve some permissive effect on either necrosis or apoptosis induction.

Finally, the experimental evidence reported that 2.4-D can not only affect the nervous system or other hormone-sensitive organs, but also exert a very important, deleterious effect on embryonic and fetal development.

6. Acknowledgment

We thank Prof. Dr. Jose Luis Ferretti for his assistance in writing of the chapter.

7. References

Abdollahi, M., Ranjbar, A., Shadnia, S., Nikfar, S. & Rezaie, A. (2004). Pesticides and oxidative stress: a review. *Medical Science Monitor* , Vol. 10, No. 6, (june 2004), pp.141-147, ISSN 1234-1010.

Adkins, D.L. & Jones, T.A. (2005). D-Amphetamine enhances skilled reaching after ischemic cortical lesions in rats. Neuroscience Letter, Vol. 380, No. 3, (June 2005), pp. 214–218, ISSN 0304-3940.

Agarwal, A., Gupta, S. & Sharma, R.K. (2005). Role of oxidative stress in female reproduction. Reproduction Biology and Endocrinology, Vol. 3, No. 28, (july 2005), pp , ISSN 1477-7827.

Agarwal, A., Gupta, S. & Sikka, S. (2006). The role of free radicals and antioxidants in reproduction. Current Opinion in Obstetrics and Gynecology, Vol. 18, No. 3 (june 2006), pp. 325-332, ISSN 1040-872X.

Bagchi, M., Hassoun, E.A., Bagchi, D. & Stohs, S.J. (1992). Endrin-induced increases in hepatic lipid peroxidation, membrane microviscosity, and DNA damage in rats. Archives of Environmental Contamination and Toxicology, Vol. 23, No. 1, (july 1992), pp. 1-5, ISSN 0090-4341.

Barnekow, D.E., Hamburg, A.W., Puvanesarajah, V. & Guo, M. (2001). Metabolism of 2,4-dichlorophenoxyacetic acid in laying hens and lactating goats. Journal of Agriculture and Food Chemistry, Vol. 49, No. 1, (January 2001), pp. 156-163, ISSN 0021-856.

Behrman, H. R., Kodaman, P. H., Preston, S. L. & Gao, S. (2001). Oxidative stress and the ovary. Journal of Society for Gynecology and Investigation, Vol. 8, (1 Suppl Proceedings), (january 2001), pp. 40-42, ISSN 1071-5576.

Belizário, J.E., Alves, J., Occhiucci, J.M., Garay-Malpartida, M. & Sesso, A. (2007). A mechanistic view of mitochondrial death decision pores. Brazilian Journal of Medical and Biological Research, Vol. 40, No. 8, (August 2007), pp. 1011–1024, ISSN 1678-451.

Bendell, J. J. & Dorrington J. (1991). Estradiol-17 beta stimulates DNA synthesis in rat granulosa cells: action mediated by transforming growth factor-beta. Endocrinology, Vol. 128, No. 5, (may 1991), pp. 2663-2665, ISSN 0013-7227.

Blerkley M.C. & Magee, K.R. (1963). Neuropathy following exposure to a dimethylamine salt of 2,4-D. Archives of Internal Medicine, Vol. 111, (March 1963), pp. 351-352, ISSN 0003-9926.

Bongaerts, P.A. (2008).What of apoptosis is important: the decay process or the causative origin? Medical Hypotheses, Vol. 70, No. 3, (February 2008), pp. 482-487, ISSN 0306-9877.

Bongiovanni, B., De Lorenzi, P., Ferri, A., Konjuh, C., Rassetto, M., Evangelista de Duffard, A.M., Cardinali, D.P. & Duffard, R. (2007). Melatonin decreases the oxidative stress produced by 2,4-dichlorophenoxyacetic acid in rat cerebellar granule cells. Neurotoxicity Research, Vol. 11, No. 2, (February 2007), pp. 93-99, ISSN 1029-8428.

Bongiovanni, B., Ferri, A., Brusco, A., Rassetto, M., Lopez, L.M., Evangelista de Duffard, A.M. & Duffard, R. (2011) Adverse effects of 2,4-dichlorophenoxyacetic acid on rat cerebellar granule cell cultures were attenuated by amphetamine. Neurotoxicity Research. 2011, Vol. 19, No. 4 (May 2011), pp. 544-555, ISSN 1029-8428.

Bortolozzi, A.; Duffard, R. & Evangelista de Duffard, A. (1999). Behavioral alterations induced in rats by a pre- and postnatal exposure to 2,4-dichlorophenoxyacetic acid. Neurotoxicolology and Teratology, 21(4), (Jul-August 1999), pp. 4514-65, ISSN: 0892-0362.

Bortolozzi, A., Evangelista de Duffard, A.M., Dajas, F., Duffard, R., Silveira, R. (2001). Intracerebral administration of 2,4-diclorophenoxyacetic acid induces behavioral and neurochemical alterations in the rat brain. Neurotoxicology, Vol. 22, No. 2, (April 2001), pp. 221–232, ISSN 0161-813X.

Bortolozzi, A.; Duffard, R. & Evangelista de Duffard, A. (2003). Asymmetrical development of the monoamine systems in 2,4-dichlorophenoxyacetic acid treated rats. *Neurotoxicology*,24(1), (Jan 2003), pp. 149-157, ISSN: 0161-813X.

Bradberry, S.M., Watt, B.E., Proudfoot , A.T. & Vale, J.A. (2000). Mechanisms of toxicity, clinical features, and management of acute chlorophenoxy herbicide poisoning: a review. *Journal of Clinical Toxicology*, Vol. 38, No. 2, () pp. 111–122, ISSN 0731-3810.

Brusco,A., Pecci Saavedra, J., Garcia, G., Tagliaferro, P., Evangelista de Duffard, A.M. & Duffard, R. (1997). 2,4-Dichlorophenoxyactic acid through lactation induces astrogliosis in rat brain. Molecular and Chemical Neuropatology, Vol. 30, No. 3, (April 1997), pp. 175-185, ISSN 1044-7393.

Bukowska, B. (2003). Effects of 2,4-D and its metabolite 2,4-dichlorophenol on antioxidant enzymes and level ofglutathione in human erythrocytes. *Comparative Biochemistry and Physiology (Part C)*, Vol.135, No. 4, (August 2003), pp.435–441, ISSN 1532-0456-

Carmody, R.J. & Cotter, T.G. (2001). Signalling apoptosis: a radical approach. Redox Report, Vol. 6, No. 2, (April 2001), pp. 77–90, ISSN 1351-0002.

Casey, P.H. & Collie, W.R. (1984). Severe mental retardation and multiple cogenital anomalies of uncertain cause after extreme parental exposure to 2,4-D, *Journal of Pediatrics, Vol.* 104, No. 2, (February 1984), pp. 313-315, ISSN 0022-3476.

Cazenave, J., Bistoni Mde, L. Pesce, S.F. & Wunderlin, D.A. (2006). Differential detoxification and antioxidant response in diverse organs of Corydoras paleatus experimentally exposed to microcystin-RR. *Aquatic Toxicology,* Vol. 76, No. 1, (january 2006), pp.1-12, ISSN 0166-445X.

Celik, I. & Tuluce Y. (2007). Determination of toxicity of subacute treatment of some plant growth regulators on rats. Environmental Toxicology, Vol. 22, No. 6, (december 2007), pp. 613-619, ISSN 1520-408.

Chatterjee, A. & Chatterjee, R. (2009). How stress affects female reproduction: and overview. *Biomedical Research,* Vol. 20, No. 2, (august 2009), pp.79-83, ISSN 0970938X.

Crews, D., Willingham, E., & Skiper, JK. (2000). Endocrine disruptors: present issues, future directions. *The Quarterly Review of Biology,* Vol. 75, No. 3, (september 2000), pp. 243-260, ISSN 0033-5770.

De Moliner, K.L., Evangelista de Duffard, A.M., Soto, E., Duffard, R. & Adamo, A.M. (2002). Induction of apoptosis in cerebellar granule cells by 2,4-dichlorophenoxyacetic acid. Neurochemical Research, Vol. 27, No. 11, (November 2002), pp. 1439–1446, ISSN 0364-3190.

Diaz, J. & Samson, H. (1980). Impaired brain growth in neonatal rats exposed to ethanol. Science, Vol. 208, No. 4445, (May 1980), pp. 751–753, ISSN 0036-8075.

Dringen, R. (2000). Metabolism and funcions of glutathione in brain. *Progress in Neurobiology,* 62(6), (December 2000), pp. 649–671, ISSN: 0301-0082.

Duchnowicz, P., Koter, M. & Duda, W. (2002). Damage of erythrocyte by phenoxyacetic herbicides and their metabolites. Pesticide Biochemistry and Physiology, Vol. 74, No.1, (September 2002), pp.1–7, ISSN 0048-3575.

Emerit, J., Edeas, M.& Bricaire, F. (2004). Neurodegenerative diseases and oxidative stress. *Biomedicine and Pharmacotherapy, Vol.* 58, No. 1, (January 2004), pp. 39-46, ISSN 0753-3322.

Ferri, A., Bortolozzi, A., Duffard, R. & Evangelista de Duffard, AM. (2000). Monoamine levels in neonate rats lactationally exposed to 2,4-dichlorophenoxyacetic acid. *Biogenic Amines*, Vol. 16, No. 1, (January 2000), pp. 73-100, ISSN 0168-8561.

Ferri, A., Duffard, R., Stürtz, N. & Evangelista de Duffard, A.M. (2003) Iron, zinc and copper levels in brain, serum and liver of neonates exponed to 2,4-dichlorophenoxyacetic acid. *Neurotoxicology and Teratology*, Vol. 25, No. 5 ,(September – October 2003), pp. 607–613, ISSN0892-0362.

Ferri, A., Duffard, R. & Evangelista de Duffard, A.M. (2007) Selective oxidative stress in brain areas of neonate rats exposed to 2,4-dichlorophenoxyacetic acid through mother's milk. Drug and Chemical Toxicology, Vol. 30, No. 1, (January 2007), pp.17–30, ISSN 0148-0545.

Garcia-Segura, L. M., Cardona-Gomez, P. , Naftolin, F. & Chowen, J.A. (1998). Estradiol upregulates Bcl-2 expression in adult brain neurons. Neuroreport , Vol. 9, No. 4, (march 1998) pp. 593-597, ISSN 0959-4965.

Goldstein , L.B. (2000). Effects of amphetamines and small related molecules on recovery after stroke in animals and man. Neuropharmacology, Vol. 39, No. 5, (April 2000), pp. 852–859, ISSN 0028-3908.

Gonzalez, M., Soloneski, S., Reigosa, M.A. & Larramendy, M.L. (2005). Genotoxicity of the herbicide 2,4-dichlorophenoxyacetic and a commercial formulation, 2,4-Dichlorophenoxyacetic acid dimethylamine salt. I. Evaluation of DNA damage and cytogenetic endpoints in Chinese Hamster ovary (CHO) cells. Toxicology In Vitro, Vol. 19, No. 2, (march 2005), pp. 289-297, ISSN 0887-2333.

Gupta, A., Agarwal, R. & Shukla, G.S. (1999). Functional impairment of blood-brain barrier following pesticide exposure during early development in rats. *Human Experimental Toxicology*, Vol. 18, No. 3, (March 1999), pp. 174-179, ISSN 0960-3271.

Gupta, R. K., Schuh, R. A., Fiskum, G. & Flaws, J.A. (2006). Methoxychlor causes mitochondrial dysfunction and oxidative damage in the mouse ovary. Toxicology and Applied Pharmacology, Vol. 216, No. 3, (november 2006) pp. 436-445, ISSN 0041-008X.

Halliwell, B. & Gutteridge, J.M.C. (1998). *Free Radicals in Biology and Medicine. (3rd Edition)*, Oxford University Press, ISBN 9780198500445, Oxford.

Huang, X., Atwoog, C.S., Hartshorn, M.A., Multhaup, G., Goldstein, L.E., Scarpa, R.C., Cuajungco, M.P., Lim, D.N., Moir, R.D., Tanzi, R.E. & Bush, A.I. (1999). The Aβ peptide of Alzheimer's disease directly produces hydrogen peroxide through metal ion reduction. Biochemistry, Vol. 38, No. 24, (May 1999), pp. 7609–7616, ISSN 0006-2960.

IARC. (1986). IARC Monographs on the Evaluation of the Carcinogenic Risk of Chemicals to Humans; Occupational Exposures to Chlorophenoxy Herbicides. In IARC Monographs on the Evaluation of Carcinogenic Risks to Humans (Lyon, France: IARC, World Health Organization), pp. 357-407.

Ishige, K., Chen, Q., Sagara, Y. & Schubert, D. (2001).The Activation of Dopamine D4 Receptors Inhibits Oxidative Stress-Induced Nerve Cell Death. *The Journal of Neuroscience*, Vol. 21, No. 16, (August 2001), pp. 6069–6076, ISSN 0270-6474.

Jurado, A., Fernandes, M.A., Videira, R., Peixoto, F. & Vicente J. (2011) *Herbicides: The Face and the Reverse of the Coin. An in vitro Approach to the Toxicity of Herbicides in Non-Target Organisms*. In: *Herbicides and Environment*, Andreas Kortekamp, pp. 1-44, InTech, ISBN 978-953-307-476-4,

Kaioumova, D., Kaioumov, F., Opelz, G. & Süsa,l C. (2001a) Toxic effects of the herbicide 2,4-dichlorophenoxyacetic acid on lymphoid organs of the rat. *Chemosphere*, Vol. 43, No. 4-7, (May- June 2001), pp. 801–805, ISSN 0045-6535.

Kaioumova, D., Süsal, C. & Opelz, G. (2001b) Induction of apoptosis in human lymphocytes by the herbicide 2,4-dichlorophenoxyacetic acid. *Human Immunology*, Vol. 62, No. 1, (January 2000), pp. 64–74, ISSN 0198-8859.

Kaur, P., Kaur, G. & Bansal, M.P. (2006). Tertiary-butyl hydroperoxide induced oxidative stress and male reproductive activity in mice: role of transcription factor NF-kappaB and testicular antioxidant enzymes. *Reproductive Toxicology*, Vol. 22, No. 3 (october 2006) pp. 479-484, ISSN 0890-6238.

Kim, H.J., Park, Y.I. & Dong, M.S. (2005). Effects of 2,4-D and DCP on the DHT-induced androgenic action in human prostate cancer cells. Toxicological Science, Vol. 88, No. 1, (August 2005), pp. 52-59, ISSN 1096-6080.

Kolb, B. & Wishaw, Y.K. (1989). Plasticity in the neocortex mechanisms underlying recovery from early brain damage. *Progress in Neurobiology*, Vol. 32, No. 4, (September 1989), pp. 235-276, ISSN 0301-0082.

Konjuh, C., García, G., López, L., de Duffard, AM, Brusco, A & Duffard, R.(2008). Neonatal hypomyelination by the herbicide 2,4-dichlorophenoxyacetic acid. Chemical and ultrastructural studies in rats. *Toxicological Science*, Vol. 104, No. 2, (August 2008), pp.332-340, ISSN 1096-6080.

Lee, J.Y., Baw, C.K., Gupta, S., Aziz, N. & Agarwal, A. (2010). Role of Oxidative Stress in Polycystic Ovary Syndrome. *Current Women's Health Reviews*, Vol. 6, No. 2, (april 2010), pp.96-107 , ISSN 1573-404.

Leist, M., Single, B., Castoldi, A.F., Kuhnle, S. & Nicotera, P.(1997). Intracellular adenosine triphosphate (ATP) concentration: a switch in the decision between apoptosis and necrosis. *The Journal of Experimental Medicine*, Vol. 185, No. 8, (April 1997), pp.1481-1486, ISSN 0022-1007.

Martinsson, L. Eksborg, S. (2004). Drugs for stroke recovery: the example of amphetamines. *Drugs Aging*, Vol. 21, No. 2, (January 2004), pp. 67-79, ISSN 1170-229X.

Milton, N.G. (2004). Role of hydrogen peroxide in the aetiology of Alzheimer's disease: implications for treatment. Drugs Aging, Vol. 21, No. 2, (January 2004), pp. 81-100, ISSN 1170-229X.

Moroz, I.A., Peciña, S., Schallert, T. & Stewart, J. (2004). Sparing of behavior and basal extracellular dopamine after 6-hydroxydopamine lesions of the nigrostriatal pathway in rats exposed to a prelesion sensitizing regime of amphetamine. Experimental Neurology, Vol. 189, No. 1, (September 2004), pp. 78-93, ISSN 0014-4886.

Mueller, D., Chapman, C.A. & Stewart, J. (2006). Amphetamine induces dendritic growth in Ventral Tegmental area dopaminergic neurons in vivo via basic fibroblast growth factor. *Neuroscience*, Vol. 137, No. 3, (February 2006), pp. 727-735, ISSN 03064522.

Mukherjee, S., Koner, B. C., Ray, S. & Ray, A. (2006). Environmental contaminants in pathogenesis of breast cancer. *Indian Journal of Exprimental Biology*, Vol. 44, No. 8, (august 2006), pp. 597-617, ISSN 0019-5189.

Munro, I.C., Carlo, G.L., Orr, J.C., Sund, K.G., & Wilson, R.M. (1992). A comprehensive integrated review and evaluation of the scientific evidence relating to the safety of the herbicide 2,4–D. *International Journal of Toxicology*, Vol. 11, No. 5, (October 1992), pp. 559-664, ISSN 1091-5818.

Nicotera, P. & Leist, M. (1997). Mitochondrial signals and energy requirement in cell death. Cell Death and Differentiation, Vol. 4, No. 6, (August 1997), 516-516, ISSN 1350-9047.

Nieminen, A.L., Saylor, A.K., Tesfal, S.A., Herman, B. & Lemasters, J.J. (1995). Contribution of mitochondrial permeability transition to lethal injury after exposure of hepatocyte to tbutylhydroperoxide. Biochemical Journal , Vol. 307, No. 1, (April 1995), pp. 99–106, ISSN 0264-6021.

Nieminen, A.L. (2003) Apoptosis and necrosis in health and disease: role of mitochondria. *International Review of Cytology*, Vol. 224, pp. 29–55, ISSN 0074-7696.

Padros, J., Chun, D., Chen, L., Rigolli, P., Flarakos, T. & Jurima-Romer, M. (2003). A cell-based assay for screening the uridine 5(')-diphosphate-glucuronosyltransferase 1A inhibitory potential of new chemical entities. *Analitical Biochemistry*, Vol. 320, No. 2, (september 2003), pp. 310-312, ISSN 0003-2697.

Pandi-Perumal, S.R., Srinivasan, V., Maestroni, G.J.M., Cardinali, D.P., Poeggeler, B. and Hardeland, R. (2006). Melatonin: nature's most versatile biological signal?, *Federation of European Biochemical Societies Journal*, Vol. 273, No. 13, (July 2006), pp. 2813- 2838, ISSN 1742-4658.

Park, J.S., Wang, M., Park, S.J. & Lee, S.H. (1999). Zinc finger of replication protein A, a non-DNA binding element, regulates its DNA binding activity through redox. Journal of Biological Chemistry, Vol. 274, No.41, (October 1999), 29075-29080, ISSN 0021-9258.

Pelzer, T., Schumann, M., Neumann, M., deJager, T., Stimpel, M., Serfling, E. & Neyses, L. (2000). 17beta-estradiol prevents programmed cell death in cardiac myocytes. *Biochemical and Biophysical Research Communications*, Vol. 268, No. 1, (February 2000), pp. 192-200, ISSN 0006-291X.

Pietsch, C., Wiegand, C., Ame, M. Nicklisch, A., Wunderlin, D. & Pflugmacher, S. (2001). The effects of a cyanobacterial crude extract on different aquatic organisms: evidence for cyanobacterial toxin modulating factors. *Environmental Toxicology*, Vol. 16, No. 6, (november 2001), pp. 535-542, ISSN 1520-4081.

Pochettino, A.A., Bongiovanni, B., Duffard, R.O. & Evangelista de Duffard, A.M. (2011). Oxidative stress in ventral prostate, ovary, and breast by 2,4-dichlorophenoxyacetic acid in pre- and postnatal exposed rats. *Environmental Toxicology*, doi: 10.1002/tox.20690, ISSN 1522-7278.

Qian, T., Herman, B. & Lemasters, J. (1999). The mitochondrial permeability transition mediates both necrotic and apoptotic death of hepatocytes exposed to Br-A23187. Toxicology and Applied Pharmacology, Vol.154, No. 2, (January 1999), pp.117-125, ISSN 0041-008X.

Richards, J. S., Jonassen, J. A. & Kersey, K (1980). Evidence that changes in tonic luteinizing hormone secretion determine the growth of preovulatory follicles in the rat. *Endocrinology*, Vol. 107, No. 3, (September 1980), pp. 641-648, ISSN 0013-7227.

Rodier, P.M. (1995). Developing brain as a target of toxicity. *Environmental Health Perspectives*. Vol. 103, No. 6, (September 1995), pp. 73–76, ISSN 0091-6765.

Rosso, S.B., Garcia, G.B., Madariaga, M.J., Evangelista de Duffard, A.M. & Duffard, R.O. (2000a). 2,4-Dichlorophenoxyacetic acid in developing rats alters behaviour, myelination and regions brain gangliosides pattern. *Neurotoxicology*, Vol. 21, No. 1-2, (February – April 2000), pp. 155-63, ISSN 0161-813X.

Rosso, S.B., Cáceres, A.O., Evangelista de Duffard, A.M., Duffard, R., Quiroga, S. (2000b). 2,4-Dichlorophenoxyacetic acid disrupts the cytoskeleton and disorganizes the Golgi apparatus of culture neurons. Toxicological Science, Vol. 56, No. 1, (July 2000), pp. 133–140, ISNN 1096-6080.

Sanders, C.A. & Rogers, L.J. (1981). 2,4,5-Trichlorophenoxyacetic acid causes behavioral effects in chickens at environmentally relevant doses. *Science, Vol.* 211, No. 4482, (February 1981), pp. 593-595, ISSN 0036-8075.

Saradha, B., Vaithinathan, S. & Mathur, P.P. (2008). Lindane alters the levels of HSP70 and clusterin in adult rat testis.*Toxicology*, Vol. 243, No.1-2, (january 2008), pp.116-123, ISSN 0300-483X.

Sjoden, P.O. & Soderberg, U. (1972). Sex-dependent effects of prenatal 2,4,5 trichlorophenoxyacetic acid on rats open field behavior. *Physiology and Behavior, Vol.* 9, No. 3, (September 1972), pp. 357-360, ISSN 0031-9384.

Spyridopoulos, I., A., Sullivan, B., Kearney, M. Isner, J.M. & Losordo, D.W. (1997). Estrogen-receptor-mediated inhibition of human endothelial cell apoptosis. Estradiol as a survival factor. *Circulation*, Vol. 95, No. 6, (march 1997), pp. 1505-1514, ISSN 0009-7322.

Stürtz, N.; Evangelista de Duffard, A. & Duffard, R. (2000). Detection of 2,4-dichlorophenoxyacetic acid (2,4-D) residues in neonates breast-fed by 2,4-D exposed dams. *Neurotoxicology*, 21(1-2), (Feb-April 2000), pp. 147-154, ISSN: 0161-813X.

Stürtz, N., Bongiovanni, B., Rassetto, M., Ferri, A., Evangelista de Duffard, A.M. & Duffard, R. (2006). Detection of 2,4-dichlorophenoxyacetic acid in rat milk of dams exposed during lactation and milk analysis of their major components. *Food and Chemical Toxicology*, Vol. 44, No. 1, (January 2006), pp. 8–16, ISSN 0278-6915.

Stürtz, N., Jahn, G.A., Deis, R.P., Rettori, V., Duffard, R.O. & Evangelista de Duffard AM. (2010). Effect of 2,4-dichlorophenoxyacetic acid on milk transfer to the litter and prolactin release in lactating rats. *Toxicology*, Vol. 271, No.1-2, (April 2010), pp. 13-20, ISSN 0300-483X.

Sulik, M., Kisielewski, W., Szynaka, B., Kemona, A., Sulik, A., Sulkowska, M. & Baltaziak, M. (1998). Morphological changes in mitochondria and lysosomes of hepatocytes in acute intoxication with 2,4-dichlorophenoxyacetic acid (2,4-D). *Materia Medica Polona*, Vol. 30, No. 1-2, (January-June 1998), pp.16-19, ISSN 0025-5246.

Symonds, D. A., Merchenthaler, I. & Flaws, J. A. (2008). Methoxychlor and estradiol induce oxidative stress DNA damage in the mouse ovarian surface epithelium. *Toxicological Science*, Vol. 105, No. 1, (), pp.182-187, ISSN 1096-0929.

Tilson, H.A. (1998). Developmental neurotoxicology of endocrine disruptors and pesticides: identification of information gaps and research needs. *Environmental Health Perspectives*. Vol. 106, No. 3, (June 1998), pp. 807–811, ISSN 0091-6765.

Tomkinson, A., Reeve, J., Shaw, R.W. & Noble, B.S. (1997). The death of osteocytes via apoptosis accompanies estrogen withdrawal in human bone. *Journal of Clinical Endocrinology and Metabolism*, Vol. 82, No. 9, (september 1997), pp. 3128-3135, ISSN 0021-972X.

Tuschl, H. & Schwab, S. (2003) Cytotoxic effects of the herbicide 2,4-dichlorophenoxyacetic acid in HepG2 cells. *Food and Chemical Toxicology*, Vol. 41, No. 1, (March 2003), pp. 385–393, ISSN 0278-6915.

U.S.E.P.A. Edwards, D. (2006). Reregistration Eligibility Decision for 2,4-D. Prevention, Pesticides EPA 738-R-05-002. Environmental Protection and Toxic Substances (June 2005). Agency (7508C).

Wüllner, U., Seyfried, J., Groscurth, P., Beinroth, S., Winter, S., Gleichmann, M., Heneka, M., Löschmann, P., Schulz, J.B., Weller, M. & Klockgether, T. (1999). Glutathione depletion and neuronal cell death: the role of reactive oxygen intermediates and mitochondrial function. *Brain Research*, Vol. 826, No. 1, (April 1999), pp-53–62, ISSN 00068993.

Yama, O. E., Duru F.I.O., Oremosu A.A. & Noronha C.C. (2011). Testicular oxidative stress in Sprague-Dawley rats treated with bitter melon (Momordica charantia): the effect of antioxidant supplementation. *Bangladesh Journal of Medical Science*, Vol.10, No. 2, (april 2011), pp 104-111, ISSN 2223-4721.

Yutaka, E., Shigeomi, S. & Yoshihide, T. (1997). Intracellular ATP levels determine cell death fate by apoptosis or necrosis. Cancer Research, Vol. 57 (10), (May 1997), pp. 1835-1840, ISSN 0008-5472.

6

Effects of Herbicide Atrazine in Experimental Animal Models

Grasiela D.C. Severi-Aguiar[2] and Elaine C.M. Silva-Zacarin[1]
*[1]Laboratório de Biologia Estrutural e Funcional (LABEF),
Universidade Federal de São Carlos – UFSCAR, Sorocaba, São Paulo;
[2]Programa de Pós-Graduação em Ciências Biomédicas, Centro Universitário Hermínio
Ometto, UNIARARAS, Araras, São Paulo
Brazil*

1. Introduction

Atrazine (2-chloro-4-ethylamino-6-isopropylamino-1,3,5-triazine) is a widely used herbicide in many countries for the control of broadleaf and grassy weeds in agricultural crops. The State of São Paulo, located in Brazil Southeast, is an important sugarcane, soybean and corn producing area with high use of chemicals in agriculture and potential risk of environmental contamination because of the pesticide dissemination, among them the atrazine (ATZ) leaching to groundwater (Cerdeira et al., 2005).

The prolonged use of ATZ and its persistence involves the risk of its retention in crops and soils; moreover, these compounds may also pass from surface to ground waters (Figure 1). In this way, ever-increasing agriculture has caused contamination of natural water sources (Mundiam et al., 2011). The maximum contaminant levels for the most of triazines in drinking water are 3 parts per billion ($\mu g/L$) (Costa Silva et al., 2010).

Some European countries have included ATZ on the list of pesticide residues to be controlled because it is a potential contaminant due to its chemical characteristics, including lipophilicity, slow hydrolysis, moderate to low water solubility, and high solubility in organic solvents with high absorption by organic matter, clay, and fat tissues (Ross et al., 2009). The lack of data about the effects of ATZ metabolites has prompted the U.S. Environmental Protection Agency (U.S. EPA) to state that the toxicity of atrazine's metabolites is equivalent to that of its parent compound and that exposure to these metabolites should be taken into account for risk assessment purposes (Ralston-Hooper et al., 2009).

Mohammad & Itoh (2011) presented the relative risk of various scenarios of exposure and recovery with an aquatic test organism submitted a long-term exposure to herbicides and demonstrated the toxicity of isolated or mixtures of ATZ at different concentrations. However, the patterns of accumulation of xenobiotics vary depending on the organism, characteristics of the chemical compound, quantity of this substance present in the environment, and the balance between assimilation and metabolic rates (Nwani et al., 2011).

It is necessary to study the effects of atrazine exposure in a great variety of experimental animal models in order to understand its action in the organisms and their target organs. In this context, it is very important to verify the effect of high concentration of herbicides in animal tests as positive controls. Saal & Welshons (2006) related the importance of positive controls in toxicological research to determine whether conclusions from experiments that report no significant effects in low-dose of the toxicant are valid or false.

Since in the last decade many efforts have resulted in intensive research about action of ATZ in various organisms, it could be necessary to identify morphological, molecular, biochemical or physiological biomarkers that detect biological effects of this triazine herbicide on the organisms (Campero et al., 2007). So we present in this chapter an extensive bibliographical review about this herbicide in animal tissues, focusing some target-organs, in order to gain insight into its cellular mechanisms, highlighting the results obtained by our research group.

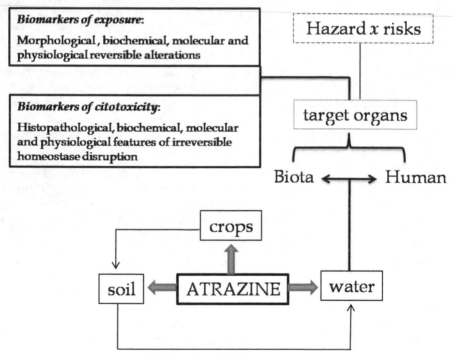

Fig. 1. Environmental contamination way of atrazine. Studies about biota and human health hazard are the basis for risk assessment and, in this context, cellular markers in target organs from organisms exposed to this herbicide could be used in monitoring programs.

2. Morphological and molecular alterations caused by ATZ

2.1 Hepatotoxicity

In liver, organ responsible for detoxification process, our researches with Wistar rats orally exposed to 400 mg/kg body weight (bw) of ATZ for 14 days, showed reduced accumulation of hepatic glycogen and early symptoms of cytotoxicity. This event is attributed to the hepatotoxic effect of ATZ, which inhibits the activity of key enzymes of glyconeogenesis such as hexokinase, glycogen synthase, and glucokinase (Glusczak et al., 2006) and it can explain decrease in animals' body weight observed in our study. This finding agrees with Curic et al. (1999), who studied fish exposed to low doses of ATZ (2 mg/kg) for two weeks and observed the decrease of glycogen and the increase of lipids in the liver.

On the other hand, no differences in glycogen or lipid storage were noted in livers of *Xenopus laevis* tadpoles exposed to both atrazine concentrations 200 and 400µg/L (Zaya et al., 2011). Livers of ATZ-exposed tadpoles were significantly smaller and those from 400µg/L-exposed tadpoles had higher numbers of activated caspase-3 immuno-positive cells suggesting increased rates of apoptosis. The changes noted in body and organ size at 200 and 400µg/L ATZ indicated that exposure throughout development compromised the tadpoles. Additionally, fat body size decreased significantly after exposure to 200 and 400µg/L of ATZ, although this organ still contained some lipid and lacked any pathology. Zaya et al. (2011) suggested that significant reductions in fat body size could potentially decrease their ability to survive the stresses of metamorphosis or reduce reproductive fitness as frogs rely on lipid storage for these processes.

In male Japanese quail (*Coturnix japonica*), vacuolar degeneration in liver was observed at high doses of ATZ (500 mg/kg bw) ingested orally by 45 days (Hussain et al., 2011). Additionally, biliary hyperplasia and mild renal tubular necrosis were observed in these quails. In our studies with Wistar rats orally exposed to 400 mg/kg body weight (bw) of ATZ for 14 days, similar data were observed in liver and renal tubular necrosis was also observed.

Zebrafish (*Danio rerio*) is other model organism that presented histopathological effects in liver, which were induced by atrazine exposure. Yuanxiang et al. (2011) found seven proteins that were upregulated >2- fold, whereas 6 protein were downregulated >2-fold, after 10 and 1000 µg/l ATZ exposures in zebrafish for 14 days. They suggested that these changes in protein regulation were associated with a variety of cellular biological processes, such as response to oxidative stress, oncogenesis and others.

Another example of cellular biological process that could be changed in response to atrazine exposure is the lipid metabolism and insulin resistance. Study performed in Sprague-Dawley rats treated for 5 months with vehicle or ATZ (30 or 300 µg kg^{-1} day^{-1}), supplied in drinking water, showed prominent accumulation of lipid droplets in the livers of ATZ-treated rats (Lim et al., 2009). By means of transmission electron microscopy, some liver mitochondria from the ATZ-treated group showed partially disrupted cristae. Despite the fact that mitochondrial morphology was altered in liver and, additionally, in muscle, protein expression levels of mitochondrial OXPHOS complex subunits in liver and muscle tissues were not changed significantly by ATZ administration. Since no treatment-related changes in food or water intake or physical activity were observed at any point during the study, Lim et al. (2009) believe that the development of insulin resistance by ATZ might be related to energy metabolism and they suggest that long-term exposure to the herbicide ATZ might contribute to the development of insulin resistance and obesity, particularly where a high-fat diet is prevalent.

2.2 Reproductive toxicity

In review of Sifakis et al. (2011) about pesticide exposure and health, related issues in male and female reproductive system have been presented and they showed that ATZ seems to have estrogenic and anti-androgenic properties.

Our research group evaluated histopathological effects of low and high doses of ATZ in ovary and testicles from exposed Wistar rats and the compilation of data are presented in the Table 1.

Testicular lesions observed in our studies (Table 1) also be detected, associated with reduced germ cell numbers, in teleost fish, amphibians, reptiles, and mammals; and induces partial and/or complete feminization in fish, amphibians, and reptiles (Hayes et al., 2011). Then, ATZ is an endocrine disruptor that demasculinizes and feminizes the gonads of male vertebrates by means of the reduction in androgen levels and the induction of estrogen synthesis - demonstrated in fish, amphibians, and reptiles - that represent plausible and coherent mechanisms to explain these effects, according to Hayes et al. (2011). ATZ reduce testicular testosterone in male rats and it was associated with poor semen quality (Sifakis et al., 2011).

Morphological alterations induced by atrazine oral exposure			
Ovaries		Testis	
0,75mg/kg	400mg/kg	0,75mg/kg	400mg/kg
Primordial follicles without alterations	Primordial follicles without alterations	Normal histoarchitecture	Disorganized histoarchitecture
Primary follicles without alterations	Primary follicles without alterations	Absence of degeneration in seminiferous epithelium	Degeneration in some areas of the seminiferous epithelium
Presence of some multioocytic Preantral follicles	Preantral follicles with disorganized granulose layer and/or a degenerating oocyte	Germinative cells keep their typical morphology	Some germinative cells presented apoptotic or necrosis features
Antral follicles without alterations	Presence of some Antral follicles with a degenerating oocyte	Germinative cells were not released to tubular lumen	Releasing of germinative cells to the tubular lumen
Atretic antral follicles with intensification of apoptosis in granulose cells	High intensity of apoptosis in the granulose cells of Atretic antral follicles	Intertubular tissue around seminiferous tubules remained intact	Intertubular tissue around seminiferous tubules remained intact

Table 1. Histopathological analysis of ovaries and testis from Wistar rats submitted to subchronic oral exposure: 0,75mg atrazine/kg/day during 30 days; and subchronic oral exposure: 400mg atrazine/kg/day during 14 days.

A study developed by Hussain et al. (2011) that intended to determine the pathological and genotoxic effects of ATZ in male Japanese quail (*Coturnix japonica*) demonstrated that testis from ATZ treated birds were comparatively smaller in size and seminiferous tubules in group treated with 500 mg/kg bw exhibited decreased number of spermatocytes, necrotic nuclei of spermatids, and lesser number or absence of spermatozoa.

A significant dose dependent induction in the levels of mRNA expression of genes of steroidogenic acute regulatory protein (STAR), cytochrome P450-11A1, 3β-hydroxysteroid dehydrogenase (3β-HSD), and other steroidogenic proteins were observed in cells exposed to ATZ. These data suggest the applicability of these selected marker genes of steroidogenesis as an indicator of short term exposure of ATZ induced rat testicular toxicity in interstitial Leydig cells (ILCs) (Abarikwu et al., 2011).

Pogrmic-Majkic et al. (2010) examined Leydig cells treated for 24 h with the concentrations 0.001, 1, 10, 20, and 50µM of ATZ and they observed increased basal and human chorion gonadotropin-stimulated testosterone production and accumulation of cAMP in the medium of treated cells. The stimulatory action of atrazine on androgen production but not on cAMP accumulation was abolished in cells with inhibited protein kinase A. They observed that Leydig cells obtained from rats treated with 200 mg ATZ/kg body weight, by gavage, during the first 3 days of treatment, stimulated the expression of mRNA transcripts for steroidogenic factor-1, steroidogenic acute regulatory protein, cytochrome P450(CYP)-17A1, and 17b-hydroxysteroid dehydrogenase (HSD), as well as the activity of CYP17A1 and 17bHSD and cAMP accumulation and androgen production. However, this behavior is followed by a decline during further treatment (6 days). These results indicate that ATZ has a transient stimulatory action on cAMP signaling pathway in Leydig cells, leading to facilitated androgenesis.

ATZ exposure (120 or 200 mg/kg body weight ATZ orally for 7 and 16 days) has a dose-dependent adverse effect on the testicular and epididymal sperm numbers, motility, viability, morphology, and daily sperm production in rats (Abarikwu et al., 2009). Although the testis of the ATZ -treated animals appear normal, few tubules had mild degeneration with the presence of defoliated cells, similar to observed in our research group for rat testis (Table 1). Likewise, no perceptible morphological changes were observed in the epididymis. The results suggest that ATZ impairs reproductive function and elicits a depletion of the antioxidant defense system in the testis and epididymis, indicating the induction of oxidative stress. Glutathione (GSH) and glutathione-S-transferase (GST) activities were elevated in the high-dose group, whereas the activity of superoxide-dismutase (SOD), catalase (CAT); ascorbate (AA), and malondialdehyde (MDA) levels and hydrogen peroxide production were unchanged in the testis during the 7-day exposure protocol. When ATZ treatment was increased to 16 days, GSH levels remained unchanged, but lipid peroxidation levels were significantly increased in both the testis and epididymis. This corresponded to the significant diminution in the activities of GST and SOD. CAT activities were unaffected in the testis and then dropped in the epididymis. These experiments performed by Abarikwu et al. (2009) was important to understand the antioxidant defense system in the testis and epididymis; and it is interesting to note that ATZ can also affect mitochondrial electron transport and oxidative stress in the insect *Drosophila melanogaster* (Thornton et al., 2010).

Eggs of alligator *Caiman latirostris,* at stage 20 of embryonic development, were exposed to 0,02ppm of ATZ and incubated at 33°C resulted in male hatchlings. Tortuous seminiferous tubules with increased perimeter, disrupted distribution of peritubular myoid cells (desmin positive), and emptied tubular lumens characterized the testis of pesticide-exposed *Caiman* (Rey et al., 2009).

ATZ is a known ovarian toxicant which increase progesterone (P4) secretion and induce luteal cell hypertrophy following repeated administration. The aim of Taketa (2011) study group was to define the pathways by which these compounds exerted their effects on the ovary and hypothalamic-pituitary-gonadal (HPG) axis. They demonstrated that 300 mg ATZ/kg were orally given daily from proestrus to diestrus in normal cycling rats resulted in significant increased serum P4 levels, upregulation of the follow steroidogenic factors: scavenger receptor class B type I, steroidogenic acute regulatory protein, P450 cholesterol side-chain cleavage and 3β-hydroxysteroid dehydrogenase (HSD), and so downregulation of luteolytic gene 20α-HSD. ATZ may directly activate new corpora lutea by stimulating steroidogenic factor expressions and the authors suggest that multiple pathways mediate its effects the HPG axis and luteal P4 production in female rats *in vivo*.

One of the molecular events that may be triggered by stressful conditions, like pesticide exposure, is the synthesis of heat shock proteins (HSP). Additionally to histophatological analysis of rat ovaries (Table 1), our studies also emphasized the immunohistochemical labeling of 90 KD heat shock protein (HSP 90) in order to evaluate the role played by this protein in the ovary, under stressed conditions induced by ATZ exposure. Our results indicated that atrazine induced impaired folliculogenesis, increased follicular atresia and HSP90 depletion in female rats submitted to subacute treatment, while the subchronic treatment with the lowest dose of ATZ could compromise the reproductive capacity reflected by the presence of multioocytic follicle and stress-inducible HSP90 (Juliani et al., 2008).

Experiments developed by our research group also showed that low doses of ATZ, which does not affect estrous cyclicity, induced a higher HSP70 expression in cells of the oviduct when compared to the control group, indicating that HSP70 may be acting in the tissue response to stress caused by chronic exposure to the herbicide. In subacute exposure, with the dose that disrupts the estrous cycle, the expression of HSP70 was higher than the control group and the subchronic treatment (Figure 2), probably indicating a major protective function of HSP70 in addition to the estrogen receptor baseline level maturation. In literature, HSP70 is related to the maturation of the estrogen receptor in the oviduct (Mariani et al., 2000). We concluded that the increased expression of HSP70 induced by ATZ is mainly related to the protective effect of these chaperones in response to chemical stress generated by exposure to this herbicide.

2.3 Glandular alterations

Due to the adrenal gland is reported to be the most common endocrine organ associated with chemically induced lesions, our research group also evaluated adrenal glands of adult rats submitted to subacute and subchronic treatment with this herbicide, respectively. The morphological and histochemical analyses were performed and the results indicated that the subacute treatment induced drastic alterations in the cortex of the adrenal glands as well as

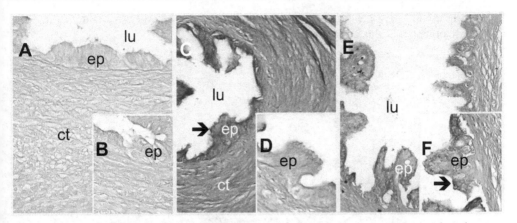

Fig. 2. Immunohistochemical detection of HSP70 (Heat Shoch Protein - 70) in oviduct from Wistar rats. A–B) Control Group; C-D) Experimental group submitted to subchronic oral exposure: 0,75mg atrazine/kg/day during 30 days; E-F) Experimental group submitted to subchronic oral exposure: 400mg atrazine/kg/day during 14 days. Oviducts present different degrees of immunopositive reaction (brown color) in epithelium (ep) and connective tissue (ct). Arrows indicate areas with high imunolabeling of HSP70. A, C, E – Magnification = 200x; B, D, F – Magnification = 400x.

in the medullar region. The subchronic treatment with the low dose of ATZ caused slight morphological alterations in the cortex of adrenal glands, but not in the medullar region. The histochemical analyses showed abnormal accumulations of lipid droplets mainly in the Reticularis Zona of the adrenal cortex suggesting alteration in the steroidogenesis process that occur in this region (Figure 3).

Foradori et al. (2011) demonstrated that high doses of ATZ (200mg/kg), administered for 4 days, suppress luteinizing hormone (LH) release and increase adrenal hormones levels. Considering the known inhibitory effects of adrenal hormones on the hypothalamo-pituitary-gonadal axis, the authors investigated the possible role that the adrenal gland has in mediating ATZ inhibition of LH release and observed that adrenolectomy had no effect on ATZ inhibition of the LH surge but prevented the ATZ disruption of pulsatile LH release. These data indicate that ATZ selectively affects the LH pulse generator through alterations in adrenal hormone secretion. Adrenal activation does not play a role in ATZ's suppression of the LH surge and therefore, ATZ may work centrally to alter the preovulatory LH surge in female rats.

2.4 Genotoxicity

Although the toxic properties of ATZ are well known, there is not a consensus about the genotoxic effects of ATZ. On aquatic organisms they are rather scarce. To evaluate the genotoxic effects of ATZ and an ATZ-based herbicide (Gesaprim®) on a model fish species *Carassius auratus* L., 1758, (Pisces: Cyprinidae) using the micronucleus test and the comet assay in peripheral blood erythrocytes, fish were exposed to 5, 10 and 15 μg/L ATZ and to its commercial formulation for 2, 4 and 6 days (Cavas, 2011). The results revealed significant increases in the frequencies of micronuclei and DNA strand breaks in erythrocytes of C.

auratus, following exposure to commercial formulation of ATZ and thus demonstrated the genotoxic potential of this pesticide on fish.

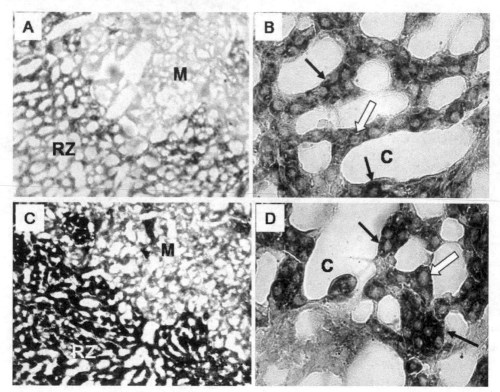

Fig. 3. Cryosections of adrenal gland from Wistar rats, stained with Sudam Black. The dark-brown color indicates the presence of lipids in the Reticularis Zone (RZ) but not in the Medullar region (M). A–B) Control Group; C-D) Experimental group submitted to subchronic oral exposure: 0,75mg atrazine/kg/day during 30 days. In (B) and (D), lipids in cells of Reticularis Zone are indicated with arrows (black arrow= the strongest intensity; white arrow= strong intensity). A, C – Magnification = 200x; B, D – Magnification = 400x

Significantly longer comet tails of DNA damage in leukocytes and isolated hepatocytes of male Japanese quail (*Coturnix japonica*) were recorded with 500 mg/kg bw ATZ (Hussain et al., 2010).

In our results with rats treated with 400mg ATZ/kg bw too have been observed a significant increase of micronucleated polycromatic erythrocytes (data not published), corroborating that authors and suggesting a possible genotoxic potencial of ATZ in mammals, which have to make its use highly controlled.

2.5 Mutagenicity and cancer

Chronic studies of ATZ and simazine and their common metabolites show an elevated incidence of mammary tumors only in female Sprague Dawley (SD) rats. On the basis of the clear tumor increase in female SD rats, ATZ was proposed to be classified as a likely human carcinogen by US Environmental Protection Agency (EPA) in 1999. With Fischer rats, all strains of mice, and dogs, there was no evidence of increased incidence of ATZ -associated tumors of any type. Evidence related to the pivotal role of hormonal control of the estrus cycle in SD rats appears to indicate that the mechanism for mammary tumor induction is specific to this strain of rats and thus is not relevant to humans. In humans the menstrual cycle is controlled by estrogen released by the ovary rather than depending on the LH surge, as estrus is in SD rats. However, the relevance of the tumors to humans continues to be debated based on endocrine effects of triazines. No strong evidence exists for ATZ mutagenicity, while there is evidence of clastogenicity at elevated concentrations. ATZ does not appear to interact strongly with estrogen receptors α or β but may interact with putative estrogen receptor GPR30 (G-protein-coupled receptor). A large number of epidemiologic studies conducted on manufacturing workers, pesticide applicators, and farming families do not indicate that triazines are carcinogenic in these populations. A rat-specific hormonal mechanism for mammary tumors has now been accepted by US EPA, International Agency for Research on Cancer, and the European Union. Chlorotriazines do influence endocrine responses, but their potential impact on humans appears to be primarily on reproduction and development and is not related to carcinogenesis (for revision, see Jowa & Howd, 2011).

According an extensive review, epidemiology studies do not provide consistent, scientifically convincing evidence of a causal relationship between exposure to ATZ or triazine herbicides and cancer in humans. Based upon the assessment studies, there is no scientific basis for inferring the existence of a causal relationship between triazine exposure and the occurrence of cancer in humans (Sathiakumar et al., 2011).

A study developed by NIEHS (National Institute of Environmental Health Sciences) that extended analysis of cancer risk associated with occupational hazard of ATZ showed that there was no strong or consistent evidence of an association between ATZ and any cancer. There was a non-statistically significant increased risk of ovarian cancer related to occupational hazard for female who reported to use ATZ compared to those who did not; however, this observation was based on a small number of cases among ATZ users. The authors found an elevated risk of thyroid cancer, has not been previously reported, for the highest versus lowest category of intensity weighted ATZ use, but the trend was not monotonic and not statistically significant when lifetime days of use was considered as the exposure metric. In contrast, they observed little evidence for an association between ATZ occupational use and other cancers previously reported in the literature, such as NHL non-Hodgkin lymphoma) and leukemia, or with cancers of the breast or prostate, for which ATZ has been hypothesized to be a risk factor because of its hormonal properties (Freeman et al., 2011).

Although there is conflicting information about relationship between ATZ and cancer some researches have been demonstrated preoccupation with this aspect and they highlighted the importance of many studies to confirm or not this supposition.

3. Conclusion

We concluded that:

-   In adult model animals, lower doses of atrazine generally induce accumulation of lipids in hepatocytes, otherwise higher doses induce hepatotoxicity with degree variation according to animal. Amphibian tadpole's liver presents morphological response pattern, which is different those from the adult model animals.

- In rat and bird testis, atrazine has a dose-dependent adverse effects varying from no perceptible morphological changes to degeneration of seminiferous epithelium because ATZ impairs reproductive function and induces a depletion of the antioxidant defense system, according to the dose and time of exposure. Otherwise, in testis of teleost fish, amphibian and reptile, atrazine has a demascunilization/feminization effect that can be partial or complete what depends on the dose and time of exposure.

- High concentrations of atrazine induce morphological alterations in rat ovarian follicles, but not in oviduct. Induction of HSP70 in oviduct (low and high doses) could be used as exposure cell marker, as well as HSP90 depletion (high dose) or HSP90 increasing (low dose) could indicate the degree of ATZ exposure.

- In adrenal glands of rats, atrazine exposure induced varied degree of morpho-physiological alterations, which is observed in a dose-dependent way due its endocrine disruptor property.

- There is not a consensus about the genotoxic effects of atrazine, and then it is necessary further studies in experimental animal models.

- Although high doses of atrazine induce clastogenicity, there is not consistent evidence that associate mutagenecity with cancer in humans.

4. Acknowledgment

We thank to Fundação Hermínio Ometto for financial support; to Dra. Patrícia Aline Boer for to start these researches with us and her important contribution about the kidneys; to the students Franco D. C. Pereira, Kelly L. Calisto, Patrícia M. Dias, Rafael Muniz and Silvia Takai for join us to get knowledge about effects of atrazine, and Renata Barbieri for technical assistance.

5. References

Abarikwu S. O., Adesiyan A. C., Oyeloja T. O., Oyeyemi M. O., Farombi e. O. (2009). Changes in Sperm Characteristics and Induction of Oxidative Stress in the Testis and Epididymis of Experimental Rats by a Herbicide, Atrazine. *Arch Environ Contam Toxicol.*, Vol. 58, pp. 874–882

Abarikwu S.O., Farombi E.O., Kashyap M. P., Pant A. B. (2011). Atrazine induces transcriptional changes in marker genes associated with steroidogenesis in primary cultures of rat Leydig cells. *Toxicol In Vitro*, DOI:10.1016/j.tiv.2011.06.002.

Campero M., Ollevier F., Stoks R. (2007). Ecological relevance and sensitivity depending on the exposure time for two biomarkers. *Environ Toxicol., Vol.* 22, No. 6, pp. 572-81.

Cavas T. (2011). In vivo genotoxicity evaluation of atrazine and atrazine-based herbicide on fish *Carassius auratus* using the micronucleus test and the comet assay. *Food Chem Toxicol.*, Vol. 49, No 6, pp. 1431-5.

Cerdeira, A. L., Dornelas-DeSouza, M., Bolonhezi, D., Queiroz, S. C. N., Ferracini, V. L., Ligo, M . A. V., Pessoa, M. C. P. Y., Smith Jr, S. (2005). Effects of Sugar Cane Mechanical Harvesting Followed by No-Tillage Crop Systems on Leaching of Triazine Herbicides in Brazil. *Bull. Environ. Contam. Toxicol.*, Vol. 75, pp. 805-812.

Costa Silva, R. G. C., Vigna, C. R. M, Bottoli, C. B. G., Collins, C. H., Augusto, F. (2010). Molecularly imprinted silica as a selective SPE sorbent for triazine herbicides. *J. Sep. Sci.*, Vol. 33, pp. 1319–1324.

Curic, S.; Gojmerac, T.; Zuric, M. (1999). Morphological changes in the organs of gilts induced with low-dose atrazine. *Vet. Archiv.*, Vol. 69, pp. 135-148.

Foradori C. D., Hinds L. R., Quihuis A. M., Lacagnina A. F., Breckenridge C. B., Hand R. J. (2011). The Differential Effect of Atrazine on Luteinizing Hormone Release in Adrenalectomized Adult Female Wistar Rats. DOI:10.1095/biolreprod.111.092452

Freeman L. E. B., Rusiecki J. A., Hoppin J. A., Lubin J. H., Koutros S., Andreotti G., Zahm S. H., Hines C. J., Coble J. B., Barone-Adesi F., Sloan J., Sandler D.P., Blair A., Alavanja M. C. R. (2011). Atrazine and Cancer Incidence Among Pesticide Applicators in the Agricultural Health Study (1994-2007). DOI: 10.1289/ehp.1103561

Gluzczak, L. Miron, D.S., Moraes, B.S., Simoes, R.R. Schetinger, M.R.C., Morsch, V.M.,Loro, V.R. (2007). Acute effect of glyphosate herbicide on metabolic and enzymatic parameters of silver catfish (Rhamdia quelen). *Comparative Biochemistry and Physiology.*, Vol. 146, pp. 519-524.

Hayes T. B., Anderson L. L., Beasley V. R., de Solla S. R., Iguchi T., Ingraham H., Kestemont P., Kniewald J., Kniewald Z., Langlois V. S., Luque E. H., McCoy K. A., Muñoz-de-Toro M., Oka T., Oliveira C. A., Orton F., Ruby S., Suzawa M., Tavera-Mendoza L. E., Trudeau V. L., Victor-Costa A. B., Willingham E. (2011). Demasculinization and feminization of male gonads by atrazine: Consistent effects across vertebrate classes. *J Steroid Biochem Mol Biol.*, DOI:10.1016/j.jsbmb.2011.03.015.

Juliani, C.C.; Silva-Zacarin, E.C.M.; Carvalho, D., Boer, P. A. (2008). Effects of atrazine on female Wistar rats: morphological alterations in ovarian follicles and immunocytochemical labeling of 90 kDa heat shock protein. *Micron*, Vol. 30, pp. 607-616.

Jowa L., Howd R. (2011). Should atrazine and related chlorotriazines be considered carcinogenic for human health risk assessment? *J Environ Sci Health C Environ Carcinog Ecotoxicol Rev.*, Vol. 29, No 2, pp. 91-144.

Lim S., Ahn S. Y., Song I. C., Chung M. H., Jang H. C., Par K. S., Lee K., Pak Y. K., Lee H. K. (2009). Chronic Exposure to the Herbicide Atrazine Causes Mitochondrial Dysfunction and Insulin Resistance. *PLOS ONE*, Vol. 4, No 4, pp. 1-11

Mariani, M. L.; Souto, M.; Fanelli, M. A., Ciocca, D. R. (2000). Constitutive expression of heat shock proteins hsp25 and hsp70 in the rat oviduct during neonatal development, the oestrous cycle and early pregnancy. *Journal of Reproduction and Fertility*, Vol. 120, No 2, pp. 217-223.

Mudiam, M. K. R., Pathak, S. P., Gopal, K., Murthy, R. C. (2011). Studies on urban drinking water quality in a tropical zone. *Environ. Monit. Acess.*, DOI 10.1007/s10661-011-1980-3.

Mohammad M., Itoh K. (2011). New Concept for Evaluating the Toxicity of Herbicides for Ecological Risk Assessment, In: *Herbicides and Environment.*, Andreas Kortekamp, pp. 561-582, InTech Open Access Publisher, Retrieved from www.intechweb.org/books/show/title/herbicides-and-environment, ISBN 978-953-307-476-4.

Nwani C.D., Lakra W.S., Nagpure N.S., Kumar R., Kushwaha B and Srivastava S.K. (2010). Toxicity of the herbicide atrazine: effects on lipid peroxidation and activities of antioxidant enzymes in the freshwater fish *Channa punctatus* (Bloch). *Int J Environ Res Public Health, Vol.* 8, pp. 3298-312.

Pogrmic-Majkic K., Fa S., Dakic V., Kaisarevic S., Kovacevic R. (2010). Upregulation of Peripubertal Rat Leydig Cell Steroidogenesis Following 24 h *In Vitro* and *In Vivo* Exposure to Atrazine. *Toxicol.Sci., Vol.* 118, No 1, pp. 52-60.

Ralston-Hooper, K., Hardy, J., Hahn, L., Ochoa-Acuña, H., Lee, L.S., Mollenhauer, R., Maria S. Sepúlveda, M.S. (2009). Acute and chronic toxicity of atrazine and its metabolites deethylatrazine and deisopropylatrazine on aquatic organisms. *Ecotoxicol., Vol.* 18, No 7, pp. 899-905.

Rey F., González M., Zayas M. A., Stoker C., Durando M., Luque e. H., Muñoz-de-Toro M. (2009). Prenatal exposure to pesticides disrupts testicular histoarchitecture and alters testosterone levels in male *Caiman latirostris. General and Comparative Endocrinology, Vol.* 162, pp. 286–292

Hussain R., Mahmood F., Khan M. Z., Khan A., Muhammad F. (2011). Pathological and genotoxic effects of atrazine in male Japanese quail (*Coturnix japonica*). *Ecotoxicology, Vol.* 20, pp. 1–8

Ross, M. K., Jones, T. L., Filipov, N. M. (2009). Disposition of the Herbicide 2-Chloro-4-(ethylamino)-6-(isopropylamino)-s-triazine (Atrazine) and Its Major Metabolites in Mice: A Liquid Chromatography/Mass Spectrometry Analysis of Urine, Plasma, and Tissue Levels. *Drug Metabolism and Disposition, Vol.* 37, pp. 776-786.

Saal F. S. and Welshons W. V. (2006). Large effects from small exposures. II. The importance of positive controls in low-dose research on bisphenol A. *Environ Res.,*Vol. 100, No. 1, pp. 50-76.

Sathiakumar N., MacLennan P. A., Jack Mandel, Delzell E. (2011). A review of epidemiologic studies of triazine herbicides and cancer. *Critical Reviews in Toxicology, Vol.* 41, No. 1, pp. 1–34

Sifakis S., Mparmpas M., Soldin O. P., Tsatsakis A. (2011). Pesticide Exposure and Health Related Issues in Male and Female Reproductive System. In: *Pesticides - Formulations, Effects, Fate.* Margarita Stoytcheva, pp. 495-526, InTech Open Access Publisher, Retrieved from www.intechweb.org/books/show/title/pesticides-formulations-effects-fate, ISBN 978-953-307-532-7.

Taketa Y., Yoshida M., Inoue K., Takahashi M., Sakamoto Y., Watanabe G., Taya K., Yamate J., Nishikawa A. (2011). Differential stimulation pathways of progesterone secretion from newly formed corpora lutea in rats treated with ethylene glycol monomethyl ether, sulpiride, or atrazine. *Toxicol Sci., Vol.* 121, No. 2, pp. 267-78.

Thornton B. J., Elthon T. E., Cerny R. L., Siegfried B. D. (2010). Proteomic analysis of atrazine exposure in *Drosophila melanogaster* (Diptera: Drosophilidae). *Chemosphere, Vol.* 81, No. 2, pp. 235-41.

Yuanxiang J., Xiangxiang Z., Dezhao L., Zhengwei F. (2011). Proteomic Analysis of Hepatic Tissue in Adult Female Zebrafish (*Danio rerio*) Exposed to Atrazine. *Arch Environ Contam Toxicol.,* DOI 10.1007/s00244-011-9678-7

Zaya R. M., Amini Z., Whitaker A. S., Kohler S. L., Ide C. F. (2011). Atrazine exposure affects growth, body condition and liver health in *Xenopus laevis* tadpoles. *Aquat Toxicol., Vol.* 104, No. 3-4, pp. 243-53.

7

Weed Population Dynamics

Aurélio Vaz De Melo[1], Rubens Ribeiro da Silva[1],
Hélio Bandeira Barros[1] and Cíntia Ribeiro de Souza[2]
[1]*Federal University of Tocantins*
[2]*Federal Institute of Education, Science and Technology of Pará*
Brazil

1. Introduction

Clearly, the growing infestation of weeds in agricultural systems causes damage to crops, with sharp declines in productivity, either by direct competition for factors of production, whether by allelopathic compounds released into the soil (MARTINS and PITELLI, 1994). There are many factors related to population dynamics of these plants. However this chapter will be referred to those who, according to research, seem to be the most important.

2. Factors influencing the population dynamics of weed

In various cultures was observed the influence of farming system, being of fundamental importance to understand these dynamics through studies on the floristic composition and phytosociological structure of the same. The cultivation of maize intercropped with tropical forages in the system of direct planting can reduce the incidence of weeds due to the high biomass production and allelopathy provided by surface deposition of straw on the soil. The presence of *B. brizantha* intercropping reduced weed density. Therefore, the use of intercropping maize with *B. brizantha* provides control rate of 95% of the weeds in the soil (BORGHI et al., 2008).

A survey of weeds in conventional farming sunflower family Poaceae was the most representative among species (SILVA et al., 2010). In the experiment carried out by MARQUES et al. (2010) the plants originate in poultry farming sprouts in cowpea had the highest rates of importance values. However, they are dependent on the season and the continuity of the system.

Mechanized harvesting of raw cane enables the maintenance of the layer of straw on the surface, so by reducing the movement of soil and alter the dynamics of herbicides. These changes promote changes in microclimatic conditions, which in turn affect the composition of specific weeds. In this culture, the population dynamics of weeds in no-tillage system reduces up to 531% the incidence of weeds compared to the conventional system after treatment with herbicides. This provides 27% reduction in the productivity of cane sugar in conventional tillage soil (DUARTE JR et al., 2009).

The results obtained by VAZ-DE-MELO et al. (2007) showed that the practices adopted in growing organic green corn under no-tillage system, provide the appropriate management of weeds while with adoption of soil cover with oat straw. Among the weed species, *B. pilosa* was the one that had the highest relative importance, that due to higher efficiency of this species to produce biomass in the absence of competition. In addition, this species shows high capacity for regrowth after mowing adopted in the organic system. The maize cultivars also interfere with the relative importance of *C. rotundus*, indicating the importance of knowledge of the floristic composition and the cultivar of corn to be adopted for the proper management of weeds.

Another important factor in weed control is to define the ability of weed species to compete for water, light and nutrients that are the factors responsible for reduced productivity of the main crop.

According GAZZIERO et al. (2004), features such as growth rate, efficiency of space occupation of the soil, shading, release of toxic chemicals to weeds, crop residues produced different and the specific methods of control used in each culture are also considered important features of competition between the crop and weeds.

2.1 Humic substances in the dynamics and allelopathy and weed control

The herbicides have specific mechanisms of interaction with organic compounds in humic matter in soil. This interaction interferes with the dynamics of molecules of herbicides in soil as well as implementing the recommendations.

The humic substances present in agricultural systems are caused by the biological degradation of plant and animal remains in the soil. The maintenance of straw on the soil surface and the permanence of the root system of crops harvested in the soil increases, the medium and long term, its organic matter content. This enables the maintenance of temperature and soil moisture at adequate levels, favoring the perfect physiological functioning of plants, ensuring the survival of a wide variety of organisms such as fungi and bacteria, which are primary decomposers of crop residues and serve as food for the small animals (MATZENBACHER, 1999).

In addition, the products then formed associate themselves into complex structures more stable, dark colored, high molecular weight, separated on the basis of solubility characteristics. Classified into: humin, humic acids, fulvic and hymatomelâmicos.

The humic substances increase the CTC CTA and soil, protecting and providing the cations and anions for the plants. This property of ion exchange of soil humic substances, when properly managed, enhances the efficiency of pesticides and fertilizers, and reduce the contaminant action (FOLONI e SOUZA, 2010, 2010).

The cultures used as soil cover in general have the ability to recycle nutrients, promote decompression of the soil, increase organic matter content and suppress weeds (THEISEN et al., 2000). The suppression occurs through the production of secondary metabolites, called allelochemicals, which accumulate in various organs of plants and are released with important ecological function. The main forms of release into the environment occur through the processes of volatilization, exudation from roots, leaching and decomposition of waste (DURIGAN and ALMEIDA, 1993).

The allelopathic action, both during vegetative growth and during the decomposition process, interspecific inhibition exerts on other species. The inhibition is linked primarily to reduced availability of light and to allelopathic effects, which have potent phytotoxic and can act as inhibitors of photosystem II (CZARNOTA et al. 2003; KADIOGLU et al., 2005).

The biochemical production of inhibitors, the remains of crops or the soil microorganisms can inhibit the germination and emergence of some species (MATEUS, 2004), as well as reduce the initial growth of plants.

Sorghum and millet are C4 plants, which have fast growth and good ability to cover soil. Furthermore, sorghum has allelopathic compound that is exuded by their roots, sorgoleone, which is able to differentially suppress the growth of various weeds and crops (NIMBAL et al., 1996).

Another factor that may alter the production of allelopathic compounds and modify the intensity of the effects found in the field, are the characteristics of the environment where these allelochemicals are produced. According to MARTINS et al. (1999) in the system of production of sugarcane, if allelopathic compounds are present in the straw, without fire, will be released larger amounts of these substances in the soil and may promote weed control or cause reductions shoots in culture due to autointoxication, similar to that observed in the cultivation of *Brachiaria brizantha* (RODRIGUES and REIS, 1994).

Thus, determining the nature of the effects of straw on the seed germination process and may lead to the adoption of different techniques of behavior control of invasive plants. In addition to the species and numbers of viable weed seeds that are dependent on culture, the presence of humic substances in the soil and in the application solution interferes with the population dynamics of weeds.

2.2 Weed control through trash on the soil

In Brazil, the adoption of production systems where crops are planted on some kind of plant waste / trash has increased in several regions. The layer of straw on the soil is essential to the success of no-tillage system. It creates an extremely favorable environment for the improvement of physical, chemical and biological soil, contributing to weed control, stabilization and recovery of production or maintenance of soil quality. The system of crop rotation and succession must be suitable for the maintenance of a minimum cover the soil with straw.

The weed control by vegetation can occur by physical effect, preventing the incidence of light, which can be maximized by reducing the spacing between rows, favoring rapid soil cover (BARROS et al 2009), and by allelopathic effects (KREMER, 1998; THEISEN and VIDAL, 1999; FÁVERO et al., 2001 and MESCHEDE et al., 2007).

According to RODRIGUES and ALMEIDA (2005), species such as *Galinsoga parviflora* (buttercup) and *Sonchus oleraceus* (milkweed) did not germinate in soil covered, while *Raphanus raphanistrum* (wild radish) sprouts normally. The greater the amount of straw, the greater the physical barrier that will influence the germination of weed seeds and the higher the amount of allelochemicals produced.

The amount of straw depends on the source material, soil and climatic conditions of the region and the management system. The decomposition of the mulch depends directly on the relationship between the levels of carbon and nitrogen in each material. C / N ratio indicates a high content of high cellulose and lignin. In places where climatic conditions favor the rapid decomposition of the mulch should be preferred to species with high C / N ratio, for example, grains.

According to ALVARENGA et al. (2002), can be considered that 6 t ha [-1] of residue on the soil surface constitutes an adequate amount to no-tillage system, with which it can proper rate of soil cover. However, for this same author, the quantity and quality of straw on the soil surface depends largely on the type of plant cover and management practices is given. Therefore, this amount can vary greatly depending on the ease or difficulty of biomass production or the rate of decomposition. In this case, one must consider the permanence of straw on the soil surface. It is known that the C / N ratio becomes larger as the plant grows, and the C / N ratio around 40 seems satisfactory when the objective is to collect straw.

A plant of adequate coverage is one that maintains or improves soil conditions. Grasses have fasciculate roots, making them useful in reconstructing the structure of the soil, improving water infiltration and controlling erosion. Since legumes are the most efficient in the process of biological nitrogen fixation, rapidly decomposing waste by the lower C / N ratio (PECHE-FILHO et al., 1999). In choosing these plants, is a decisive factor to know their adaptation to the region and its ability to grow in an environment less favorable, since the crops are laid down in auspicious times. Therefore the straw to form the Brazilian Cerrado conditions, can be sown corn, sorghum, millet, pigeon pea, sunflower and crotalaria after cultivation of main crop (second crop).

The cultivation of oats and other species to cover the soil, either alone or in consortium, within a system of crop rotation, promotes significant increases in the yield of subsequent crops, and make them more lucrative by reducing the use of mineral fertilizers (MATZENBACHER, 1999). DERPSCH et al. (1985), reported that oat winter covering produced larger amounts of dry matter (8670 kg ha [-1]) and high levels of total N (147 kg ha [-1]), while reducing the amplitude of variation of temperature and soil humidity. To VIDAL et al. (1998), the mulch originated from oat straw to reduce weed infestations.

Results of experiments conducted in five tillage in Nebraska, USA, indicated that five seven t ha [-1] of wheat straw residue on the soil reduced the biomass of weeds in 21 and 73%, respectively, compared with soil discovered (WICKS et al., 1994). CRUTCHFIELD et al. (1985) reported that five t ha [-1] of wheat residue reduced weed density by 65%, contrasted with soils without residue.

Thus, we can say that the presence of plant residue / straw affects the establishment of weeds in different ways. Among the forms of interference cites are: formation of physical barrier to be broken by the plant in emergency temperature control on the soil surface, increased microbial biomass that can reduce the seed bank in the soil, apart from possible allelopathic effects that inhibit germination (FOLONI e SOUZA, 2010).After the biochemical transformation of these residues occur in the presence of soil humic substances, which will interfere with the dynamics of molecules of herbicides in the soil, or even the recommended doses of herbicides.

Herbicides when applied to the soil come, either directly or by incorporation of vegetable crop residues. In this sequence, there is a branch of the most complex and fascinating study of environmental soil chemistry, where little is known.

3. Organic matter and sorption of herbicides in the soil

Organic Matter (OM) present a strong ability to absorb herbicides (STEVENSON, 1972) and this reduces the mobility and biological activity of chemical compounds applied to soil

(SCHEUNERT et al., 1992). The pronounced reactivity of the OM is mainly related to its high specific surface area and presence of various functional groups such as carboxyl, hydroxyl and amine, and aliphatic and aromatic structures (STEVENSON, 1972; STEARMAN et al. 1989; KUCKUK et al. 1997).

CHEFETZ et al. (2004) observed greater adsorption of ametryn in sediment with 1.25 dag kg $^{-1}$ of organic carbon in relation to the other with 1.63 dag kg $^{-1.}$ The authors attributed this behavior to the fact that most of the sediment showed higher adsorption of aromatic compounds in the organic fraction, followed by a smaller number of polysaccharides, which favors its adsorption capacity.

Among the compartments of soil organic matter, humic substances are reported as the main responsible for the sorption of herbicides (PUSINO et al. 1992; CELIS et al., 1997). Most humic substances, especially in tropical regions, is in the form of clay-organic complexes (52 to 98%), according to STEVENSON (1994), whose total binding energy depends on the different forms of interaction promoted by functional groups of components organics. In this condition, neutralize their functional groups with loads of clay minerals, which reduces its sorption capacity of herbicides (PROCOPIO et al., 2001).

3.1 Quality of OM and HS on the sorption of herbicides

The quality of organic matter because of their functional groups determine the sorption of atrazine in soils, since different humic substances show different sorption intensities (PICCOLO et al., 1992). Among the constituents of organic matter, humic acid is responsible for about 70% of the sorption capacity of atrazine (BARRIUSO et al., 1992). Another important physical and chemical sorption of pesticides in soil is pH. In this sense, TRAGETTA et al. (1996) observed maximum sorption of atrazine in humic and fulvic acid near pH 3, while for pH conditions normally found in soils (5-7) sorption is lower.

PROCOPIO et al. (2001) observed that the sorption of atrazine by humic acids isolated was approximately nine times greater than that found sorption to kaolinite, goethite and ferridrita also examined separately. This indicates the high affinity between the herbicide and humic acids (MARTIN-NETO et al., 1994).

According MARTIN-NETO et al. (1999), the high intensity of hydrophobic sites to which the fraction of atrazine coupled (non-ionic behavior) can bind would be the main mechanism involved in sorption of atrazine with humic substances.

FERRI et al. (2005), in his study of sorption of herbicides on different substrates and found that the sorption ability of acetochlor was higher in the soil under no-tillage compared to conventional tillage.This behavior was explained only partially by higher content of C in the treatment of the sample of direct seeding.

When isolated, humin and humic acids showed acetochlor ability to absorb higher than when contained in the soil. Humin had higher ability to absorb acetochlor than the humic acids (FERRI et al., 2005).

Spectroscopic studies performed by PICCOLO et al. (1996) demonstrated that the main mechanism linking glyphosate and humic substances can be hydrogen bonds. The sorption of this herbicide may vary depending on the macromolecular structure and size of humic substances. The less aromatic C, the greater the sorption of the molecule.

3.2 Humic substances on the recommendation of herbicides

The humic acids in general allow a better exchange of chemicals, so that some authors suggest a decrease in those with the maintenance of efficacy was observed that the FOLONI and SOUZA (2010) working with cane sugar, concluding in his work that the use of humic acid in doses of 3.0 and 6.0 L ha $^{-1}$ did not cause phytotoxicity apparent effect on the culture of sugar cane plant. The addition of the use of humic acid with Dual Gold in different combinations, even with dose reduction of 25% allowed an excellent control of the main weeds present, equaling or surpassing traditional treatments until 120 DAT.

In assessing the remobilization of bound residues of 14C-anilazina fulvic acids in two soils of Germany, LAVORENTI et al. (1998) observed a small percentage of this waste in the humic and humin fractions and that the microorganisms were stimulated by applying the treatments corn stover (1.5 g 100 g $^{-1}$ dry soil) and glucose + peptone (0.2 g + 0.2 g 100 g^{-1} dry soil).

4. Allelochemicals release and control weed

4.1 Concept and production of allelopathy

Allelopathy is said to have any effect that plants have on the production of other chemical compounds released into the environment (RICE, 1984). Plants are able to produce chemicals with properties that affect beneficial or harmful in other plant species phenomenon called allelopathy, which is of Greek origin meaning *allelon* (from one to another) and *pathos* (suffering) (MOLISCH, 1937). Currently, this conceptual definition has become broader, expanding into the animal kingdom, since the interaction can occur between them and the plants and between animals and plants (GARDEN OF FLOWERS, 2001).

The chemicals responsible for allelopathy are called allelochemicals, whose function is primarily protective (SORIANO, 2001). The natural compounds with phytotoxic properties may have a high potential to control weeds (SOUZA-FILHO, 2006). According to MORALES et al. (2007), these compounds tend to have low toxicity to non-target organisms of control, as a potential source in the discovery of new molecules of herbicides less harmful to the ecosystem.

Unlike the common occurrence of compounds with allelopathic properties in higher plants, the amount and composition of these may vary depending on the species, age of the organ of the plant, temperature, light intensity, nutrient availability, microbial activity of the rhizosphere and composition of the soil in which they are the roots (PUTNAM, 1985; EINHELLIG and LEATHER, 1988).

Many are organic compounds produced by higher plants or microorganisms that have been identified as allelochemicals, which are: terpenes, steroids, organic acids, soluble in water, aliphatic aldehydes, ketones, long chain fatty acids, polyacetylenes, naphthoquinones, anthraquinones and complex quinones, originate from the mevalonate metabolic pathway of acetate (REZENDE and PINTO, 2003). Already the simple phenols, benzoic acids and derivatives, cinnamic acids and derivatives, coumarins, amino acids, polypeptides and sulfides and glycosides, alkaloids, cianidrina, flavonoids, and purine nucleoside derivatives, quinones and hydrolysable and condensed tannins are derived from the acid metabolic pathway shikimic (REZENDE and PINTO, 2003).

These compounds can be released from allelopathic plants through leaching and volatilization, root exudation and decomposition of plant residues (WEIR et al., 2004). A large number of allelopathic compounds such as oxalic acid, the amygdalin, coumarin and transcinâmico acid, are released into the rhizosphere and can act directly or indirectly in plant-plant interactions and the action of microorganisms.

The allelopathic effects may occur in the forms of auto toxicity and hetero toxicity (MILLER, 1996). The autotoxicidade occurs when the plant produces toxic substances that inhibit seed germination and growth of plants of the same species. Research has shown that alfalfa plants contain water-soluble phytotoxic compounds that are released into the soil environment by means of fresh leaves, stems and crown tissues as well as dry material, decaying roots and seeds (HALL and HENDERLONG, 1989). The phytotoxic hetero toxicity occurs when substances are released by leaching and root exudation and decomposition of waste in any type of plant on seed germination and growth of other plants (NUÑEZ et al., 2006). This second form is more potential to be explored by science, as a subsidy for the control of weeds in organic farming systems, or even as a tool to reduce costs with herbicides in conventional systems management.

4.2 Mode of action of allelochemicals

The action of allelochemicals and modification involves inhibition of growth or development of plants. According to SEIGLER (1996), the allelochemicals can be selective in their actions and plants can be selective in their response, which is why it is difficult to clarify the mode of action of these compounds. Several mechanisms of action of allelochemicals can affect the processes of respiration, photosynthesis, enzyme activity, water relations, stomatal opening, level of hormones, mineral availability, division and cell elongation, structure and permeability of membranes and cell wall (REZENDe et al. 2003). The same authors reported that many of these processes occur as a result of oxidative stress. One of the many effects of allelochemicals in plants is to control the production and accumulation of reactive oxygen species (ROS), which accumulates in cells in response to the allelochemical, thereby being responsible for damaging the cells causing their death (TESTA, 1995). Among them, blocking chain that carries electrons, where electrons are free and easily react with O_2 to form superoxide. Another known mechanism in the formation of ROS is the activity of allelochemicals on the NADPH oxidase, an enzyme that transfers electrons from the NDPH and donates to an acceptor (O_2) forming superoxide (FOREMAN et al., 2003).

Some allelochemicals rapidly depolarize cell membranes, increasing permeability and inducing lipid peroxidation, causing a generalized cell disorder that leads to cell death (YU et al., 2003).

4.3 Weed control by allelopathic effect

Seeds of *Coronilla L varies.* showed reduced germination rate when exposed to aqueous extracts of *Eucalyptus camaldulensis* and *Juglans regia* (ISFAHAN and SHARIAT, 2007). Soils planted with species *E.grandis, E.* and *E. urophylla grandis x urophylla* contains water-soluble phenolic compounds that inhibit the germination and early growth of black beans *(Phaseolus vulgar)* (ESPINOSA-GARCIA et al., 2008). According BURGOS et al. (2004), the allelochemicals produced by *Secale cereale* L. reduces root growth of *Cucumis sativus* L. causing changes in cellular structures of the roots. Thus, a large part of allelochemicals acts on oxidative stress by producing reactive oxygen species, which act directly or as flags to the processes of cell degradation, thus preventing the germination and early development, as well as physiological processes of plants.

Another effective technique for controlling weeds, mainly due to physical and allelopathic effects is the use of green manure. FONTANÉTTI et al. (2004) observed that species velvetbean *(Stizolobium aterrimum)* and pork-bean *(Canavalia ensiformis)* significantly reduced the number of nutsedge when incorporated into the soil before planting romaine lettuce and cabbage. Already CAVA et al. (2008) found that plants such as *C. juncea, C. spectabilis, M. aterrima* and *M. pruriens* have high competitive capacity by reducing the production of dry mass and number of weeds. SEVERINO and CHRISTOFFOLETI (2001) found that the use of green manures *Arachis pintoi, C. juncea* and *Cajanus cajan* have significantly reduced the seed bank of species *Brachiaria decumbens, Panicum maximum* and *Bidens pilosa.*

Thus, it is clear that the population dynamics of weed species is a function not only of the main crop, soil and planting season, but also because of the tillage system, the quantity and quality of dry matter present in the soil surface. This is due to the fact that each species used in land cover has production of metabolites that interfere with specific control of weeds. The association provides interaction with herbicides and microorganisms in the soil fauna, as well as decreasing the dose of herbicides due to the increased amount of straw on the soil surface.

5. References

ALVARENGA, R.C.; JUCKSH, I.; NOLIA, A.; ANDRADE, C.L.I.; CRUZ, J.C. Adubação verde como fonte exclusiva de nutrientes para a cultura do milho orgânico. In: Congresso Nacional de Milho e Sorgo: Meio Ambiente e o Agronegócio para o Milho e Sorgo, 24., 2002, Florianópolis. *Resumos...* Florianópolis, 2002. CD.

BARRIUSO, E.; FELLER, C.; CALVET, R.; CERRI, C. Sorption of atrazine, terbutryn and 2,4-D herbicides in two Brazilian Oxisols. *Geoderma,* v.53, n.1/2, p.155-167, 1992.

BARROS, H.B; SILVA, A.A; SEDIYAMA, T. (2009). Manejo de plantas daninhas. In: SEDIYAMA, T. *Tecnologias de produção e usos da soja.* Londrina: Mecenas, 314p.

BORGHI, E.; COSTA, N.V.; CRUSCIOL, C.A.C. (2008). Influência da distribuição espacial do milho e da *Brachiaria brizantha* consorciados sobre a população de plantas daninhas em sistema plantio direto na palha. *Planta Daninha,* Viçosa-MG, v. 26, n. 3, p. 559-568.

BURGOS, N.R., R.E. TALBERT, K.S. KIM AND Y.I. KUK. 2004. Growth inhibition and root ultrastructure of cucumber seedlings exposed to allelochemicals from rye *(Secale cereale). Journal of Chemical Ecology,* 30(3):671-689.

CAVA, M.G.B. et al. Adubos verdes para a renovação de canaviais do sudoeste goiano. In: CONGRESSO INTERNACIONAL DE TECNOLOGIA NA CADEIA PRODUTIVA DA CANA, 2, 2008, Uberaba. *Anais...*Uberaba, 2008. (CD-ROM)

CELIS, R.; COX, L.; HERMOSIN, M.C.; CORNEJO, J. Sorption of triazafluron by iron and humic acid-coated montmorillonite. *Journal of Environmental Quality,* v.26, p.472-479, 1997.

CHEFETZ, B.; BILKIS, Y.I.; POLUBESOVA, T. Sorption-desorption behavior of triazine and phenylurea herbicides in Kishon river sediments. *Water Research,* v. 38, p. 4383-4394, 2004.

CRUTCHFIELD, D.A.; WICKS, G.A.; BURNSIDE, O.C. Effect of winter wheat (*Triticum aestivum*) straw mulch level on weeds control. *Weed Science,* v.34, p.110-114, 1985.

CZARNOTA, N.A.; RIMANDO, A.M. Evaluation of root exudates of seven sorghum accessions. *Journal of Chemical Ecology,* v. 29, n. 9, p. 2073-2083, 2003.

DERPSCH, R.; SIDIRAS, N.; HEINZMANN, F.X. Manejo do solo com coberturas verdes de inverno. *Pesquisa Agropecuária Brasileira,* v. 20, p. 761-773, 1985

DUARTE JÚNIOR, J.B.; COELHO, F.C.; FREITAS, S.P. (2009). Dinâmica de populações de plantas daninhas na cana-de-açúcar em sistema de plantio direto e convencional. *Semina: Ciências Agrárias,* Londrina, v. 30, n. 3, p. 595-612.

DURIGAN, J. C.; ALMEIDA, F. L. S. *Noções sobre alelopatia.* Jaboticabal: FUNEP, 1993. 28 p.

EINHELLIG, F.A. AND G.R. LEATHER. 1988. Potetials for exploiting allelopathy to enhance cropo production. *Journal of Chemical Ecology,* 14(10):1829-1844.

ESPINOSA-GARCIA, F.J.; MARTINEZ, H.E.; QUIROZ, F.A. 2008. Allelopathic potential of Eucalyptus spp plantations on germination and early growth of annual crops. *Allelopath. J.* 21(1):25-37.

FAVERO, C. M.; JUCKSCH, I. ; ALVARENGA, R. C.; COSTA, L. M. Modificações na população de plantas espontâneas na presença de adubos verdes. *Pesquisa Agropecuária Brasileira,* v. 36, n. 11, p. 1355–1362, 2001.

FERRI, M.V.W., GOMES, J., DICK, D.P., SOUZA, R.F., VIDAL, R.A.R. Sorção do herbicida acetochlor em amostras de solo, ácidos húmicos e huminas de Argissolo submetido à semeadura direta e ao preparo convencional. *Revista Brasileira de Ciência do Solo,* 29:705-714, 2005.

FOLONI, L.L; SOUZA, E.L.C. Avaliação do uso de ácido húmico na redução do uso de herbicidas pré-Emergentes na cana planta. Ribeirão Preto, 2010. *Anais.*

FONTANÉTTI, A.; CARVALHO, G. J.; MORAIS, A. R.; ALMEIDA, K.; DUARTE, W. F. Adubação verde no controle de plantas invasoras nas culturas de alface-americana e de repolho. *Ciência e Agrotecnologia,* v.28, n.5, p.967-973, 2004.

FOREMAN, J., V. DEMIDCHIK, J.H.F. BOTHWELL, P. MYLONA, H. MIEDEMA, M.A. TORRES, P. LISTEAD, S. COSTA, C. BROWNLEE, J.D.G. JONES, J.M. DAVIES AND L. DOLAN. 2003. Reactive oxygen species produced by NAPH oxidase regulate plant cell growth. *Nature.* 422(6930):422-445.

GAZZIERO, D.L.P.; VARGAS, L.; ROMAN, E.S. (2004). Manejo e controle de plantas daninhas em soja. In: VARGAS, L.; ROMAN, S.R. (Org.). *Manual de manejo e controle de plantas daninhas.* Bento Gonçalves: Embrapa Uva e Vinho, p. 595-636.

HALL, M. H.; HENDERLONG, P. R. Alfafa autotoxic fraction characterization and initial separation. *Crop Science.,* v. 29, n. 2, p. 425-428, 1989.

ISFAHAN, M.N. AND M. SHARIATI. 2007. The effect os some allelochemicals on seed germination of *Coronilla varia* L. seeds. *American-Eurasian Journal of Agricultural & Environmental Science.* 2(5):534-538.

JARDIM DE FLORES. Alelopatia: a defesa natural das plantas. [S.I.: s.n.], 2000. Disponível em: <*http://www.jardimdeflores.com.br/ESPECIAIS/A09alelopatia.html #topo1*>. Acesso em: 11 maio 2001.

KADIOGLU, I.; YANAR, Y.; ASAV, U. Allelopathic effects of weeds extracts against seed germination of some plants. *Journal of Environmental Biology*, v. 26, n. 2, p. 169-173, 2005.

KREMER, R.J. Integration of biological methods for weed management. *WSSA Abstracts*, v.38, p. 46, 1998.

KUCKUK, R.; HILL. W.; NOLTE. J. & DAVIES, A.N. Preliminary investigations into the interactions of herbicides with aqueous humic substances. *Pesticide Science*, 51:450-454, 1997.

LAVORENTI, A.; BURAUEL, P.; WAIS, ANDREAS; FUHR, F. Remobilization of bound anilazine residues in fulvic acids. *Pesticidas: Revista de Ecotoxicologia e Meio Ambiente*, Londrina, v.8, 1998.

MARQUES, L.J.P.; SILVA, M.R.M.; ARAÚJO, M.S.; LOPES, G.S.; CORRÊA, M.J.P.; FREITAS, A.C.R.; MUNIZ, F.H. (2010). Composição florística de plantas daninhas na cultura do feijão-caupi no sistema de capoeira triturada. *Plantas Daninhas*, v.28, p.953-961.

MARTIN-NETO, L.; FERREIRA, J.A.; NASCIMENTO, O.R.; TRAGHETTA, D.G.; VAZ, C.M.P.; SIMÕES, M.L. Interação herbicidas e substâncias húmicas: estudos com espectroscopia e polarografia. In: ENCONTRO BRASILEIRO SOBRE SUBSTÂNCIAS HÚMICAS, 3, 1999, Santa Maria, RS. *Anais...* Santa Maira: 1999.

MARTIN-NETO, L.; VIEIRA, M.E.; SPOSITO, G. Mechanism of atrazine sorption by humic acid: a spectroscopy study. *Environmental Science & Technology*, v.28, p.1867-1873, 1994.

MARTINS, D.; PITELLI, R. A. (1994). Interferência das plantas daninhas na cultura do amendoim das águas: efeitos de espaçamentos, variedades e períodos de convivência. *Planta Daninha*, v. 12, n. 2, p. 87-92.

MARTINS, D.M.; VELINI, E.D.; MARTINS, C.C.; SOUZA, L.S. (1999). Emergência em campo de dicotiledôneas infestantes em solo coberto com palha de cana-de-açúcar. *Planta Daninha*, v. 17, n. 1.

MATEUS, G. P.; CRUSCIOL, C. A. C.; NEGRISOLI, E. Palhada do sorgo de guiné gigante no estabelecimento de plantas daninhas em área de plantio direto. *Pesquisa Agropecuária Brasileira*, v. 39, n. 6, p. 539-542, 2004.

MATZENBACHER R. G. Manejo e Utilização da Cultura. In: MATZENBACHER, R.G. (Eds). *A Cultura da aveia no sistema plantio direto*. Cruz Alta: FUNDACEP/ FECOTRIGO, 1999, p. 22-54

MESCHEDE, D.K.; FERREIRA, A.B.; RIBEIRO JR, C.C. Avaliação de diferentes coberturas na supressão de plantas daninhas no cerrado. *Planta Daninha*, Viçosa-MG, v. 25, n. 3, p. 465-471, 2007.

MILLER, D.A. 1996. Allelopathy in forage crop systems. *Agronomy Journal*. 88(6):854-859.

MOLISCH, H. 1937. Der Einfluss einer pflanze auf die andere. *Allelopathy*. Gustav Fischer Verlag, Jena. 106 p.

MORALES, F.F., M.I. AGUILAT, B.K. DIÁZ, J.R. DE SANTIAGO-GOMÉZ AND B.L. HENNSEN. 2007. Natural diterpenes from *Croton ciliatoglanduliferus* as photsystem II and photosystem I inhibitors in spinach chloroplasts. *Photosynthesis Research*. 91(1):71-80, 2007.

NIMBAL, C.I.; PEDERSON, J.; YERKES, C.N. et al. Phytotoxicity and distribution of sorgoleone in grain sorghum germplasm. *Journal of Agricultural And Food Chemistry*, v. 44, n. 5, p. 1343-1347, 1996.

NUÑEZ, L.A., T. ROMERO, J.L. VENTURA, V. BLANCAS, A.L. ANAYA AND R.G. ORTEGA. 2006. Allelochemical stress causes inhibition of growth and oxidative damage in *Lycopresicon esculentum* Mill. *Plant, Cell and Environment.* 29(11):2009-2016. p.13-20.

PECHE-FILHO, A. GOMES, J.A., BERNARDI, J.A. Manejo de Fitomassa: considerações técnicas. In: AMBROSANO, E. (Eds). *Agricultura Ecológica.* Guaíba: Agropecuária, 1999. 398p.

PICCOLO, CELANO, A. G.; CONTE, P. Adsorption of glyphosate by humic substances. *Journal of Agriculture and Food Chemistry,* v.44, p.2442-2446, 1996.

PICCOLO, A.; CELANO, G.; SIMONE, C. de. Interactions of atrazine with humic substances of different origins and their hydrolysed products. *The Science of the Total Environment,* v.117-118, p.403-412, 1992.

PROCÓPIO, S.O., PIRES, F.R., WERLANG, R.C., SILVA, A.A., QUEIROZ, M.E.L.R., NEVES, A.A., MENDONÇA, E.S., SANTOS, J.B. e EGREJA FILHO, F.B. Sorção do herbicida atrazine em complexos organominerais. *Planta Daninha,* Viçosa-MG, v.19, n.3, p.391-400, 2001.

PUSINO, A.; LIU, W.; GESSA, C. Influence of organic matter and its clay complexes on metolachlor adsorption on soil. *Pesticide Science,* v.36, p.283-286, 1992.

PUTNAM, A.R. Weed allelopathy. *In:* DUKE, S.O. (Ed.). *Weed phisiology,* Flórida, p. 131-155, 1985.

REZENDE, C.P., J.C. PINTO, A.R. EVANGELISTA E I.P.A. SANTOS. 2003. Alelopatia e suas interações na formação e manejo de pastagens. Boletim Agropecuário, Universidade Federal de Lavras, MG. (54):1-55.

RICE, E.L. 1984. Allelopathy. 2nd ed. Academic Press, New York. 422 p.

RODRIGUES, B.N.; ALMEIDA, F.S. (2005). *Guia de herbicidas.* 5. ed. Londrina-PR: 591p.

RODRIGUES, L.R.A.; REIS, R.A. Estabelecimento de outras forrageiras em áreas de *Brachiaria* spp. IN: SIMPÓSIO SOBRE MANEJO DA PASTAGEM, 11, 1994. *Anais...* Piracicaba: FEALQ, 1994. p. 299-325.

SCHEUNERT, I.; MANSOUR, M. & ANDREUX, F. Binding of organic pollutants to soil organic matter. *International Journal of Environmental Analytical Chemistry,* 46:189-199, 1992.

SEIGLER, D.S. 1996. Chemistry and mechanisms of allelopathy interactions. *Agronomy Journal,* 88(6):876-885.

SEVERINO, F. J.; CHRISTOFFOLETI, P. J. Banco de sementes de plantas daninhas em solo cultivado com adubos verdes. *Bragantia,* v.60, n.3, p.201-204, 2001.

SILVA, H. P.; GAMA, J.C.M.; NEVES, J.M.G.; BRANDÃO JÚNIOR, D.S.; KARAM, D. (2010). Levantamento das plantas espontâneas na cultura do girassol. *Revista Verde,* v.5, n.1, p.162 – 167.

SORIANO, U. M. Alelopatia. 1996. Disponível em: http://mailweb.pue.udlap.mx/aleph/alephzero6/alelopatia.htm . Acesso em: 10 maio 2001.

SOUZA-FILHO, A.P.S. 2006. Proposta metodologica para análise de ocorrência de sinergismos e efeitos potencializadores ente aleloquímicos. *Planta Daninha* 24(3):607-610.

STEARMAN, G.K.; LEWIS, R.J.; TORTEROLLI, L.J. & TYLER, D.D. Herbicides reactivity of soil organic matter fractions in no-tilled and tilled cotton. *Soil Science Society of America Journal*, 53:1690-1694, 1989.

STEVENSON, F.J. *Humus chemistry*. Genesis, composition, reactions. 2.ed. New York: John Willey & Sons, 1994. 443p.

STEVENSON, F.J. Organic matter reactions involving herbicides in soil. *Journal of Environmental Quality*, 1:333-343, 1972.

TESTA, B. 1995. The metabolism of drugs and other xenobiotcs. Academic Press, New York. 475p.

THEISEN, G.; VIDAL, R. A.; FLECK, N. G. Redução da infestação de*Brachiaria plantaginea* em soja pela cobertura do solo com palha de aveia preta. *Pesquisa Agropecuária Brasileira*, v. 35, p. 753-756, 2000.

THEISEN, G; VIDAL, R. A. (1999). Efeito da cobertura do solo com resíduos de aveia preta nas etapas do ciclo de vida do capim-marmelada. *Planta Daninha*, v. 17, n.2, p.189-196.

TRAGETTA, D.G.; VAZ, C.M.P.; MACHADO, S.A.S.; VIEIRA, E.M.; MARTIN-NETO, L. Mecanismos de sorção da atrazina em solos: estudos espectroscópicos e polarográficos. *Comunicado Técnico*, São Carlos, n.14, p. 1-7, dez. 1996.

VAZ DE MELO, A. GALVÃO, J.C.C., FERREIRA, L.R. , MIRANDA, G.V. , TUFFI SANTOS, L.D., SANTOS, I.C. e SOUZA, L.V. (2007). Dinâmica populacional de plantas daninhas no cultivo de milho-verde no sistema de plantio direto orgânico e tradicional. *Planta Daninha*, Viçosa, MG, v. 25, n.3, p. 521-527.

VIDAL, R.A.; THEISEN, G.; FLECK, N.G.; BAUMAN, T.T. Palha no sistema de semeadura direta reduz a infestação de gramíneas anuais e aumenta a produtividade da soja. *Revista Científica Rural*, v.28, p. 373-377, 1998.

WEIR, T.L., S.W. PARK AND J.M VIVANCO. 2004. Biochemical and physiological mechanisms mediated by allelochemicals. *Current Opinion* in *Plant Biology*, 7(4):p.472-479.

WICKS, G.A.; CRUTCHFIELD, D.A.; BURNSIDE, O.C. Influence of *wheat (Triticum aestivum)* straw mulch and metalachlor on corn (*Zea mays*) growth and yield. *Weed Science*, v.42, p.141-147, 1994

YU, J.Q., S.F. YE, M.F. ZHANG AND W.H. HU. 2003. Effects of root exudates and aqueous root extracts of cucumber (*Cucumis sativus*), and allelochemicals on photosynthesis and antioxidant enzymes in cucumber. *Biochemical Systematics and Ecology*. 31(2):129-139.

8

Ecological Production Technology of Phenoxyacetic Herbicides MCPA and 2,4-D in the Highest World Standard

Wiesław Moszczyński and Arkadiusz Białek
The Institute of Industrial Organic Chemistry,
Poland

„Selectivity is a major goal in modern synthetic chemistry"
Bartman W., Trost B. M.

1. Introduction

Herbicides MCPA (4-chloro-2-metylphenoxyacetic acid) and 2,4-D (2,4-dichloro-phenoxyacetic acid), which belong to the group of chlorophenoxyacetic acids, have been produced in Poland since the break of the sixties in the scale of many thousands of tons per year, which constitutes 5-7% of the world's production. Acids and their salts are exported to all continents. An advantage of herbicides within this group is their harmlessness for man and environment in doses used in agriculture. The condition is the high content and purity of the active substance in utility preparations. Unfortunately, classic technologies based on the reaction of phenols chlorination or their derivatives used until today all over the world do not ensure high purity, and significant quantities of highly toxical chloroorganic waste compounds originate in the production process. The main cause lies in the low selectivity of the reaction of chlorination of phenols' aromatic ring. Numerous producers enrich the purity of chlorophenols with the method of rectification. As our research has shown, in the process of rectification and while burning post-rectification wastes, dioxines and dibenzofurans can originate.

Below we present research conducted in the Institute of Industrial Organic Chemistry (IPO) in Warsaw with strict cooperation with production plants „Rokita Agro" in Brzeg Dolny and „Organika Sarzyna" in Nowa Sarzyna over the development of technologies of chlorophenoxyacetic herbicides. Commonly used „classical" technologies of production of MCPA and 2,4-D are based on two subsequent reactions – chlorination of phenol or 2-methylphenol and the reaction of the obtained chlorophenol with chloroacetic acid (MCAA), commonly called condensation. The first stage – chlorination reaction – is critical for the nature of the entire reaction, i.e. the number of operations, selection of equipment, purification methods, wastes. Electrophilic reaction of chlorination of aromatic phenol ring is non-selective. Isomers and polychlorinated compounds are always originated. In order to increase the selectivity of reaction and the purity of the the final product, various technologies of chlorination and purification of chlorophenols are used all over the world, including regioselective catalysts, replacement of chlorine with sulfuryl chloride, rectification of chlorophenol and reversed technologies, i.e. condensation first, and then chlorination of the obtained phenoxyacetic acid. Also in our country technologies of production of MCPA and 2,4-D have been changed many times, achieving higher and higher selectivity of the chlorination reaction. The current production of MCPA (Chwastoks) is based on a unique technology of chlorination of 98% selectivity, while in the 2,4-D technology 96% selectivity has been achieved, and after crystallization it ensures the purity of the product equal to 98%. Technologies minimize or eliminate waste chloroorganic compounds. The final products have the highest quality standard.

2. MCPA

In the years 1959-1961 in IPO-Warszawa an experimental production MCPA in the scale of 5 tons per year of 30% „Chwastoks R-30" preparation was started. It made possible the conduct of vast agricultural research, including the research at 15500 linum plantations. In 1962 in Nowa Sarzyna was started the first technical installation of 1000 tons of Chwastoks R-30 (Moszczyński et al, 1963). The synthesis was based on chlorination of 2-methylphenol with gaseous chlorine. Technical 4-chloro-2-methylphenol without purification was exposed to condensation with monochloroacetic acid (MCAA). The diagram of the reaction is presented below.

2-methylphenol   4-chloro-2-methylphenol   2-methyl-4-chlorophenoxyacetic acid
                                                                    (MCPA)

The reaction of chlorination of melted 2-methylophenol had a very low selectivity – 60 to 70%. Technical 4-chloro-2-methylphenol was a mixture of chlorophenols of the following composition:

4-chloro-2-methylphenol	60-70%
6-chloro-	12-20%
4,6-dichloro-	3-9%
2-methylphenol-	4-10%

After the condensation of technical 4-chloro-2-methylphenol with chloroacetic acid there was derived a mixture of chloromethylphenoxyacetic acids containing 60 to 70% of MCPA. Below such a product is referred to as „MCPA 70". In the process of condensation the efficacy varied within broad limits and there always remained several per cent of non-reacted chloromethylphenols.

The final product Chwastoks R-30 was separated from the post-condensation mass without purification, the preparation contained up to 6% of free chloromethylphenols. A similar MCPA technology was applied by many producers, e.g. in Leuna-Werke plants, GDR, where the raw post-chlorination mixture was exposed to vacuum rectification. Together with that 40% of waste chloromethylphenols was originated. The preparation Chwastoks R-30 had a good opinion among its users. It did not freeze in winter in unheated warehouses, it perfectly dissolved in water. It was cheap. In IPO a comparative research of the activity of MCPA-70 with pure MCPA was conducted, and the activity of each chloromethylphenoxyacetic acids separately, and also their mixtures. It appeared that isomeric 6-chloro-2-methylphenoxyacetic acid had a high biological activity, and in the mixtures the synergism of activity was observed. After 10 years of MCPA-70 production in a small installation in 1972 a new production facility of 4500 tons per year capacity was started without substantial changes in technology. In the discussed decade in the European market there appeared MCPA preparations of high purity, without free chloromethylphenols containing 80-90% of the pure active component. In the technologies sulfuryl chloride was used as a more selective chlorinating agent, and sometimes additionally vacuum rectification of technical 4-chloro-2-methylphenol. Chemische Werke Schwarzweide in GDR started the production of MCPA basing on sulfuryl chloride on the license purchased in Great Britain. The technology based on sulfuryl chloride is currently commonly used by chemical companies all over the world.

Skeeters was the first to describe in 1956 the MCPA synthesis through chlorination of 2-methylphenoxyacetic acid (MPA) in 1,2-dichloroethane (Skeeters et al., 1956). In the seventies such a method was used in industry. It was a new generation technology, below referred to as the reverse method, and the obtained product – „MCPA 90". The diagram of the reaction is presented below.

| 2-methylphenol | 2-methylphenoxyacetic acid (MPA) | MCPA |

The aromatic ring of MPA acid is less prone to electrophilic substitution, and the side chain is a steric hindrance for creating the isomer 6-chloro-. In the reaction of MPA, chlorination selectivity of 80-90% is achieved, i.e. higher than in the classical method, where sulfuryl chloride without purification of the product by rectification is used. In the reverse technology a barrier which is difficult to lift is the high melting temperature of MPA (157°C) and the low solubility (7%) in organic solvents. Chlorination of MPA acid without solvents in hot water cannot be performed to the end and there remains 20% of non-chlorinated MPA. In Bratysława, the chlorination of 20-30% of water solution of MPA sodium salt MPA in the temperature of 90-100 °C was used on the industrial scale. The final product, containing 90% of the active substance, contained 5% of isomer 6-chloro- and 3-6% of free MPA (Rapoš et al, 1960). Chimzawod in Ufa, USSR, launched a multi-ton installation of MPA acid chlorination in 7% solution in solvent. The license with the full equipment was purchased from the Fisons company (Great Britain). The MCPA purity was 85-90%, the product contained 1-3% of free MPA, efficiency about 92% (Moszczyński, 1971).

In Poland, in 1976, after the complete redevelopment of MCPA-70 system (Moszczyński et al., 1975), a new generation reverse technology was developed and used in industry. MPA acid heated with tetrachloroethylene (TCE) in post-condensation mass (brine) forms an eutectic of melting temperature of 100°C. It enables the separation of the liquid organic phase MPA/TCE from brine effluent from the post-condensation mass. The ogranic phase of MPA/TCE without drying was chlorinated in the temperature of 90-100%°C. TCE was removed from the post-chlorination mass with the use of destillation with water vapour. The obtained product was of 90% purity, with a mix of 7-9% of isomer 6-chloro-, and up to 3% of free chloromethylphenols. With the evaluation of efficiency in precise balance tests, 3 to 5% of the product was missing, and the quantity of free chloromethylphenols was higher that the quantity derived from MPA. In the technology used in Ufa, MPA before chlorination was effectively dephenoled by the extracting method, and in spite of that, in the final product chloromethylphenols were present. In patent literature the low efficiency of MPA chlorination attracts attention, as well as the lack of material balance of reaction. We have recognized that as a normal known phenomenon of breaking MPA and MCPA ether bonding in the presence of hydrogen chloride according to the following diagram:

The hydrolysis research which we performed did not confirm the presence of such a reaction in the conditions of MCPA synthesis, in the MPA post-chlorination mass there was found and separated several per cent and a compound with spirolactone structure was identified. An identical compound was depicted earlier by H. Lund (Lund, 1958).

H. Lund suggests the mechanism of spirolactone origination, a derivative of MPA, based on joining 2 molecules of Cl_2 to the aromatic ring before the formation of cyclic ketal (spirolactone) according to the diagram:

He refers to known phenols reactions, which in the process of long-lasting chlorination, after saturation of all the places with chlorinum, form cyclohexanones. It is difficult to agree with such a mechanism, where in the chlorination of MPA with one chlorinum atom, in mild conditions, with the help of hypochlorite with the known susceptibility of phenol aromatic ring to replacement of o- and p-, there might occur such a significant addition of chlorinum atoms. The life of cyclohexanone ring should not be conditioned by the presence of lactone side ring. Meanwhile, the reconstruction of the side ring transforms the cyclic structure into the aromatic one. W. Moszczyński suggests a completely different mechanism (Moszczyński, 1998). In the process of MPA chlorination after binding chlorinum atom in the electrophilic reaction there is formed a transitory sigma complex, which is subject to preservation in the intramolecular nucleophile trap according to the following diagram:

sigma complex   sigma complex preserved in the intramolecular nucleophile trap

The preserved sigma complex is easily disintegrated into chloromethylphenol and glycolic acid. The discussed reaction is of a general nature, we repeated it while chlorinating 2,4-dichlorophenoxyacetic acid and chloromethylphenoxypropionic acid, obtaining appropriate preserved sigma complexes. Our discovery that chlorophenoxyalkanecarboxylic acids are traps of sigma complexes in electrophilic reaction of chlorination in the environment of organic solvents explains the causes of appearance of free chloromethylphenols and losses of the final product in the synthesis with the reverse method.

In MCPA production with the reverse method, the first stage was the synthesis of 2-methylphenoxyacetic acid (MPA) in the Williamson's reaction according to the following diagram:

The main reaction is always accompanied by the side reaction of hydrolysis of chloroacetic acid into glycolic acid.

$$ClCH_2COOH \xrightarrow[\text{water}]{\text{NaOH}} HOCH_2COOH + NaCl$$

In the domestic industrial production of MPA and 2,4-D, for many years the efficiency of condensation calculated to chloroacetic acid varied within broad limits, achieving on average 85%. Research conducted by W. Moszczyński with the use of regression function explained the causes of changeable efficiency (Moszczyński, 1999). Repetitively the efficiency of MPA synthesis of about 92% was achieved. It was established that the first 15 minutes of reaction are critical for the efficiency of the main reaction. In that time about 90% of MCAA reacts. The nucleophilic reaction of synthesis of MPA with Williamson's method proceeds according to the bimolecular mechanism according to the scheme:

$$V = k[ArO^-] [ClCH_2COO^-]$$

In the temperature <90°C, in the water alkaline environment its speed is low, while in the temperature of 90-110°C full substrate reaction occurs after 1 hour (Moszczyński, 1994). pH has very clear influence on this speed, and the concentration of ions of phenoxyl ArO $^-$ and hydroxyl OH- depends on pH. Both mentioned nucleophilic agents compete with each other in the reaction with substrate – apart from the main reaction there occurs hydrolysis of chloroacetic acid into glycolic acid. Hydrolysis of substract makes the reaction's kinetics more complicated, and the formula of speed of reaction of synthesis of MPA needs to be corrected.

Hydrolysis of chloroacetic acid has a course which is clearly distinct from the main reaction. According to Berhenke and Britton, it occurs both in acidic environment and in alkaline environment and in the range of pH 3 – 10 it has similar speed. (Berhenke, Britton, 1946). Occurrence of hydrolysis both in the acidic and lightly alkaline environment as well as acceleration of speed of hydrolysis in the final stage of MPA synthesis along with sodium chloride accumulation proved by this research confirm that the reaction proceeds according to the mechanism S_N1. With pH = 11 - 11,5 MCAA hydrolysis is 5,5%, and with pH = 12 - 12,5 it increases to 42%, which suggests that with a greater concentration of OH- the reaction proceeds according to the bimolecular mechanism type S_N2 with the participation of OH- ions. This research confirms the findings of Dawson and Pycock (Dawson, Pycock, 1936) that the reaction of MCAA hydrolysis is a combination of mechanisms S_N1 and S_N2.

For research on the optimalization of parametres of MPA synthesis with the use of mathematical model the following parametres were accepted as critical: reagent molar ratio (X_1), reagent molar ratio (X_2), pH (X_3). Optimum values of variables X_1 and X_2 marked by experiment and calculated for the synthesis of MPA are respecively 0,8-1, and 3,3-3,7 mol/dm³.

pH of reaction environment has a critical influence on the efficiency of MPA synthesis. With pH <9,5 MPA synthesis does not occur, while with pH >11, ion ArO- loses in the competition with ion OH-. The optimum value of pH indicated on the mathematical model is 10-10,7.

In the period of MCPA production with the reverse method in „Organika Sarzyna" in the seventies, countries classical technologies of chlorination of 2-methylphenol with sulfuryl chloride used in different were modernized. New regioselective catalysts combined with vacuum rectification of technical 4-chloro-2-methylphenol were commonly used. It meant an improvement of the purity of preparations with over 90 to 96% of active substance.

IPO for the third time has joined the competition in the market, overtaking, in terms of modernity, all technologies based on sulfuryl chloride. MCPA of 98% purity was offered directly from the synthesis, without purification and wastes, with an outstanding simplification of technology.

After several years of research on regioselective catalysts of chlorination of MPA with chlorinum, sodium hypochlorite and *t*-butyl, a laboratory synthesis „MCPA 98" of 98% of purity of active substance was developed. In May 1992 a combined research-implementation team, composed of 25 specialists from IPO was assigned, W. Moszczyński, I. Górska et al., and 19 from „Organika Sarzyna", including. T. Jakubas, J. Peć to conduct technical tests, supervising redevelopment of construction and to conduct the launch of production. Synthesis of MCPA 98 consisted in one-stage condensation of MCAA with o-cresol to MPA and chlorination of MPA in water in room temperature with the use of sodium hypochlorite against amine catalyst (Moszczyński et al., 1992). A simple laboratory technology appeared to be unrepeatable in industrial conditions. For two years of ½-technical and technical tests not a single manufactured unit of product of >94% purity was obtained.

In the laboratory in glass all syntheses without exception had selectivity and efficiency of 98-99%. Meanwhile, with any multiplication of scale in glass and apparatuses from 50 l to 2,5 m³ after exceeding the stage of about 2/3 pure chlorination *para-* suddenly broke down and a product of 90-94% purity of isomer *para-* was originated. Only the research of W. Moszczyński on the mechanism of reaction showed that we deal wigh a new, unknown in this group of compounds, aromatic free radicals reaction of 99% selectivity (Moszczyński, 1998). Free radicals reactions, as opposed to ion reactions, are highly selective, but sensitive to the conditions of reactions and dozens of external agents, and anytime there may occur a break of chain generation of radicals and return to the ion mechanism of a lower selectivity. Phenol reactions occupy a special place in this group. They are subject to reactions with such free radicals, which in other cases are inactive. There are proofs that with the presence of unpaired electrones delocalized into the aromatic ring, the replacement only occurs into the location *orto-* or *para-* (Dermer & Edmison 1957). The presence of free radicals was proved by EPR method at the University of Wrocław (Jezierski et al., 1999). It allowed to discover and eliminate radicals inhibitors and easily repeat the synthesis on an industrial scale. The installation launch took place in 1995. For the third time a new MCPA production installation was assembled with the full automatic control of the technological process and continuous chlorination and educing of MCPA acid. In the table below a standard of active substance of domestic MCPA was shown, according to the technology from 1976 and the last from 1996 with the purity declared in that time by leading western producers (Moszczyński et al. 2010). MCPA 98 from „Organika Sarzyna" is the best in all parametres.

No.	Parametres	MCPA 90 „Organika Sarzyna" 1976	MCPA AK 20 The Netherlands prospectus 1992	MCPA BASF Germany prospectus 1992	MCPA 98 „Organika Sarzyna" 1995
		[%]			
1	MCPA	85-90	96.0	94.0	97-98
2	6-chloro-2-methylphenoxyacetic acid	8-14	1.0	1.5	0.7
3	2-methylphenoxyacetic acid (MPA)	2	1.0	1.5	0.2
4	4,6-dichloro-2-methylphenoxyacetic acid	2	1.5	2.5	0.2
5	chlorocresols	1-3	0.5	0.5	traces

Table 1. Comparison of MCPA quality standards.

MCPA 98 authors have won a series of awards in national competitions for the technical and ecological level of technologies.

3. 2,4-D

In Poland in a short time after the publication of Hamner and Tukey in Science (Hamner, Tukey, 1944) about the discovery of selective chemical herbicide 2,4-D there was started research on production and agricultural use of 2,4-D. 2,4-dichlorophenoxyacetic acid and its sodium salt are marked with the symbol 2,4-D. The dynamic development of technology and production of 2,4-D and its derivatives in our country is a constant keeping pace with the growing needs of agriculture. Sodium salt 2,4-D („Pielik"), the first modern chemical selective herbicide in the domestic market, quickly became a strategical item in agricultural cultivation. In the years 1952-53 in the Chemical Plant „Rokita", an experimental production of 20-30 tons per year according to the technology developed in the Intitute of Industrial Organic Chemistry was launched (Hirszowski, Moszczyński, 1952).

The technology was based on the reaction of dichlorophenol (2,4-DCP) with chloroacetic acid (MCAA):

2,4-dichlorophenol

88-90% of isomer 2,4-

MCAA   2,4-D 88-90% izomer 2,4-

96% 2,4-D after crystalization from water

In 1960 in the Chemical Plants „Rokita", a production installation of 500 tons per year was launched. One year later ½-technique, and in 1964 a technical installation of esters 2,4-D, so called Pieliki Płynne. In 1968 in IPO there was developed the technology of technical enrichment of 90% 2,4-DCP with the help of vacuum rectification. The process turned out to be highly energy-consuming and costly. Because of similar boiling temperatures of 4-chloro-2,4- and 2,6-dichlorophenols within the limits of 210-220°C, strongly corrosive for the equipment and generating 15% of chlorophenols in the form of cube residue. In that time we did not have the knowledge about the possibility of forming dioxines in the process of chlorophenols rectification.

The turning point in the production and usage of 2,4-D occurred after the appearance and launching production of dimethylamine salt, so called Aminopielik in 1976. Aminopieliks are liquid preparations soluble in water, easy to use and prepare working liquids. Previously used sodium salt 2,4-D, so called Pielik, in the form of powder was poorly soluble in water, and in hard water residues of magnesium, calcium and iron salt were precipitated. In 1968 a brand new production installation 3000 tons per year was launched. As was shown in the diagram above, the synthesis of 2,4-D is based on two subsequent reactions of chlorination and condensation. This method is used by producers of 2,4-D in all countries, invariably since the first synthesis published in 1941 by J. Pokorny (Pokorny, 1941). Below, this technology is referred to as „classical". This method is based on easily available resources and the standard prodution equipment, which makes it generally accessible. A disadvantage to this method is the low quality of the product, burdened with 10-12% of undesired chlorophenoxyacetic acids, as well as free chlorophenols.

In the process of production as a result of standard purification of the product with the method of crystallization and washing there are originated 15 tons of wastes per 1 ton of 2,4-D, burdened with chloroorganics, chlorophenoxyacetic acids, chlorophenols, glycolic acid and sodium chloride. With multi-thousand scale of production, it means a serious environmental problem. For several thousand tons of annual production for many years there were created many thousand tons of wastes in the solid form and in liquid wastes. In some manufactured units there was confirmed the presence of nanogram quantities of highly toxic dioxines. Technology 2,4-D required research and solutions eliminating dioxines. Thermal processing, including burning of chloroorganic compounds, particularly chlorophenols, conduces originating dioxines. For some period waste chlorophenols were separated from the liquid wastes and chlorinated up to pentachlorophenol PCF, a known fungicide used mainly for the preservation of railway sleepers. In the first period in the installation 500 tons of 2,4-D per year, chlorophenol wastes were oxygenated with sodium hypochlorite on separated uncovered ground fields, a technology of Ostrowska J. IPO. The breakthrough occurred only after condensation was mastered at the beginning of the nineties, thanks to using several metre long reactors with stirrers for the reaction of disintegration of chlorophenols with sodium hypochlorite and additional cleaning of liquid wastes on biological treatment plant.

The border of purity of 2,4-D in the classical technology is determined by 90% of isomer 2,4-DF in the process of phenol chlorination and 96% of the active substance in acid and preparations 2,4-D after crystallization and washing.

Composition of 2,4-D obtained in the classical technology [% GC]	
chlorophenols	0.3
o-chlorophenoxyacetic acid	2.0
p- chlorophenoxyacetic acid	0.4
2,6-di chlorophenoxyacetic acid	0.9
2,4- di chlorophenoxyacetic acid	95.6
2,4,6-tri chlorophenoxyacetic acid	0.3

Table 2. 2,4-D obtained in the classical technology. (Białek, Moszczyński, 2009).

For many years of production of 2,4-D on bigger and bigger installations, apart from the cumulation of wastes, a weak link was the poor efficiency of condensation reaction, maximum 90% of calculated on 2,4-dichlorophenol. In particular manufactured units, it varied up to several per cent. The reacting mass, initially thin, sometimes as a result of a violent reaction and salting out the product, was getting thick so quickly that it could not be stirred and it was thrown out of the hatchway from the reactor. Never ending experiments with a change of concentration and order of introducing products, molar ratio, temperature and height of pH gave no results. During the synthesis, in the course of reacting of chloroacetic acid pH was changing, which required adding sodium lye. The measurement and adjustment of pH was done by hand and as a rule – delayed in relation to the speed of reaction. As it was mentioned above, in „Organika Sarzyna" was produced since 1962 on IPO technology 4-chloro-2-methylphenoxyacetic acid (MCPA, Chwastoks). In the reaction of condensation of 4-chloro-2-methylphenol with chloroacetic acid there occurred similar difficulties. They were mastered only after partial continuity of the process and using automatic adjustment of pH with the help of pH-metre controlling the dispensing of MCAA and sodium lye.

The causes of difficulties occurring at MCAA and phenols condensation were only explained by the research of Moszczyński over the mechanism of the main reaction and the competitive MCAA hydrolysis reaction depicted before (Moszczyński, 1999). The research concerned the reaction of o-krezol with chloroacetic acid, but they are of a general nature and they can be related to 2,4-dichlorophenol (2,4-D) and other alkilo- and chlorophenols. Particular phenols depending on the substituents and solubility of the final chlorophenoxyacetic acid react in a slightly different way. The maximum industrial capacity of condensation in MPA synthesis is 92%, while in 2,4-D synthesis 97% is achieved.

The composition of mass after condensation of raw 2,4-dichlorophenol with MCAA acid was shown below in table 3.

Composition of post-condensation mass	[%GC]
o-chlorophenoxyacetic acid	2.5
p- chlorophenoxyacetic acid	0.5
2,6-dichlorophenoxyacetic acid	7.4
2,4- dichlorophenoxyacetic acid	84.6
2,4,6-trichlorophenoxyacetic acid	1.2
o-chlorophenol	0.0
2,4-dichlorophenol	2.2
p-chlorophenol	0.0
2,6-dichlorophenol	1.1
2,4,6-trichlorophenol	0.4
NaCl	
glycolic acid	
water	

Table 3. Composition of reaction mass after condensation of raw 89% 2,4-DCP with MCAA.

Purification of acid with crystallization method has a limited efficiency. Isomer 2,6 is easily eliminated, while 2-chlorophenoxyacetic acid is hard to dissolve and remains in 2,4-D. 2,4-D of 96% purity of active substance is obtained. Rectification of about 89% raw 2,4-dichlorophenol is non-economical and threatened with originating dioxines in cube residue. The difficulty in mastering the technology of selective chlorination, purifying raw 89% 2,4-

dichlorophenol or technical 2,4-D acid is soundly confirmed by the EU position regarding purity of commercial 2,4-D and its derivatives.

In the countries of the European Union, after the process of verification and re-registration according to the directive 91/414/EWG, only preparations of the highest quality were allowed into the market, broadly tested, man friendly and natural environment friendly. In the nineties, over one half of active substances used in preparations of plant protection was recalled from sales. With allowing a plant protection agent into the market full chemical specification and toxicity tests of all pollutions of the active substance which were found in the quantity of over 0,1% were required.

In the meantime, the working team for toxicological and ecotoxicological evaluation appointed under directive 91/414 EEC (over 200 of open and closed sorts of research) 2,4-D, consisting of 12 producers and suppliers of preparation 2,4-D in the European Union, including DowElanco Europe UK, Nufarm-Agrolinz Austria, Rhone-Poulenc France and AH Marks Co Ltd. UK submitted to the European Committee for registration 2,4-D of the quality of 96% of active substance. This standard was approved for UE producers by the directive 2001/103/EC of 28.11.2001. Maintaining the traditional standard 96% of active substance 2,4-D, in spite of a great pressure of environmentalists, shows difficulties in obtaining product of a higher purity.

Moszczyński and Białek have conducted a vast research on selectivity of reaction of phenol chlorination to 2,4-dichlorophenol (Białek, Moszczyński, 2009).

Phenol is a substrate which easily undergoes electrophylic substitution in aromatic ring. Additional pair of electrons from hydroxyl group after delocalization and because of messomerism shifts to moves in aromatic ring which results in higher density of electrons in *ortho*- and *para*- positions.

Substitution in the ring with an electrophilic agent is fast but not regio- and chemoselective. Just after using 1 M of chlorine for 1 M of phenol a mixture of six products is formed (Watson, 1985).

The following conclusion can be drawn from the scheme above: obtaining *p*-chlorophenol as an intermediate product in synthesis of 2,4-dichlorophenol is much more advantageous than obtaining *o*-chlorophenol which is followed by forming 2,4-dichlorophenol and 2,4,6-trichlorophenol or even 2,6- dichlorophenol.

It is commonly known that many factors are included in the selectivity of reaction. Electron density in the substrat molecule – active phenol centres are located in the position *ortho-*, *para-* with preference for *para-* . The dissociation of hydroxyl group in phenol in different conditions of pH influences the orientation of electrophilic substitution. The activity of electrophilic agent is of crucial significance has. Chlorine molecules are a strong electrophilic agent reacting violently with active aromatic substrats. The decisive significance should be ascribed to steric effect and catalysts. Intermolecular hydrogen bondings through hydroxyl group are formed in melted phenol. The association of molecules makes it more difficult to replace in the position *ortho-*. A similar effect is achieved at phenol chlorination in strong acids and polar proton dissolvents. The presence of hydrogen bondings in phenols was confirmed by IR methods and proton NMR method.

Our research mostly concerned catalysts and chlorinating agents, including high-molecular ones.

It is not necessary to use catalyst during the synthesis of 2,4-dichlorophenol from melted phenol and gaseous chlorine. The process of chlorination stops at 2,4,6-trichlorophenol. Substitution of next chlorine atoms is only possible when Friedel-Crafts catalysts are used. It was stated that Lewis acids used at chlorination of less active substrates do not influence the selectivity of substitution of highly reactive phenols. Catalysts from amines group direct substituents into *ortho-* position. Catalysts used by Moszczyński in MPA production appeared to be totally inactive. Sulphur catalysts have a limited directing effect. The combination of Lewis catalysts with sulphur compounds resulted with a high regioselectivity into *para-* position. Probably in the reaction a high-molecular complex is originated, where the catalytic effect interferes with the steric effect.

There were also performed tests of chlorination with the reverse method, i.e. phenoxyacetic acid (PA). Przondo, in cooperation with fellow workers in the installation ½-technical in Chemical Plants „Rokita", did research on chlorination of PA acid with chlorine in alkaline water solution. Satisfactory results were not achieved. The process required significant dilutions of the reaction mass. The technology appeared to be highly waste-generating and the selectivity of chlorination not much better than at phenol chlorination (Dudycz et al. 1985).

Our team IPO and Rokita continued research on the reversed process. Using *tert*-butyl hypochlorite in the place of chlorine (t-BuOCl) of spatially developed molecule. A high selectivity of chlorination was achieved (98%) and purity of 2,4-D up to 99% of active substance. A very high selectivity of chlorination reaction is the result of the interference of two steric effects – side chain and volumetric molecule t-BuOCl (Moszczyński et al. 2008). A negative side to the technology is the necessity of operating flammable t–BuOCl and t-BuOH.

A technology based on sulfuryl chloride was developed and confirmed in the ½ technical scale results presented by Watson (Watson, 1974). The process is easy to implement and control. A negative side is the necessity of SO_2 recycling.

Three new methods of 2,4-D synthesis were developed, each of them of > 95% selectivity. The barrier of 90% selectivity of isomer 2,4-D in phenol chlorination was lifted. Results were shown below in table 4.

Chlorinating agent, catalyst,	Cl_2	Cl_2 catalytic complex, Lewis acids, sulphur compounds	SO_2Cl_2 catalytic complex	tert-BuOCl phenoxyacetic acid, reverse method
Selectivity of chlorination [%]	90	96	98	98
Production scale	industrial	industrial (Moszczynski et al., 2002)	experimental (Moszczyński et al., 2009)	laboratory

Table 4. Progress in selectivity of the process of chlorination in technology 2,4-D.

While discussing the technology of production of 2,4-dichlorophenol, the problem of trace pollutions must be mentioned, which includes remains of catalyst, present always in industrial water iron, polychlorinated compounds, cyclohexenones derivatives. There may also appear polychlorodibenzodioxines and polychlorodibenzofurans. Trace pollutions in 2,4-DCP are undesired, because they influence the quality and stability of 2,4-D. Iron compounds may worsen the hue, remains of sulphur catalysts – odour, form during storing liquid forms slimy residues and also, as an effect, decrease the durability of 2,4-D preparations. Used by some 2,4-D producers, the vacuum rectification of technical 2,4-DCP is efficient, but it threatens with dioxines forming. In the tests of vacuum rectification of technical 2,4-dichlorophenol there was formed about 15% of post-destillation tars, containing significant quantities of dioxines, including the most toxic 2,3,7,8-TeCDD and TeCDF, which was shown below in table 5.

2,3,7,8-TCDD

Authors have developed the technology of removing catalysts and trace pollutions (Moszczyński et al. 2005).

2,4-dichlorophenol obtained with the method of phenol chlorination in the presence of catalysts is washed off with mineral acid, it is neutralized to pH >10,5 and dilutes with water to 50%. A suspension of mineral sorbents is introduced to chlorophenolate obtained in this way. Chlorophenolate with sorbents is stirred for 30 minutes and the introduced sorbents are filtered, which are then destroyed thermally.

Congener PCDD/PCDF	Toxicity equivalency factor (TEF)	Congener content in an specimen, m_i [ng/g]	Toxicity TEQ m_i x TEF [ng-TEQ/g]
2,3,7,8-TeCDD	1	2.32	2.3200
1,2,3,7,8-PeCDD	1	51.71	51.7100
1,2,3,4,7,8-HxCDD	0.1	0.28	0.0280
1,2,3,6,7,8-HxCDD	0.1	16.22	1.6220
1,2,3,7,8,9-HxCDD	0.1	6.63	0.6630
1,2,3,4,6,7,8-HpCDD	0.01	0.76	0.0076
OCDD	0.0001	3.11	0.0003
2,3,7,8-TeCDF	0.1	10.69	1.0690
1,2,3,7,8-PeCDF	0.05	1.45	0.0725
2,3,4,7,8-PeCDF	0.5	1.50	0.7500
1,2,3,4,7,8-HxCDF	0.1	0.55	0.0550
1,2,3,6,7,8-HxCDF	0.1	4.58	0.4580
1,2,3,7,8,9-HxCDF	0.1	3.16	0.3160
2,3,4,6,7,8-HxCDF	0.1	0.17	0.0170
1,2,3,4,6,7,8-HpCDF	0.01	5.90	0.0590
1,2,3,4,7,8,9-HpCDF	0.01	0.00	0.0000
OCDF	0.0001	67.75	0.0068
Result in ng TEQ/g			59.15 ± 0,05

Table 5. Contents of PCDD/F in the cube residue from the destillation of 2,4-DCP (Białek, 2009).

Authors have developed the technology of removing catalysts and trace pollutions (Moszczyński et al. 2005).

2,4-dichlorophenol obtained with the method of phenol chlorination in the presence of catalysts is washed off with mineral acid, it is neutralized to pH >10,5 and dilutes with water to 50%. A suspension of mineral sorbents is introduced to chlorophenolate obtained in this way. Chlorophenolate with sorbents is stirred for 30 minutes and the introduced sorbents are filtered, which are then destroyed thermally.

Trace pollutions in purified 2,4-dichlorophenol are given below in table 6.

Iron contents [ppm]	Derivatives DPS [%]	PCDD/F [TEQ ng/g]
<5	<0.001	0.027

Table 6. Trace pollutions in purified 2,4-dichlorophenol.

The product contains trace pollutions PCDD/F in the amount allowed by technical producers norms.

The world production of 2,4-dichlorophenol without Russia, China and India is estimated for 44 000 tons per year. The current production of 2,4-DCP in Poland is 6 to 8 000 tons per year. It places our country in the strict world lead. After using in 2004 a new technology of selective chlorination of phenol, two barriers were lifted: chlorination selectivity increased from 89 to 96% and the purity of commercial acid 2,4-D from 96 to 98%. Waste chloroorganic compounds were decreased by one half in the production process, also organic pollutions introduced to the soil during agricultural operations were decreased by one half. Recommendations of directive 91/414 EEC on the improvement of ecological conditions in the production and usage of plant protection agents were implemented.

The author team W. Moszczyński, A. Białek, E. Makieła, B. Rippel, Listopadzki E., Okulewicz, Z. Dancewicz G. for the technology and product won prizes in national and world competitions, including:

-   Złoty Orbital (Gold Orbital) of the monthly Rynek Chemiczny (Chemical Year), year 2004
-   The main prize in the competition „Polish Product of the Future" („Polski Produkt Przyszłości") in category „Technology of the Future" („Technologia Przyszłości") organized by the Polish Agency for Enterprise Development, year 2004
-   The Gold Medal with Mentions in the World Inventions, Research and Innovations Exhibition „Brussels Eureka", year 2005.

4. References

Berhenke, L. F. Britton, E. C. (1946) Effect of pH on Hydrolysis Rate of Chloroacetic Acid. *Ind. Eng. Chem.* 38 (5), pp 544–546.

Białek, A. (2009). *Technology of 2,4-dichlorophenol. PhD Dissertation*, Warsaw University of Technology, Warsaw.

Białek, A. Moszczyński, W. (2009). Technological aspects of the synthesis of 2,4-dichlorophenol. *Polish Journal Of Chemical Technology*, 11, 2, pp 21-30.

Dermer, O. C. Edmison M.T. (1957) Radical Substitution In Aromatic Nuclei. *Chem. Rev.* 57 (1), pp 77–122.

Dudycz, R. Przondo, J. Strzyż, B. Marszal, Z. (1985). Pat. PL 243770. *Metoda wytwarzania kwasu 2,4-dichlorofenoksyoctowego.* Nadodrzańskie Zakłady Przemysłu Organicznego, Organika-Rokita.

Hamner, C. L. Tukey, H. (1944) The Herbicidal Action Of 2,4 Dichlorophenoxyacetic And 2,4,5 Trichlorophenoxyacetic Acid On Bindweed. *Science*, 100 (2590), pp 154-155.

Hirszowski, J. Moszczyński W. (1952) Dokumentacja technologiczna otrzymywania 2,4-D oraz Aneks Nr 2, Archive of The Institute of Industrial Organic Chemistry, Warsaw.

Jezierski, A. Zakrzewski, J. Moszczyński, W. (1999) Increased intensity of *tert*-butyloxyl radical emission in 4-chloro-2-methylphenoxy acetic acid (MCPA) synthesis. *Pestic. Sci.*, 55, 1229-1232.

Lund, H. (1958), Chlorination of o-Cresoxyacetic Acid. A Peculiar Side Reaction. *Acta Chim. Scand.* 12, 793-796.

Moszczyński, W. Ostrowski, T. Tomasik W. (1963). Pat PL 49289.

Moszczyński, W. (1971) Sprawozdanie konsultacji w Ufijskim Chimiczeskim Zawodzie, ZSRR, Archive of The Institute of Industrial Organic Chemistry, Warsaw.

Moszczyński, W. Budziar, M. Jach, J. (1975) Pat. PL 100642

Moszczyński, W. Górska, I. Jakubas, T. Piłat, W. Peć, J (1992) Pat. PL 167510

Moszczyński, W. Górska, I. Jakubas, T. Piłat, W. Peć, J (1994) Pat. PL 174569

Moszczyński, W. Górska, I. Jakubas, T. Peć, J. Piłat, W. (1997) Pat. PL 185459

Moszczyński, W. (1998) Prace Naukowe Instytutu Przemysłu Organicznego „Organika", pp 57-60 .

Moszczyński, W. (1999) Synthesis of 2-methyl-phenoxyacetic acid by the Williamson's reaction. Optimisation of reaction parameters by application of regression function *Przem. Chem.* 78/11, pp 398-400.

Moszczyński, W. Białek, A. Makieła, E. Rippel, B. Listopadzki, E. (2002) Pat. PL 354552, Method of the manufacture of 2,4-dichlorophenol.

Moszczyński, W. Białek, A. Makieła, E. Rippel, B. Okulewicz, Z. Dancewicz, G. (2005) Pat. PL 377671 Method for the manufacture of 2,4 - dichlorophenoxyacetic acid and its salts.

Moszczyński, W. Białek, A. Makieła, E. Rippel, B. (2008). Development of 2,4-D technology in Poland, *Przem. Chem.* (87), 7 pp 757-761.

Moszczyński, W. Białek A. Maliszewski, T. Wyrzykowska, U. (2009) Apl. Pat. PL 389029. Process for the preparation of 2,4-dichlorophenoxyacetic acid.

Moszczyński,W. Białek, A. Jakubas, T. Peć, J. (2010) Production of 4-chloro-2-methylphenoxyacetic acid in Organika-Sarzyna at Nowa Sarzyna. An example of the cooperation between an R&D institute and chemical company. *Przem. Chem.* 89/2, pp 124-127.

Pokorny R. (1941) New Compounds. Some Chlorophenoxyacetic Acids. *J. Am. Chem. Soc.*, 63 (6), 1768.

Skeeters, M. (1956). Pat. US 2740810, Preparation of 2-methyl-4-chlorophenoxyacetic acid.

Rapos, P. Kovac, J. Batora, V. (1960), *Vyskumny Ustav Agrochemickiej Technologie*, Sbornik prac I , Slovenska Akademie Vied, p 63, Bratislava.

Watson, W. D. (1975). Pat. US 3920757. Chlorination With Sulfuryl Chloride.

Watson, W. D. (1985). Regioselective para-chlorination of activated aromatic compounds. J. Org. Chem., 1985, 50 (12), pp 2145-2148

9

Adverse Effects of Herbicides on Freshwater Zooplankton

Roberto Rico-Martínez[1], Juan Carlos Arias-Almeida[2], Ignacio Alejandro Pérez-Legaspi[3], Jesús Alvarado-Flores[1] and José Luis Retes-Pruneda[4]
[1]Departamento de Química, Centro de Ciencias Básicas,
Universidad Autónoma de Aguascalientes, Aguascalientes,
[2]Limnología Básica y Experimental, Instituto de Biología,
Universidad de Antioquia, Medellín,
[3]División de Estudios de Posgrado e Investigación.
Instituto Tecnológico de Boca del Rio, Boca del Rio, Veracruz,
[4]Departamento de Ingeniería Bioquímica, Centro de Ciencias Básicas, Universidad
Autónoma de Aguascalientes, Aguascalientes,
[1,3,4]México
[2]Colombia

1. Introduction

The use of herbicides to control weeds is a part of agricultural management throughout the world. Unfortunately, the indiscriminate use of these herbicides may have impacts on non-target organisms (Sarma et al., 2001; Nwani et al., 2010). The long persistence of many herbicides in freshwater suggests that they are capable of producing adverse effects on freshwater zooplankton. Dalapon persist in water for 2 to 3 days, paraquat and diquat persist more than dalapon, and 2,4-D amine salt persist for 4 to 6 weeks; chlorthiamid breaks down into dichlobenil that stays for three months in water. On the other hand, terbutryne and diuron persist for more than three months in the water. These periods of time in the water show that most herbicides will cause serious adverse effects in the populations of freshwater zooplankton (Newbold, 1975). The herbicide n-chloridazon (n-CLZ) is degraded to desphenyl-chloridazon (DPC). This transformation product is more toxic than n-CLZ, and can last more than 98 days in surface water. Maximum concentrations of 7.4 µg/L DPC have been found in Germany (Buttiglieri et al., 2009). Atrazine (2-chloro-4-ethylamino-6-isopropylamino-s-triazine) is one of the most commonly used herbicides found in the rural environments, easily transported and one of the most detected pesticides in streams, rivers, ponds, reservoirs and ground waters (Battaglin et al., 2003; Battaglin et al., 2008). It has a hydrolysis half-life of 30 days and relatively high water solubility (32 mg/L), which aids in its infiltration into ground water. Atrazine concentrations of 20 to 700 µg/L in runoff surface waters have been reported (Nwani et al., 2010). Table 1 show some physicochemical properties of herbicides which are used to determine the toxic effects on freshwater zooplankton, as well as lethal values of some of these herbicides.

Herbicides	CAS Registry number	Molecular formula	Breakdown in water	Mobility in water	Species	LC$_{50}$ 24h	LC$_{50}$ 48h mg/l	LC$_{50}$ 96h	Reference
2,4 -D	94-75-7	$C_8H_6Cl_2O_3$	4 to 6 weeks		Pteronarcys califórnica (I)		1.8		Walker (1971)
					Daphnia pulex (C)		3.2		Walker (1971)
					Simochepalus serrulatus (C)		4.9		Walker (1971)
					Daphnia magna(C)		>100		Newbold (1975)
Dalapon	75-99-0	$C_3H_4Cl_2O_2$	2 to 3 days	very mobile	Pteronarcys califórnica (I)		100		Sanders and Cope (1968)
					Simocephalus serrulatus (C)		16		Walker (1971)
					Daphnia pulex (C)		11		Walker (1971)
					Daphnia magna (C)		6		Newbold (1975)
Dichlobenil	1194-65-6	$Cl_2C_6H_3CN$	2 to 3 months	low	Hyalella azeteca (A)		12.5	8.5	Wilson and Bond (1969)
					Callibaetis sp. (I)		15.2	12	Wilson and Bond (1969)
					Limnephilus sp. (I)		23.3	13	Wilson and Bond (1969)
					Enallagma sp. (I)		24.2	20.7	Wilson and Bond (1969)
					Pteronarcys califórnica (I)		8.4		Cope (1966)
					Daphnia pulex (C)		3.7		Cope (1966)
					Simocephalus serrulatus (C)		5.8		Cope (1966)
					Daphnia magna (C)		3.7		Newbold (1975)
Diquat	2764-72-9	$C_{12}H_{12}N_2$	8 to 11 days	immobile	Hyalella azeteca (A)		0.12	0.048	Wilson and Bond (1969)
					Callibaetis sp. (I)		65	33	Wilson and Bond (1969)
					Limnephilus sp. (I)		>100	>100	Wilson and Bond (1969)
					Enallagma sp. (I)		>100	>100	Wilson and Bond (1969)
					Daphnia magna (C)		7.1		Newbold (1975)
Paraquat	4685-14-7	$C_{12}H_{14}N_2$	7 to 14 days	immobile	Simocephalus serrulatus (C)		0.45		Walker (1971)
					Daphnia pulex (C)		0.24		Walker (1971)
					Daphnia magna (C)		3.7		Newbold (1975)
Terbutryn	886-50-0	$C_{10}H_{19}N_5S$			Daphnia magna (C)		1.4		Newbold (1975)
Diuron	330-54-1	$C_9H_{10}Cl_2N_2O$	3 months	low	Daphnia magna (C)		1.4		Newbold (1975)
Propanil	709-98-8	$C_9H_9Cl_2NO$		low	Daphnia magna (C)	43.74	5.01		Villarroel et al. (2003)
					Daphnia magna (C)		0.14		Rohm and Haas (1991)
					Ceriodaphnia dubia (C)		1.65		Moore et al. (1998)
Tebuthiuron	34014-18-1	$C_9H_{16}N_4OS$		very mobile	Daphnia magna (C)	44.2			Meyerhoff et al. (1985)

(A) = Amphipoda; (C) = Cladocera; (I) = Insecta.

Table 1. Toxicological properties of some herbicides used to determine lethal and sublethal toxicity.

2. Generalities of the adverse effects of herbicides on freshwater zooplankton

Ecological effects of herbicides in freshwater systems occur direct and indirectly. Indirect effects of herbicides are defined as observed effects on consumer populations in freshwater invertebrates that are not caused by direct toxicity but due to adverse effects on primary producers such as algae and macrophytes (Fairchild, 2011). An herbicide induced death suddenly because cuts off oxygen supply during a period when growth and reproduction by freshwater zooplankton are taking place. Individuals of *Simocephalus vetulus* (Crustacea) may have died in the diquat treated ponds because of lower oxygen supply that benefited *Daphnia longispina* because increased its populations (Brooker & Edwars, 1973).

Fairchild (2011) argues that atrazine did not produce neither direct nor indirect effects on aquatic invertebrates/vertebrates. However a recent review by Rohr and McCoy (2010) concluded that atrazine produces indirect and sublethal effects on fish and amphibians at environmentally relevant concentrations. These effects were observed in reproductive success, sex ratios, gene frequencies, populations, and communities. However, these effects remain uncertain and restricted to few species. Other authors report of many indirect effects of pesticides on freshwater zooplankton obtained through meso- and microcosm experiments (see section 8 of this chapter).

The study of the direct effects of herbicides on freshwater zooplankton results in a complex mixture of data on lethal and sublethal values obtained from standard toxicity tests assessing one species relationship with chemicals of high purity in the lab, to meso- and microcosms experiments, field studies, use of biomarkers, and DNA microarrays. However, aside from environmental health protection agencies reports, the data on the mainstream scientific literature is scarce and restricted to: a) few test species, b) models, and c) small number of herbicides. The result of this diagnosis is a scattered picture with many uncertainties, but also with many opportunities for environmental toxicology research. Perhaps the fact that many authors argue that there are no direct effects of herbicides on freshwater zooplankton at environmental concentrations (Fairchild, 2011) or that herbicides do not represent a threat to aquatic communities (Relyea, 2005; Golombieski et al., 2008) has discourage research in this area. However, these authors failed to consider a series of circumstances that might be consider while analyzing the potential of herbicides for adverse effects:

a. Many herbicides are applied as commercial formulae and the formulae can be more toxic to non-target organism than the active ingredient. That is the case of glyphosate and its different commercial formulae (Domínguez-Cortinas et al., 2008).
b. The safe standards and good application techniques for herbicides are not followed as strictly as they should in developed countries and certainly less so in underdeveloped or poorly developed countries. That means that the theoretical concentrations in which many Quantitative Structure/Activity Relationship (QSAR´s) model for herbicides are based on might not apply in many cases and true environmental concentrations might be underestimated.
c. Relyea & Hoverman (2006) argue that results have shown that some herbicides may interact with a range of different natural stressors and that synergism among herbicides and other pesticides has not been studied at all. Therefore, the interaction between herbicides and the cocktail of toxicants found in many polluted sites throughout the

world has not been analyzed, and therefore, the assumption that some herbicides do not interact with other toxicants at environmentally relevant concentrations to produce direct adverse effects on freshwater zooplankton is just unsustainable (just to put it in ecological terms).

d. Ecotoxicogenomics and the development of new and more sensitive biomarkers that are unveiling effects on freshwater zooplankton (especially on endocrine disruption) at very low environmentally relevant concentrations (see sections 8 and 9 of this chapter) might change the opinion of many researchers on adverse effects of herbicides.

e. The data (at least in the mainstream scientific literature) on potential effects of herbicides on freshwater zooplankton is extremely scarce and restricted to no more than five or six taxonomic groups and less than 30 herbicides.

Herbicides can produce bioaccumulation and biomagnification, but the data is buried in different reports and few scientific articles, that a review is greatly needed. For instance, some herbicides like benfluralin, bensulide, dacthal, ethalfluralin, oxadiazon, pendimethalin, triallate, and trifluralin have the potential to accumulate in sediments and aquatic biota (USGS, 1999).

Lethal effects of a few herbicides have been determined so far in only the following freshwater zooplankton groups: amphipods, cladocerans, copepods, malacostracans, and rotifers. The information on herbicide toxicity on freshwater zooplankton is limited and mainly focused on studies of population dynamics and effects on the biodiversity of the community. Sublethal effects of herbicides on freshwater zooplankton species have focused on demographic parameters (mainly life tables and determination of "r" values), of three groups: amphipods, cladocerans, and rotifers.

Herbicides may affect the population dynamics of freshwater zooplankton by controlling individual survival and reproduction, and by altering the sex ratio. Herbicides might also produce the following effects at the community and ecosystem levels: a) induction of dominance by small species, b) an increase of species richness and diversity, and c) elongation of the food chain and reduction of energy transfer efficiency from primary producers to top predators (Hanazato, 2001).

Biomarkers used so far to study effect of herbicides on freshwater zooplankton correspond to: a) enzyme inhibition, b) mRNA expression levels, c) gen induction, and d) grazing rate inhibition.

3. Mechanism of action of herbicides related to adverse effects on freshwater zooplankton

Herbicides represent a broad variety of chemical classes of compounds, which acts over diverse sites of metabolic functions and energy transfer in plant cells (Duke, 1990). Only a few herbicides classes have a known molecular site of action, moreover, the molecular site of action and the mechanism of several important herbicide classes is still unknown (Duke, 1990). Among known mechanisms of action of herbicides, there are herbicides that inhibit photosynthesis, those that inhibit pigments and those that inhibit seedling growth (Duke, 1990; Prostko & Baughman, 1999; Gunsolus & Curran, 1999). An undesirable side-effect of herbicides is that they may enter freshwater ecosystems by spray drift, leaching, run-off, and/or accidental spills (Cuppen et al., 1997). Surface water contaminations by herbicides have been reported to have direct toxic effects on phytoplankton, epiphyton, and macrophytes. Furthermore, herbicides have indirect effects over zooplankton and animal populations (Relyea, 2005, 2009; Cuppen et al., 1997), affecting all trophic chains in freshwater reservoirs. Several studies show that herbicides selectively decreased primary producers, leading to a bottom-up reduction in the abundance of consumers due to food limitation (Fleeger et al., 2003). Contaminant-induced changes in behavior, competition and predation/grazing rate can alter species abundances or community composition, and enhance, mask or spuriously indicate direct contaminant effects (Fleeger et al., 2003). Thus, the impacts that herbicides exerts on freshwater communities are one of the main concerns about the use of these chemical compounds. The mechanisms of action of herbicides are classified according to site or specific biochemical process that is affected and are summarized in Table 2; these mechanisms have been described in plants. Below are some examples of the adverse effects of some herbicides according to their mechanism of action in freshwater zooplankton.

3.1 Amino acid synthesis inhibitors

One of the most important herbicides in this category is glyphosate because is extensively used in the aquatic environment. Martin et al. (2003), determined the acute toxicity of technical-grade glyphosate acid, isopropylamine (IPA) salt of glyphosate, Roundup and its surfactant polyoxyethylene amine (POEA) in Microtox® bacterium (*Vibrio fischeri*), microalgae (*Selenastrum capricornutum* and *Skeletonema costatum*), protozoa (*Tetrahymena pyriformis* and *Euplotes vannus*) and crustaceans (*Ceriodaphnia dubia* and *Acartia tonsa*); generally the toxicity order of the chemicals was: POEA > Roundup® > glyphosate acid > IPA salt of glyphosate, while the toxicity of glyphosate acid was mainly due to its high acidity. In *Ceriodaphnia dubia* the LC50 = 147 mg/L to glyphosate acid and for *Acartia tonsa* was LC50 = 35.3 mg/L. Glyphosate produced adverse effects on the embryonic development on time (3 and 8 mg/L), duration of juvenile and reproductive periods, average lifespan, net reproductive rate (8.0 and 10.50 mg/L), and the intrinsic population increasing rate on the freshwater rotifer *Brachionus calyciflorus* (Chu et al., 2005).

Meyerhoff et al. (1985) observed a lower length in *D. magna* exposed to the herbicide tebuthiuron than in blank control animals when the cladocerans were exposed to 44.2 mg/L herbicide. Hanazato (1998) indicated that the neonatal body size determines the size at maturation. The reduced growth rate of neonates due to the chemicals will result in a smaller size at maturation and thus a smaller adult size, leading to smaller clutch sizes.

3.2 Cell-membrane disrupters

The way in which terbutryn exerts its toxicity to rotifers is not clear. The survival curves for all *Brachionus* sp. cultures fed with terbutryn-exposed microalgae showed a drastic mortality showed that population density decreased as terbutryn concentration increased in the microalgal cells. In fact, this species of rotifer did not survive beyond four days when fed with microalgae exposed to 500 nM terbutryn. Percentage of reproductive females in rotifer populations fed with terbutryn-exposed microalgae decreased significantly as herbicide concentration increased (Rioboo et al., 2007). Interestingly the highest concentration of herbicide tested is no toxic to the algae *Chlorella vulgaris* viability, at least after 24 h of exposure (González-Barreiro et al., 2006).

Mechanism of action	Herbicide Family Chemistry	Affected site or biochemical process
Amino Acid Synthesis Inhibitors	Sulfonylureas	Inhibition of Acetolactate synthase enzyme (ALS)
	Imidazolinone	Inhibition of 5-enolpyruvyl shikimate 3-phosphate synthase (EPSP)
	Triazolopyrimidine	
	Pyrimidinylthiobenzoate	
	Sulfonyl amino carbonyl triazolinones	
Cell-Membrane Disrupters	Bipyridiliums	Inhibition of protoporphyrinogen oxidase (PPO oxidase)
	Diphenylethers	Electron acceptors, formation of reactive oxygen species (ROS)
Growth Regulators	Triazolinone	Alteration of hormonal balance
	Oxadiazoles	
	Arsenical	
	Phenoxi acids	
	Benzoic acids	
	Pyridine acids	
	Quinilonecarboxylic	
Lipid Synthesis Inhibitors	Aryloxyphenoxyproprionates	Inhibition of Aceytl Coenzyme-A carboxylase (ACCase)
	Cyclohexanediones	

Table 2. Mechanism of action of herbicides (Plimmer et al. 2005).

The herbicide molinate was tested in *Daphnia magna*, and the reproduction was significantly reduced when molinate concentration was increased in the medium, but only this effects was higher in the parental daphnids (F0) than the F1-1st and F1-3rd offspring, seem to be adapted to the herbicide molinate, showing more longevity and reproduction than their parental (Sánchez et al., 2004). Similar result were found by Julli & Krassoi (1995) who observed a significant decreased in total young per female in three broods of *Moina australiensis* when exposed to molinate.

Paraquat was toxic to almost all compartments of the plankton community including zooplankton like: rotifers (*Brachionus calyciflorus*, *Lecane* sp., *Conochiloide* sp., *Asplanchna* sp., and *Hexarthra* sp.), copepods (*Thermocyclops decipiens*, *Mesocyclops* sp.) and cladocerans (*Diaphanosoma excisum*), leading to a reduction in biomass, numbers, and overall trophic functioning, in fact *Thamnocephalus decipiens* exhibited dose-dependent sensitivity to paraquat (Leboulanger et al., 2011). Paraquat may induce peroxidation processes in non-target animal species. Furthermore, paraquat may interfere with the cellular transport of polyamines. Cochón et al. (2007), investigate some aspects related to paraquat-induction of oxidative stress (lipoperoxidation, enzymatic activities of catalase and superoxide dismutase) and also the levels of polyamines (putrescine, spermidine and spermine) in two species of freshwater invertebrates, the oligochaete *Lumbriculus variegatus* and the gastropod *Biomphalaria glabrata*. In *L. variegatus* did not induce membrane lipoperoxidation and only a transient decrease in CAT activity was observed. After 48 h of exposure, an increase of lipoperoxidation and a decrease of SOD activity were registered in the snails. It could be hypothesized that the higher resistance of *L. variegatus oligochaetes* could be due in part to a lower ability to activate the paraquat and also to a protective role of polyamines.

3.3 Growth regulators

Sarma et al. (2001) reported that the herbicide 2,4-Dichlorophenoxy acetic acid had a negative influence on the population growth of *Brachionus patulus* when the rotifers were directly exposed via water and food. Interestingly, Relyea (2005) reported 2,4-Dichlorophenoxy acetic acid had no effect on zooplankton. But exists LC_{50} = 363 and 389 mg/L values (96 h) for the *Daphnia magna* (Johnson & Finley, 1980; Verschueren, 1983, respectively). Boyle (1980) determinate the effects on 2,4-D herbicide applied two concentration 5 and 10 kg/ha, and quantifier the planktonic invertebrates (number per liter of water) rotifers and crustaceans: with a concentration of 5 kg/ha of 2,4-D, found 320 rotifers species and 40 of crustaceans, and found 207 rotifers species and 34 crustaceans with 10.0 kg/ha.

3.4 Lipid synthesis inhibitors

Metazachlor is a frequently used herbicide with high concentrations in surface waters and effects on zooplankton caused by changes in habitat structure in species such as *Keratella quadrata*, *Lecane* spp, *Brachionus calyciflorus*, *Polyathra dolicoptera* and *Bosmia longirostris*. For species such as *K. quadrata*, *Alonella excisa*, *Acropercus harpae*, *Chydorus sphaericus* and some ostracods species with negative weights indicated a decrease in abundance after metazachlor application. In contrast, species like *P. dolichoptera* or *Ceriodaphnia quadrangula* increased in abundance in the treatments as compared to the controls as indicated by the positive weight (Mohr et al., 2008). Direct toxic effects of metazachlor were not expected since this group is generally unable to synthesize fatty acids and therefore membrane functions will not be disrupted directly. EC_{50} value of 22.3 mg/L (48h) was found for *Daphnia magna* (FAO, 1999).

Another lipid synthesis inhibitors herbicide is norflurazon and is a bleaching, preemergence. Horvat et al. (2005) found that the toxicity of norflurazon caused mortality in *Polycelis feline*, and morphological and histological changes in treated animals compared to corresponding controls. The most prominent histological changes were damage of the outer mucous layer, lack of rhabdites, damage to epidermis and extensive damage to parenchyma cells.

3.5 Pigment inhibitors

Pigments inhibitors affecting plant cell by preventing the formation of photosynthetic pigments (chlorophyll and carotenoids) localized in leaf tissues, trough interfere both the chlorophyll and terpenoid synthesis pathway, inhibiting their synthesis (Duke, 1990; Prostko & Baughman, 1999; Gunsolus & Curran, 1999). This condition cause rapid photobleaching of green tissue of leafs, due the Photosystem I (PS I) reduce a chemical group of the structure of these herbicides to a radical that reduce molecular oxygen to superoxide radical. This reaction repeats continuously to form large amounts of superoxide radical; producing lipids peroxidation and photobleaching (Duke, 1990), giving to affected plants a white or translucent appearance. Because this effect, pigment inhibitors are often called "bleaching herbicides" or "photobleachers" (Prostko & Baughman, 1999). This herbicide class includes isoxazolidinones (i.e. clomazone), pyridazinones (i.e. norflurazon), fluridone, difunone, amitrole and *m*-phenoxybenzamides (Duke, 1990).

This type of herbicides has not direct effects on freshwater zooplankton, but can have indirect negative effects on them. The mechanism of action of these herbicides is targeted to photosynthetic organisms (plants), in the case of freshwater communities, the phytoplankton are the organisms that suffers direct negative effects, which affect them drastically reducing their population. However, the reduction of phytoplankton population may cause indirect negative effects on the zooplankton due a reduction of feed availability for zooplankton, reducing their abundance and/or inducing changes in the taxa composition of zooplankton (Relyea, 2005, 2009).

3.6 Photosynthesis inhibitors

Herbicides that inhibit photosynthesis are the most common type. These herbicides disrupt the vital process of photosynthesis that allows plants to convert the solar light energy into glucose. This type of herbicides binds to the quinone-binding protein (D1 protein) of photosynthetic electron transport, blocking the electron transport. Photosynthesis inhibitors herbicides include triazines (i.e. atrazine), phenylureas (i.e. linuron), uracils, nitriles and benzothidiazoles (Duke, 1990; Gunsolus & Curran, 1999; Prostko & Baughman, 1999). Diuron blocks photosynthetic electron transfer in plants and algae, it might also affect freshwater zooplankton (Leboulanger et al., 2011).

Photosynthesis inhibitors have not direct effects on freshwater zooplankton, but can have indirect effects on them. These herbicides affect mainly to phytoplankton that suffers direct toxic effects, which entails to reducing their population. Thus, the reduction of food supply, modifications of both reproduction and feeding behavior of zooplankton may cause indirect effects on the zooplankton, resulting in decrease of the abundance of some taxa (indirect negative effect), increase of some taxa (indirect positive effect), both decrease of diversity and changes in species composition of zooplankton (Solomon et al., 1996; Cuppen et al., 1997; Hanazato, 1995; Relyea, 2005, 2009; Chang et al., 2008).

Chang et al. (2008) studied the effects of application of simetryn (20 and 80 µg/L), a methylthiotriazine herbicide, and the fungicide iprobenfos (100 and 600 µg/L), on zooplankton community composed by rotifers and cladocerans. They applied four treatments (low and high concentrations of both pesticides), and their results showed that the herbicide have less apparent direct impact on zooplankton abundance within a short period; however, they observed that the diversity and species composition changed with simetryn application, suggesting that the structure of zooplankton can be altered by the herbicide application (Chang et al., 2008).

The mode of action of atrazine is blocking electron transport in photosystem II leading to chlorophyll destruction and blocking photosynthesis (Nwani et al., 2011). Dodson et al. (1999), found that atrazine have effects on male production of *Daphnia*, changing the sex ratio, which exerts a control of *Daphnia* population dynamics.

Cuppen et al. (1997) studied the effects of a chronic application of linuron (at concentrations of 0.5, 5, 15, 50 and 150 µg/L during 28 days) on freshwater microcosms, which included phytoplankton, zooplankton and macroinvertebrates. They observed that the direct negative effect of linuron on several algae (cryptophytes, diatoms) and the positive effect on green algae *Chlamydomonas* resulted in a decrease of several Rotatoria and an increase in Copepoda, and to a lesser extent, Cladocera.

3.7 Seedling growth inhibitors

This type of herbicides includes dinitroanilines (i.e. trifluralin), acetanilides (i.e. acetochlor) and thiocarbamates (i.e. EPTC). The seedling growth inhibitors are divided into two groups: a) root inhibitors; and b) shoot inhibitors. The first group binding to tubulin protein and disrupt the cell division, which inhibit the root elongation and lateral root generation. About second group, little is known about their mechanism of action, but is believe that disrupt protein synthesis and waken cell wall (Duke, 1990; Prostko & Baughman, 1999).

These herbicides may impact indirectly on freshwater zooplankton, due the direct negative effects on phytoplankton, which may be sensitive to disruption of their cell division process, limiting the growth and multiplication of phytoplankton, reducing the feed availability for zooplankton, decreasing their reproduction rate and their population (Fleeger et al., 2003; Relyea, 2009).

Relyea (2009) examined the effect of acetochlor and metolachlor on zooplankton at low concentrations (6-16 p.p.b.); he encountered that there was no clear indication of any indirect effects from the addition of these herbicides to zooplankton, and in one zooplankton taxon (*Ceriodaphnia*) the mixture of five herbicides (acetochlor, metolachlor, glyphosate, atrazine and 2,4-D) added at concentrations of 6-16 p.p.b. caused an increase in abundance. The few studies about acetochlor and other herbicides (atrazine and 2,4-D) suggest that low concentrations of these herbicides have not effect in cladoceran survival, or may cause an increase of their population due to high reproduction rate in cladocerans (Relyea, 2009).

3.8 Other kind of herbicides whose mechanism is unknown

Only two other molecular sites of action of herbicides are known. One is the herbicide asulam, which inhibits folate synthesis by inhibiting dihydropteroate synthase, although there may also be a second site of herbicide action associated with cell division. In another hand, the herbicide dichlobenil inhibits cellulose synthesis, but its molecular site of action is unknown. Photoaffinity labeling of cotton fiber proteins with a photoaffinity dichlobenil analogue resulted in specific labeling of an uncharacterized 18 kD protein (Duke, 1990). Amoung the seedling growth inhibitors, the group that inhibits plant shoot elongation have a mode of action almost unknown until today, is believe that this inhibitors disrupt protein synthesis and waken cell wall (Duke, 1990; Prostko & Baughman, 1999). In another hand, is too believed that these inhibitors could have multiple sites of action (Gunsolus & Curran, 1999).

4. Lethal effects of herbicides on freshwater zooplankton

The information on herbicide toxicity on freshwater zooplankton is limited and mainly focused on studies of population dynamics and effects on the biodiversity of the community. Some authors claim that herbicides apparently do not pose a threat to the aquatic communities, or have a lesser adverse effect than other pesticides (Golombieski et al., 2008). Relyea (2005) argue that glyphosate and 2,4-D, have no significant adverse effect on zooplankton biodiversity. Perhaps lethal effects are not so evident. However, symetrin can cause shifts in species composition, diversity and dominance of freshwater zooplankton (Hanazato, 2001; Chang et al., 2008). Therefore, it is convenient to consider data on lethal toxicity to determine the most sensitive species which might enable us to predict the direction of indirect effects on a community (Relyea & Hoverman, 2006).

Few if any environmentally relevant concentrations have been shown to have direct effects on zooplankton, fish, or amphibians in the laboratory (Fairchild, 2011). However a recent review by Rohr & McCoy (2010) concluded that atrazine produces indirect and sublethal effects on fish and amphibians at environmentally relevant concentrations. Furthermore, Domínguez-Cortinas et al. (2008) found that both glyphosate and its commercial product Faena® produce lethal toxicity to the freshwater invertebrates *Daphnia magna* and *Lecane quadridentata* at environmental concentrations (the highest concentration of glyphosate in runoff waters, 5.2 mg/L, was found in runoff occurring 1 day after treatment at the highest rate (8.6 Kg/ha of Roundup®)) (Edwards et al., 1980).

Sublethal effects of glyphosate and its formulae could be found at protective values, like the 65 µg/L value published in the Environmental Guide for protecting aquatic life of the Canadian Government (Environment Canada, 1987) for glyphosate. This value is 6.5-fold higher than the esterase inhibition NOEC value for glyphosate and 2-fold higher than the Faena® esterase inhibition NOEC value obtained by Domínguez-Cortinas et al. (2008). On the other hand, the US EPA (1986) has established a value of 700 µg/L of glyphosate for drinking water, which according to Domínguez-Cortinas et al. (2008) esterase inhibition results may represent a risk (LOEC = 62 µg/L, EC50 = 280 µg/L) especially when we consider the ample presence of acetylcholinesterases in the test organisms (Pérez-Legaspi et al., 2011).

Herbicide	Species	Criteria	Endpoint (mg/L)	Reference
Acroleine	*Daphnia magna* (C)	48-h	LC50 = 0.051mg/L	Holcombe et al., 1987
	Pennaeus aztecus (M)	48-h	LC50 = 0.100mg/L	Eisler, 1994
Atrazine	*Daphnia pulex* (C)	3-h	LC50 > 40	Keith et al., 1995
"	"	48-h	EC50 =36 –46.5	"
"	"	48-h	LC50 = 33	"
"	*Daphnia magna* (C)	26-h	LC50 = 3.6	"
"	"	48-h	LC50 = 9.4	"
"	"	48-h	EC50 = 3.6	"
"	"	24h,48h	EC50 > 39	"
"	"	48-h	LC50 = 6.9	"
			MATC = 0.14-0.25	
"	*Daphnia macrocopa* (C)	3-h	LC50 > 40	"
"	*Ceriodaphnia dubia* (C)	7-d	LC50 = 2.0	"
"	*Daphnia carinata* (C)	48-h	EC50 = 24.6	Phyu et al., 2004
"	*Hyalella azteca* (A)	96-h 21-d	LC50 = 3.0 LC50 = 1.8	Ralston-Hooper et al., 2009
"	*Diporeia* sp (A)	96-h 21-d	LC50 > 3.0 LC50 = 0.24	"
DEA (desethylatrazine)	*Hyalella azteca* (A)	96-h 21-d	LC50 = 5.1 LC50 > 3.0	Ralston-Hooper et al., 2009
"	*Diporeia* sp (A)	96-h 21-d	LC50 > 3.0 LC50 = 0.33	"
DIA (deisopropylatrazine)	*Hyalella azteca* (A)	96-h 21-d	LC50 = 7.2 LC50 > 3.0	Ralston-Hooper et al., 2009
"	*Diporeia* sp (A)	96-h 21-d	LC50 > 3.0 LC50 = 0.3	"
Diuron	*Daphnia pulex* (C)	96-h 7-d	LC50 = 17.9 LC50 = 7.1	Nebeker and Schuytema, 1998
"	*Hyalella azteca* (A)	96-h	LC50 = 19.4	"
"	"	10-d	LC50 = 18.4	"
Glyphosate	*Daphnia magna* (C)	48-h	NOEC = 120 LOEC = 140 LC50 = 146	Domínguez-Cortinas et al., 2008
"	*Lecane quadridentata* (R)	48-h	NOEC = 120 LOEC = 140 LC50 = 150	"
Glyphosate < 74 % (Faena ®)	*Daphnia magna* (C)	48-h	NOEC = 3.3 LOEC = 6.5 LC50 = 7.9	Domínguez-Cortinas et al., 2008
"	*Lecane quadridentata* (R)	48-h	NOEC = 9.8 LOEC = 13.0 LC50 = 13.1	"
Glyphosate (IPA)	*Ceriodaphnia dubia* (C)	48-h	LC50 = 415.0	Tsui and Chu, 2003
Glyphosate (POEA)	*Daphnia pulex* (C)	96-h	EC50 = 2.0	Servizi et al., 1987
Glyphosate 48 % (RON-DO®)	*Daphnia magna* (C)	24-h 48-h	EC50 = 95.96 EC50 = 61.72	Alberdi et al., 1996
"	*Daphnia spinulata* (C)	24-h 48-h	EC50 = 94.87 EC50 = 66.18	"
Glyphosate (Roundup®)	*Phyllodiaptomus annae* (Co)	48-h	LC50 = 1.06	Ashoka Deepananda et al., 2011

Herbicide	Species	Criteria	Endpoint (mg/L)	Reference
"	*Caridina nilotica* (M)	72-h	LC50 = 107.53	Folmar et al., 1979
		96-h	LC50 = 60.97	
	Daphnia magna(C)	48-h	EC50 = 3.0	Folmar et al., 1979
	Ceriodaphnia dubia (C)	48-h	LC50 = 5.7	Tsui and Chu, 2003
	Daphnia pulex (C)	96-h	EC50 = 8.5	Servizi et al., 1987
	Gammarus pseudolimnaeus (A)	48-h	LC50 = 62.0	Folmar et al., 1979
	Hyalella azteca (A)	48-h	LC50 = 1.5	Tsui and Chu, 2004
Glyphosate (Rodeo®)	*Ceriodaphnia dubia* (C)	48-h	LC50 = 415.0	Tsui and Chu, 2004
	Hyalella azteca (A)	48-h	LC50 = 225.0	Tsui and Chu, 2004
Metribuzin (Sencor®)	*Diaptomus mississippiensis* (Co)	24-h 48-h	LC50 =205.0	Syed et al., 1981
	Eucyclops agilis (Co)		LC50 =150	
Molinate	*Brachionus calyciflorus* (R)	24-h	LC50 = 11.37	Ferrando et al., 1999
"	*Daphnia carinata* (C)	48-h	EC50 = 26.5	Phyu et al., 2004
Paraquat	*Diaptomus mississippiensis*(Co) *Eucyclops agilis* (Co)	24-h 48-h	LC50 =10 LC50 = 5.0	Syed et al., 1981
"	*Diaphanosoma excisum* (C)	24-h	LOEC = 0.057	Leboulanger et al., 2008
"	*Moina micrura* (C)	24-h	LOEC = 0.577	"
Paraquat 27.6 % (OSAQUAT)	*Daphnia magna* (C)	24-h 48-h	EC50 = 16.47 EC50 = 4.55	Alberdi et al,. 1996
"	*Daphnia spinulata* (C)	24-h 48-h	EC50 = 9.91 EC50 = 2.57	"
Paraquat + metribuzin (1:1) 91% + 9%	*Diaptomus mississippiensis* (Co) *Eucyclops agilis* (Co)	24-h 48-h	LC50 = 29 LC50 = 17	Syed et al., 1981
Pendimethalin 60%	*Daphnia magna* (C)	24-h 48-h	LC50 = 112 LC50 = 53	Kyriakopoulou et al., 2009
S-metolachlor 31.2% + Terbuthilazine 18.8%	*Daphnia magna* (C)	24-h 48-h	LC50 = 20 LC50 = 9.5	Kyriakopoulou et al., 2009
Simazine (Aquazine)	*Daphnia pulex* (C)	48-h	LC50 > 50	Fitzmayer, et al., 1982
Thiobencarb	*Brachionus calyciflorus* (R)	24-h	LC50 = 47.82	Ferrando et al., 1999
2,4-D (2,4-dichlorophenoxyacetic acid)	*Brachionus calyciflorus* (R)	24-h	LC50 = 117	Snell et al., 1991
3,4- DCA (3,4-dichloroaniline)	*Daphnia magna* (C) (adults)	48-h	LC50 = 12	Ferrando and Andreu-Moliner, 1991
"	*Brachionus calyciflorus* (R)	24-h	LC50 = 61.47	"
"	*Daphnia magna,* larva (C)	24-h 48-h 96-h 7-d 14-d 3-w	LC50 = 0.40 LC50 = 0.23 LC50 = 0.16 LC50 = 0.12 LC50 = 0.10 LC50 = 0.10 EC50 = 0.01	Adema and Vink, 1981

Herbicide	Species	Criteria	Endpoint (mg/L)	Reference
"	*Daphnia magna*, adult	48-h	LC50 = 12	"
	(C)	96-h	LC50 = 1.0	
		7-d	LC50 < 0.58	
"	*Brachionus calyciflorus*	24-h	LC50 = 62	Snell et al., 1991
	(R)			

Abbreviations. (C) Cladocerans, (R) Rotifers, (Co) Copepods, (A) Amphipod, (M) Malacostracan. LC50 = Median Lethal Concentration, EC50 = Concentration where 50% inhibition occurs, MATC = Maximum Acceptable Toxicant Concentration, LOAEL = Lowest Observed Adverse Effect Level, NOAEL = No Observed Adverse Effect Level, LOEC = Lowest Observed Effect Concentration, NOEC = No observed effect concentration.

Table 3. Lethal toxicity values of herbicides with different species of freshwater zooplankton. Criteria of mortality include different exposure time to herbicide in hours (h), days (d) or weeks (w).

Lethal toxicity tests with freshwater invertebrates are based on standard protocols which are simple, reproducible, and with certain ecological relevance. They are valuable tools to estimate the adverse effect of single chemicals in short periods of exposure (usually 24 and 48 h), with or without food. The most common evaluation parameter is the death or immobility which is represented by the median lethal toxicity (LC50) or the median effect concentration (EC50) (Sarma et al., 2001; Pérez-Legaspi et al., 2011). The cladocerans (*Daphnia* sp., *Ceriodaphnia* sp. and *Moina* sp.) and the rotifer genus *Brachionus*, are among the most used freshwater organisms in toxicity tests (Table 3), mainly due to their great availability, high sensitivity towards many toxicants, ease of handling and culture and high rates of growth and reproduction (Snell & Janssen, 1998; Sancho et al., 2001; Sarma & Nandini, 2006). The amphipod (*Hyalella* sp.) and copepods have also been used (Table 3). Some of these protocols have been recognized by International Standard Organizations (ISO), USEPA, OECD, ASTM, Standard Methods (Snell & Janssen, 1995; Persoone et al., 2009).

Among herbicides, the most studied with freshwater zooplankton are atrazine (Table 3) and glyphosate (Pérez et al., 2011; Table 3). However, the most toxic herbicides are: acroelin (LC50 = 0.051 and 0.100 mg/L), the commercial formula of glyphosate, Faena® for the cladoceran *Daphnia magna* (48h-LC50 = 7.9 mg/L), Roundup® for the copepod *Phyllodiaptomus annae* (48h-LC50 = 1.06 mg/L), and 3,4- DCA (24h-LC50 = 0.40 mg/L) for *D. magna*. On the other hand, glyphosate the active ingredient is less toxic for *D. magna* (48h-LC50 = 146 mg/L) and the freshwater rotifer *Lecane quadridentata* (48h-LC50 = 150 mg/L) than its herbicide formula Roundup®; which suggests that in this particular case the substances present in the commercial formula contribute through synergistic effects to increase the toxicity towards non-target organisms (Domínguez-Cortinas et al., 2008). The 24 and 48 h exposure periods are the most common in the lethal tests, but some tests might last several days. In the case of 3,4-Dichloroaniline (3,4-DCA) the range of *D. magna* LC50 values (0.40 – 0.10 mg/L) decrease as the exposure time increases. Presence of food (microalgae) is a factor that decreases the toxicity of the herbicide as test animals are better fed; they seem to be more resistant (Sarma et al., 2001). In general among freshwater zooplankton the most sensitive model organisms to herbicides are amphipods and crustaceans. However, more toxicity testing with freshwater zooplankton are necessary because data on different species and toxicant are scarce making predictions of herbicide toxicity on zooplankton an unexplored area, and some herbicides have the potential to alter the dynamics and structure of aquatic communities.

5. Chronic effects of herbicides on freshwater zooplankton

Lethal toxicity data is considered by many environmental health protection agencies in world as reliable and significant, because comes from standard and simplified protocols. However, mortality or immobility is a parameter of lesser sensitivity in estimating adverse effects on freshwater zooplankton. Chronic tests are usually more sensitive because are based on growth, reproduction, physiological, biochemical and genetic characteristics in lower concentrations and longer exposure periods (Table 4). In other words, they assess the first responses (stress, physiological, behavioral and reproductive) to toxicants (Nimmo & McEwen, 1994). Chronic toxicity is usually expressed as the median effective concentration (EC50) or the concentration in which 50% of a specific effect is determined. Many chronic tests rely on life tables that examine demographic parameters (r, Ro, Vx, T and e_o) in freshwater invertebrates. Some chronic tests focus only on growth inhibition arguing that this is an outstanding parameter since involves all steps of a life cycle (embryos, juveniles and adults) during the test period, which makes these tests rapid, sensitive, and relevant ecologically (Snell & Moffat, 1992; Sancho et al., 2001). Besides demographic parameters, tests of chronic effects of herbicides on freshwater zooplankton also involve ingestion rate, enzymatic inhibition and behavioral parameters (Table 4). The most commonly used species belong to cladocerans, rotifers, and one species of amphipod (Table 4). Atrazine is the most studied herbicide regarding chronic effects on freshwater zooplankton; although, studies have been restricted to crustaceans. The most toxic herbicide studied so far is glyphosate, EC50 = 0.28 mg/L, for *in vivo* esterase inhibition in *L. quadridentata*, followed by thiobencarb (EC50 = 0.75 mg/L) for 21 days survival and growth inhibition tests in *D. magna*. The least toxic herbicide is 2,4-D (EC50 = 500 mg/L) for *B. patulus* and EC50 = 128 mg/L, for *B. calyciflorus* (Table 4).

As for lethal tests, the scarcity of data related to chronic effects on freshwater zooplankton becomes a research opportunity to increase the number of taxonomic groups and different herbicides studied, and to diversify the list of chronic parameters as recommended by the American Society for Testing Materials (ASTM) (Sancho et al., 2001). Such an effort would enhance our comprehension of the effects of herbicides in freshwater ecosystems (Hanazato, 2001).

6. Biomarkers assessing adverse effects of herbicides on freshwater zooplankton

The need to rely in parameters more sensitive to estimate adverse effects of toxicants in small concentrations has led to the development of biomarkers. These biomarkers detect small biochemical, cellular, genetic, physiological, morphologic and behavioral variations which can be easily and non-destructively determined in most organisms (Hagger et al., 2006; Walker et al., 2006). These small variations can led to changes in all levels of the biological organization (Hyne & Maher, 2003). These effects are usually more rapid in lower levels of biological organization and can therefore offer more sensitive responses to toxicant exposure inside the populations (Hagger et al., 2006). Therefore, Walker et al. (2006), define a biomarker as any biological response towards an environmental chemical substance

Herbicide	Test organism	Criteria	Endpoint (mg/l)	Reference
Atrazine	*Ceriodaphnia dubia* (C)	4-d	Chronic value = 6.9 NOEC = 5.0 –10 LOEC = 10–20	Keith et al. 1995
"	"	7-d	Chronic value = 3.5 NOEC = 2.5 LOEC = 5.0	"
"	"	7d	NOEC = 5.0	"
"	*Scapholeberis mucronata* (C)	F	1.0	"
"	"	ED 30 – 45-d	1.0	"
Diuron	*Daphnia pulex* (C)	R 7-d	LOAEL = 7.7 NOAEL = 4.0	Nebeker and Schuytema, 1998
"	*Hyalella azteca* (A)	S 10-d	LOAEL = 15.7 NOAEL = 7.9	"
Glyphosate	*Lecane quadridentata* (R)	cFDAam 30-m	NOEC =0.032 LOEC = 0.062 EC50 = 0.28	Domínguez Cortinas et al. 2008
"	"	PLA2 30-m	NOEC = 5.0 LOEC = 10.0 EC50 = 17.6	"
Glyphosate < 74 % (Faena ®)	*Lecane quadridentata* (R)	cFDAam 30-m	NOEC = 9.8 LOEC =13.0 EC50 = 13.1	Domínguez-Cortinas et al. 2008
"	"	PLA2 30-m	NOEC = 0.4 LOEC = 1.3 EC50 = 4.6	"
Glyphosate (Vision®)	*Simocephalus vetulus*	8-d survivorship and reproduction	0.75 mg/L	Chen et al., 2004
Molinate	*Brachionus calyciflorus* (R)	Ro T r	EC50 = 2.24 EC50 = 5.6 EC50 = 2.7	Ferrando et al. 1999
Paraquat	*Moina micrura* (C)	Population growth rate	not significant effect > 0.022	Leboulanger et al. 2008
Thiobencarb	*Brachionus calyciflorus* (R)	Ro T r	EC50 = 3.4 EC50 = 3.86 EC50 = 3.5 MATC = 3.16 NOEC = 2.0 LOEC = 5	Ferrando et al. 1999
Thiobencarb (S-4-chlorobenzyl diethylthiocarbamate)	*Daphnia magna* (C)	24-h	EC50 = 3.01	Sancho et al. 2001
"	"	R	> 0.30	"
"	"	S, r 21-d	0.75	"
2,4-D (2,4-dichlorophenoxyacetic acid)	*Brachionus calyciflorus* (R)	r 2-d	Chronic value = 70 NOEC= 58 LOEC=83 EC50= 128	Snell and Moffat, 1992

Herbicide	Test organism	Criteria	Endpoint (mg/l)	Reference
"	Brachionus calyciflorus (R)	r 2-d	NOEC = 2.5 EC10= 2.38 EC20= 4.91 EC50= 16.8	Radix et al. 1999
2,4-D (technical grade)	Brachionus patulus (R)	r	500	Sarma et al., 2001
3,4- DCA (3,4- dichloroaniline)	Brachionus calyciflorus (R)	S	5.0, 10, 20	Ferrando et al. 1993
		e_o	> 2.5	
		Ro	≥ 5.0	
		r	> 5.0	
		Vx	2.5	
		T	> 5.0	

Abbreviations. (C) Cladocerans, (R) Rotifers, (A) Amphipod. LC50 = Median Lethal Concentration, EC50 = Concentration where 50% inhibition occurs, MATC = Maximum Acceptable Toxicant Concentration, LOAEL = Lowest Observed Adverse Effect Level, NOAEL = No Observed Adverse Effect Level, LOEC = Lowest Observed Effect Concentration, NOEC = No observed effect concentration.

Table 4. Chronic toxicity of herbicides assessed to several species of freshwater zooplankton. Criteria consider a decrease or inhibition of the parameter at different exposure time to herbicide in minutes (m), hours (h) or days (d). Parameters: F = Fecundity, ED = Embryonic Development, R = Reproduction, S = Survival, cFDAam = Esterase activity, PLA2 = Phospholipase A2 activity, Ro = Net reproductive rate, T = Generation time, r = Intrinsic rate of population growth, e_o = Life expectancy, and Vx = Reproductive value.

distinct from the normal status of the individual or system health. Biomarkers are classified in three types:

1.   Effect biomarkers, which record the exposure of the organism to a toxicant or stressor without being directly related with the specific mechanism of action of the toxicant, and therefore, do not provide information on the level of adverse effect that this change causes (Hagger et al., 2006; Walker et al., 2006).
2.   Exposure biomarkers, which provide qualitative and quantitative estimations of exposure to several compounds. These biomarkers are well characterized and associated with the mechanism of action of the toxicant showing the relationship between levels of modification of the biomarker with respect to level of adverse effect (Hagger et al., 2006).
3.   Susceptibility biomarker, which provide information of the system´s health and are sensitive to toxicant exposure (Domingues et al., 2010).

There are different types of exposure biomarkers that involve important biological functions and that have been used to assess the adverse effect of many chemical substances. However, use of these biomarkers regarding aquatic invertebrates have been limited due to low availability of biological material, specificity, duration and costs (Hyne & Maher, 2003).

During a risk assessment, it is valuable to consider the range of specificity of the biomarkers. For instance, acethylcholinesterase (AChE) inhibition is consider specific for organophosphate, organochloride, and carbamate pesticides (Walker et al., 2006); and it is necessary to consider enough time to detect the presence of neurotoxic substances in the environment. Besides, AChE inhibition has been assesses in different aquatic invertebrate species. Therefore, it can be used as a good biomarker for these pesticides. The knowledge of AChE activity and its inhibition by certain herbicides can be used to relate enzymatic activity with the decrease of population densities in the field (Hyne & Maher, 2003). De Coen et al. (2001) demonstrated the relationship between parameters from carbohydrate enzymatic metabolism in *D. magna* and the specific effects of a toxicant suggesting that the activity of the piruvate kinase could potentially be the first warning sign about prolonged effects and to predict quantitative changes in the population.

Records on the use of biomarkers estimating the effect of herbicides on freshwater zooplankton are scarce. Barata et al. (2007) performed *in situ* bioassays with *D. magna*, reporting severe effects on the grazing rate, AchE, catalase, and glutathion S-transferase inhibition associated with the presence of bentazone (487 µg/L), methyl-4-chlorophenoxyacetic acid (8 µg/L), propanil (5 µg/L), molinate (0.8 µg/L), and fenitrothion (0.7 µg/L) in water. Domínguez-Cortinas et al. (2008) found that esterase and phospholipase A2 inhibition are good exposure biomarkers when the freshwater rotifer *L. quadridentata* and the cladoceran *D. magna* are exposed to the herbicide glyphosate and its commercial formula Faena (Table 1 and Table 2).

According to Barata et al. (2007) and Walker et al. (2006), the use of biomarkers is valuable to identify and assess the biological effects whenever toxicants are present in enough concentration to induce a detectable effect. Besides, Hagger et al. (2006), suggest that if the measurement of these effects shows the first responses in lower concentrations than the usual parameters of traditional toxicology, then the sensitivity of biomarker is of great use. It is important to consider that some chronic or sublethal effects can be irreversible and that can take place in ecosystems apparently healthy and where initially they were not detected (Hyne & Maher, 2003). Finally, a biomarker used as an integral parameter has the potential of establishing evidence of adverse effects caused by the presence of chemical substances in a system that can then be related with other levels of biological organization. Therefore, is fundamental to develop more research using biomarkers on freshwater zooplankton that allow to assess the adverse effect of all kind of toxicants (including herbicides), and to use these biomarkers regularly to monitor aquatic ecosystems.

7. Herbicides as endocrine disruptors of freshwater zooplankton species

Although many of the adverse physiological effects of chemicals affecting the neuroendocrine system have been known for over three decades, special attention to this issue only materialized in the early 1990s (Tackas et al., 2002). Given the high volume of use, high level of toxicity to primary producers, and long persistence in the environment, many studies have addressed the capacity of herbicides to disrupt endocrine function at concentrations that commonly occur in surface waters during application periods (Porter et al., 1999). An endocrine disruptor is defined as an exogenous agent that directly interferes with the synthesis, secretion, transport, binding action, or elimination of endogenous hormones and neurohormones, resulting in physiological manifestations of the neuroendocrine, reproductive or immune systems in an intact organism (Tackas et al., 2002).

Aquatic toxicity studies have shown that cladoceran fecundity and survival endpoints are not affected at atrazine concentrations below 100 µg/L (Takacs et al., 2002). However, Dodson et al. (1999) revealed that chronic exposure of *Daphnia pulicaria* to very low concentrations (0.5 µg/L) of atrazine induced a shift in the population sex ratio due to increased male production, indicating sex ratio is a very sensitive, ecologically-relevant endpoint. Males were produced in stress situations, in response to environmental signals such as shortening day length, reductions in food supply and pheromones produced in crowded populations (Dodson et al., 1999).

Villarroel et al. (2003) compared acute toxicity, reproductive and growth, and feeding activity alterations in *D. magna* exposed to several concentrations of propanil herbicide in a 21-days study. Some parameters analyzed were affected by herbicide: Survivorship did not decrease with increasing concentration of propanil, except with higher concentration (0.55 mg/L); number of neonates born, brood size and number of broods per female as well as the intrinsic rate of growth (r) decreased as the concentrations of propanil increased in the medium. EC50 values indicated that reproductive parameters, like the number of young per female (0.21 mg/L) and brood size (0.26 mg/L) were the most sensitive endpoints in response to propanil exposure. The filtration and ingestion rates were reduced significantly after 5-h exposure to this herbicide; this would be related with lose of coordination and paralysis caused for toxic effects of herbicide on nervous system of *D. magna* (Villarroel et al., 2003).

Other studies have shown that uptake of herbicides can directly affect survival, population growth, reproduction and feeding of rotifers. Riobbo et al. (2007) found that the *Brachionus* sp. population density decreased when females were fed with *Chlorella vulgaris* cells previously exposed to different concentrations of terbutryn, with a maximum survival of 4-days with 500 nM terbutryn in the medium. Terbutryn accumulated in *C. vulgaris* provoked a decrease in the feeding rate of *Brachionus* cultures, and a 66% reduction of the number of eggs per reproductive female compared to controls.

These results suggest that endocrine effects on zooplankton are caused by direct or indirect exposure to herbicides, where population growth rate and sex ratio can be the more sensitive parameters.

8. Field studies, mesocosms, and microcosms, involving herbicides and freshwater zooplankton

Among non-target organisms affected by herbicides in freshwater bodies, plankton and its components (bacterio-, phyto-, and zooplankton) are known to respond on short timescales to low levels of pollutants (Daam et al., 2009), mainly owing to their intrinsic sensitivity and high population turnover (Relyea, 2005). Secondary effects of herbicides on these organisms are difficult to predict since they depend on interactions between species, herbicides and the original structure of the ecosystem (Wendt-Rasch et al., 2003). For aquatic ecosystems, toxicity testing ranges from standard tests under laboratory conditions to field studies, including microcosm and mesocosm experiments (Caquet et al., 2000). These studies in enclosures are valuable tools that can help to understand how herbicides exposure may affect ecosystems as a whole, and be an aid in the assessment of the various risk scenarios resulting from the use of these chemicals (Wendt-Rasch et al., 2003).

Most of the information on the ecotoxicity of herbicides in aquatic communities is related to individual o combined effects of exposure to these chemicals at the ecosystem level (Thompson, 2006). Wendt-Rasch et al. (2003) reported no significant effects on copepod nauplii and rotifers from exposure during 14 days to metsulfuron methyl (0, 1, 5, 20 µg/L) in 24 enclosures of 80 L (height: 0.65 m, diameter: 0.4 m) in water bodies adjacent to agricultural fields. Metsulfuron methyl is a sulfonylurea herbicide that affects the synthesis of essential amino acids in plants, and hence inhibits cell division. It is highly water-soluble and has a low sorption coefficient (Tomlin, 1997). However, herbicide exposure had a significant effect on the conductivity, pH and total nitrogen in the enclosures (Wendt-Rasch et al., 2003).

Plankton communities from a tropical freshwater reservoir in Mozambique were monitored for 5 days after exposure to nominal concentrations of diuron (2.2 and 11 µg/L) and paraquat (10 and 40.5 µg/L), commonly used in the tropics for agriculture and disease vector control. Diuron blocks photosynthetic electron transfer in plants and algae, and paraquat generates superoxide O_2^- that affects all cellular components (Leboulanger et al., 2011). In general, zooplankton was slightly sensitive to diuron, and very sensitive to paraquat. Nauplii or cyclopidae copepodites and adults did not differ in microcosms inoculated with diuron relative to the controls. However, the adult stages of the copepod *Diaphanosoma excisum* were slightly reduced in high concentration compared with the control. A reduction in rotifer biomass was also noticed with a below significance level (p = 0.072). Low concentration of paraquat caused a significant reduction in *Thermocyclops decipiens* copepodite biomass relative to controls, whereas high treatments reduced the carbon biomass in all groups of zooplankton, mainly the cladocera and copepod nauplii (Leboulanger et al., 2011).

In PVC tanks of 150 L with water from the Paraná River, Gagneten (2002) evaluated the effects of paraquat (0.1, 0.2, 0.4 and 0.8 ml/L) on zooplankton community for 35 days of exposure. Contrary to what was observed with the species richness dominated by rotifers (55%), cladocerans (18%), and copepods (15%), paraquat negatively affected the zooplankton density, especially in higher concentrations. The chemical effect of the herbicide was higher on rotifers *Anuraeopsis, Lecane, Phylodina* and *Conochilus*; on the cladoceran *Ceriodaphnia*; on copepods *Eucyclops* and *Notodiaptomus*, and on thecamoebians *Arcella* and *Cucurbitella*. Dissolved oxygen, pH and water hardness did not vary significantly between controls and treatments during the experimental period. According to Pratt and Barreiro (1998), it is necessary to consider species composition, inter- and intraspecific interactions and environmental factors, such as physicochemical parameters, when analyzing the impact of herbicides on aquatic communities. This interaction between herbicides and biological and environmental factors may reduce or increase the impact of pollution on aquatic ecosystems (Gagneten, 2002).

Interactions of herbicides with others environmental stressors have also been studied. Chen et al. (2004, 2008) examined effects of interactions among pH (5.5 and 7.5), two levels of food concentrations, and the formulated products Vision® (glyphosate: 0.75 and 1.50 mg acid equivalent/L) and Release® (triclopyr) on cladoceran *Simocephalus vetulus*. Herbicide treatments resulted in significant decreases in survival, reproductive rate, and development time for *S. vetulus* at levels 5–10× below predicted worst case environmental concentrations (2.6 mg/L). High pH increased the toxic effects of the herbicide on all response variables even though it improved reproductive rate of *S. vetulus* over pH 5.5 in the absence of herbicide. Stress due to low food also interacted with pH 5.5 to diminish *S. vetulus* survival. These results support the general postulate that multiple stress interactions may exacerbate chemical effects on aquatic biota in natural systems.

Atrazine is a selective herbicide with long residual activity used on crops such as corn, sorghum, sugarcane, conifers, forestry and lawn care applications (Solomon et al., 1996). Degradation rates in water are highly variable. The DT50 in water has been estimated to range from 3-90 d or more and in sediment the range was 15-35 d (Huber, 1993). Several invertebrate community studies have been conducted with atrazine in field situations using mesocosms or whole ponds. The population density of cladocerans in ponds treated at 20 µg/L was lower than that in control ponds even one year after contamination. The most sensitive effect concentration for invertebrates in outdoor enclosures was 0.1 µg/L in which herbivorous zooplankton were reduced in abundance (Tackas et al., 2002).

Indirect effects on zooplankton were reported by Jüttner et al. (1995) during a 6 week mesocosms study. Total numbers of the cladoceran *Daphnia longispina* declined in all 7 enclosures following treatment with atrazine. This was accompanied by reduced egg ratios between day 3 and day 21. In both cases, effect concentration was 318 µg/L. Likewise, effect concentration on reduction in the density of copepod nauplii, *Synchaeta* sp. and *Polyarthra* sp was from 68, 132, and 318 µg/L atrazine, respectively. Van den Brink et al. (1995) detected only slight reductions in primary productivity over 7 weeks in multispecies microcosms exposed to 5 µg/L atrazine, and observed no significant effects on cyclopoid and cladoceran species or on the amphipod *Gammarus* and the rotifer *Keratella*.

Lozano et al. (1992) studied the temporal variation in abundance (% of control) of zooplankton following a single dose of esfenvalerate in 5 different concentrations (0.01, 0.08, 0.2, 1.0, 5.0 µg/L). Mesocosms were shallow (0.5 - 1.1 m depth), had sediment and macrophytes and ranged between 25 – 1100 m³ in volume. Dose-response curves showed that the initial impact on abundance and the subsequent recovery were dependent on the concentration: decreasing in Cladocera and Copepoda, and increasing in phytoplankton and Rotifera. Perschbacher et al. (2002) and Perschbacher and Ludwig (2004) tested the adverse impacts of common aerially applied herbicides for rice on phytoplankton, zooplankton, and water quality in 12 mesocosms (500 L, 0.7 m depth). Clomazone (0.6 kg active ingredient/ha), thiobencarb (3.4), pendamethalin (1.1), quinclorac (0.6), halosulfuron (0.07), bensulfuron methyl (0.07), triclopyr (0.4), 2,4-D-amine (1.7), and molinate (5.6) produced no measurable effects on plankton or water quality. Propanil (4.5) and diuron (1.4) significantly reduced oxygen production by 75% after their application and stimulated chlorophyll *a*, too. It was assumed to be related to compensatory action by the algae for photosynthesis inhibition. The increase in chlorophyll *a* concentration suggests an increase in food availability for zooplankton and is ultimately believed to have been responsible for the observed increase in numbers of rotifers and copepods, but not cladocerans (Perschbacher et al., 2002).

Marcial and Hagiwara (2008) determined acute toxicity of the mefenacet herbicide on the copepod *Tigriopus japonicus*, the cladoceran *Diaphanosoma celebensis* and the rotifer *Brachionus plicatilis*. Compound exposure was carried out in 6-well polystyrene plates, and mortality was evaluated after 24 h. Although species showed different sensitivities to herbicide, a dose-response relationship was consistent in all cases. *B. plicatilis* was particularly resistant to mefenacet, while *T. japonicus* and *D. celebensis* are comparatively sensitive.

Mohr et al. (2008) monitored for 140 days the effects of metazachlor (5, 20, 80, 200, and 500 µg/L) on stream and pond communities. In this study, metazachlor strongly affected mesocosms communities at all concentrations. Direct negative effects were most prominent for chlorophytes whereas diatoms and cryptophytes seemed insensitive. The effects on zooplankton were caused by changes in habitat structure due to the strong decline of macrophytes. The slow degradation of metazachlor combined with the absence of recovery in both chlorophytes and macrophytes was likely to cause long-lasting effects on aquatic ecosystems.

Jenkins and Buikema Jr. (2009) studied effects of simazine (0.1, 0.5 and 1.0 mg/L) on zooplankton and physical-chemical parameters in *in situ* microcosms for 21 days. Herbicide induced decreases in dissolved oxygen and pH, but induced increases in nitrate and ammonia levels compared to control microcosms. Rotifers dominated the zooplankton and were differentially affected by simazine. The dominant species, *Kellicottia bostomensis*, exhibited a positive response to simazine, as did *Keratella cochlearis*, due to lesser mortality in higher concentrations of simazine. *Polyarthra vulgaris* was unaffected, but *Synchaeta pectinata* was impaired by simazine at day 21.

These micro- mesocosms studies indicate that decrease in zooplankton density in the treated ponds probably was not caused by direct toxic effects of the herbicides, but to indirect effects resulting from reduced algal productivity, a change in the food source or a change in the competition for a food source.

9. Molecular genetics, DNA and protein microarrays, environmental genomics relating herbicides and freshwater zooplankton

The integration of genomic-based tools and ecotoxicology is a promising approach that may provide a broad view of how living systems respond to a given stressor (Neumann & Galvez, 2002; Robbens et al., 2007; Snape et al., 2004).

Transcription profiling using microarrays is one of the most prominent genome-wide technologies within ecotoxicogenomics since it provides an overview of changes in gene expression linked to chemical exposure (Pereira et al., 2010). Very recently, cDNA microarray-related techniques have been successfully used to address transcriptional responses of *D. magna* to different environmental toxicants, including pharmaceuticals, heavy-metals, pesticides and PAHs (Connon et al., 2008; Heckmann et al., 2008; Soetaert et al., 2006, 2007; Watanabe et al., 2007).

The evaluation of herbicides genotoxicity has been an important research line, to investigate the alterations in the molecular pathway in the organism. The most important organism for this test is *Daphnia magna*. Table 5 shows some alterations and DNA damages caused for some herbicides.

The effects of herbicides on freshwater zooplankton has been studied on molecular pathways and DNA, for example Pereira et al. (2010), to understanding the genomic responses of *D. magna* to chemical challenges, exposed to the herbicide propanil to compare phenotypic effects with changes in mRNA expression level. Propanil highly promoted synthesis of innate immunity response systems (more details in Table 3) and elicited specific up-regulation of gene transcription within neuronal pathways, including dopa decarboxylase and syntaxin 6. Atrazine induced hemoglobin genes (dhb1, dhb2 and dhb3) in *D. magna* through the hormonal pathways. This hypothesis was tested by modeling the combined effects of atrazine and the terpenoid hormone mimic pyriproxyfen on hemoglobin mRNA levels assuming the same mechanism of action (concentration addition model) and alternatively, assuming different mechanisms of action (response addition model) (Rider & Leblanc, 2006).

Herbicide	DNA alterations	Reference
Terbutryn	Cytogenetic damage Primary DNA damage	Moretti et al., 2000
Atrazine	Mutagenic and genotoxic potencial	Kaya et al., 2000 Pino et al., 1988
	DNA damage	Clements et al., 1997 Tennant et al., 2001
	Expression of haemoglobin genes	Rider and LeBlanc, 2006
Propanil	Promoted transcriptions genes of: Haemoglobin synthesis Neuronal pathways Up-regulated genes specifically related to defense mechanisms	Pereira et al., 2010

Up-regulated genes specifically related to defense mechanisms

Table 5. DNA alterations by herbicides.

10. Conclusions and future research

The study of the adverse effects of herbicides on freshwater zooplankton is an unexplored field. Studies in Quantitative Structure/Activity Relationship (QSAR's) are scarce or missing (at least from mainstream scientific literature). Ecotoxicogenomics studies are scarce and restricted to few herbicides and one species: *Daphnia magna*. Regarding biomarkers applied to herbicide exposure the small set of data available suggest that the potential of herbicides for producing adverse effects on freshwater zooplankton can be high, and warrants future research. Presently, atrazine and glyphosate are the two herbicides of great regulatory concern because of their widespread use, common detection in water having relatively long persistence in freshwater. Lethal toxicity in amphibians has been demonstrated (Reylea, 2005). Still, some authors pose serious doubts about the results suggesting direct and indirect effects of herbicides on invertebrates, amphibians and fish exposed to environmentally relevant concentrations (Fairchild, 2011). These doubts have to be clarified using well designed experiments that include effects on endocrine and immune function. Mesocosms studies will help identify and characterize the mechanisms that modify the sensitivity of zooplankton by exposure to herbicides. Compared to laboratory experiments, mixtures of herbicides combined with physical and chemical factors at the natural environment, could identify physiological, biochemical and behavioral changes more significant on zooplankton communities, mainly rotifers and copepods for which information reported is scarce. However, this chapter already includes recent data on lethal tests that suggest that at least for brief periods of time, some herbicides at environmentally relevant concentrations can produce mortality, and other relevant sublethal effects in freshwater zooplankton (for example, reduction in rate population growth).

11. Acknowledgements

Authors thank Dr. Robert L. Wallace from Ripon College, Wisconsin for critical review and fruitful comments that improved the outcome of this manuscript. R. R.-M. thanks the Fulbright Program and COMEXUS for providing the Fulbright-García Robles Scholarship. I. A. Pérez-Legaspi thanks SNI for scholarship 49351.

12. References

Adema, D.M.M. & Vink, I.G.J. (1981). A comparative study if the toxicity of 1,1,2-trichloroethane,dieldrin, pentachlorophenol and 3,4 dichloroaniline for marine and freshwater organisms. *Chemosphere*, 10, 553-554

Alberdi, J.L; Sáenz, M.E.; Di Marzio, W.D. & Tortorelli, M.C. (1996). Comparative acute toxicity of two herbicides, Paraquat and Glyphosate, to *Daphnia magna* and *D. spinulata*. *Bull Environ Contam Toxicol*, 57, 229-235

Ashoka-Deepananda, K.H.M.; Gajamange, D.; De Silva, W.A.J.P. & Wegiriya, H.C.E. (2011). Acute toxicity of a glyphosate herbicide, Roundup®, to two freshwater crustaceans. *J Natl Sci Found Sri Lanka*, 39, 169-173

Barata, C.; Damasio, J.; López, M.A.; Kuster, M.; López de Alda, M.; Barceló, D.; Riva, M.C. & Raldúa, D. (2007). Combined use of biomarkers and *in situ* bioassays in *Daphnia magna* to monitor environmental hazards of pesticides in the field. *Environ Toxicol Chem*, 26, 370-379

Battaglin, W.A.; Thurman, E.M.; Kalkhoff, S.J. & Porter, S.D. (2003). Herbicides and transformation products in surface waters of the Midwestern United States. *J Am Water Resour Assoc*, 39, 743-756

Battaglin, W.A.; Rice, C.K.; Foazio, M.J.; Salmons, S. & Barry, R.X. (2008). The occurrence of glyphosate, atrazine, and other pesticides in vernal pools and adjacent streams in Washington, DC, Maryland, Iowa and Wyoming 2005-2006. *Environ Monit Assess*, 155, 281-307

Boyle, T. (1980). Effects of the aquatic herbicide 2,4-D DMA on the ecology of experimental ponds. *Environ Pollut*, 21, 35-39

Brooker, M.P. & Edwards, R.W. (1973). Effects of the herbicide paraquat on the ecology of a reservoir. Botanical and chemical aspects. *Freshwater Biol*, 3, 157-175

Buttiglieri, G.; Peschka, M.; Frömel, T.; Müller, J.; Malpei, F.; Seel, P. & Knepper, T.P. (2009). Environmental occurrence and degradation of the herbicide n-chloridazon. *Water Res*, 43, 2865-2873

Caquet, T.; Lagadic, L. & Sheffield, S. (2000). Mesocosms in ecotoxicology: outdoor aquatic systems. *Rev Environ Contam Toxicol*, 165, 1-38

Chang, K.-H.; Sakamoto, M.; Ha, J.Y.; Murakami, T.; Miyabara, Y.; Nakano, S.I.; Imai, H.; Doi, H. & Hanazato, T. (2008). Comparative study of pesticide effects (herbicide and fungicide) on zooplankton community. In: *Interdisciplinary Studies on Environmental Chemistry – Biological Responses to Chemical Pollutants*, Murakami, Y.; Nakayama, K.; Kitamura, S.-I.; Iwata, H. & Tanabe, S. (Eds.). TERRAPUB, pp. 361-366

Chen, C.Y.; Hathaway, K.M. & Folt, C.L. (2004). Multiple stress effects of Vision® herbicide, pH, and food on zooplankton and larval amphibian species from forest wetlands. *Environ Toxicol Chem*, 23, 823-831

Chen, C.Y.; Hathaway, K.M.; Thompson, D.G. & Folt, C.L. (2008). Multiple stressor effects of herbicide, pH, and food on wetland zooplankton and a larval amphibian. *Ecotoxicol Environ Safe*, 71, 209-218

Chu, Z.; Yi, Y.; Xu, X.; Ge, Y.; Dong, L. & Chen, F. (2005). Effects of glyphosate on life history characteristics of freshwater rotifer *Brachionus calyciflorus*. *J Appl Ecol*, 6, 1142-1145

Clements, C.; Ralph, S. & Petras, M. (1997). Genotoxicity of select herbicides in *Rana catesbeiana* tadpoles using the alkaline single-cell gel DNA electrophoresis (comet) assay. *Environ Mol Mutagen*, 29, 277-288

Cochón, A.C.; Della-Penna, A.B.; Kristoff, G.; Piol, M.N.; San Martín de Viale, L.C. & Verrengia-Guerrero, N.R. (2007). Differential effects of paraquat on oxidative stress parameters and polyamine levels in two freshwater invertebrates. *Ecotoxicol Environ Safe*, 68, 286-292

Connon, R.; Hooper, H.; Sibly, R.M.; Lim, F.-L.; Heckmann, L.-H.; Moore, D.J.; Watanabe, H.; Soetaert, A.; Cook, K.; Maund, S.J.; Hutchinson, T.H.; Moggs, J.; De Coen, W.; Iguchi, T. & Callaghan, A. (2008). Linking molecular and population stress responses in *Daphnia magna* exposed to cadmium. *Environ Sci Technol*, 42, 2181-2188

Cope, O.B. (1966). Contamination of the freshwater ecosystems by pesticides. *J Appl Ecol*, 3, 517-523

Cuppen, J.G.M.; Van den Bink, P.J.; Van der Woude, H.; Zwaardemaker, N. & Brock, T.C.M. (1997). Sensitivity of macrophytes-dominated freshwater microcosms to chronic levels of the herbicide linuron. *Ecotoxicol Environ Safe*, 38, 25-35

Daam, M.A.; Rodrigues, A.M.F.; Van den Brink, P.J. & Nogueira, A.J.A. (2009). Ecological effects of the herbicide linuron in tropical freshwater microcosms. *Ecotoxicol Environ Safe*, 72, 410-423

De Coen, W.M.; Janssen, C.R. & Segner, H. (2001). The use of biomarkers in *Daphnia magna* toxicity testing V. *In Vivo* alterations in the carbohydrate metabolism of *Daphnia magna* exposed to sublethal concentrations of mercury and lindane. *Ecotoxicol Environ Safe*, 48, 223-234

Dodson, S.I.; Merritt, C.M.; Shannahan, J. & Shults, C.M. (1999). Low exposure concentrations of atrazine increase male production in *Daphnia pulicaria*. *Environ Toxicol Chem*, 18, 1568-1573

Domingues, I.; Agra, A.R.; Monaghan, K.; Soares, A.M.V.M. & Nogueira, A.J.A. (2010). Cholinesterase and Glutathione-s-transferase activities in freshwater invertebrates as biomarkers to assess pesticide contamination. *Environ Toxicol Chem*, 29, 5-18

Domínguez-Cortinas, G.; Mejía-Saavedra, J.; Santos-Medrano, G.E. & Rico-Martínez, R. (2008). Analysis of the toxicity of glyphosate and Faena® using the freshwater invertebrates *Daphnia magna* and *Lecane quadridentata*. *Toxicol Environ Chem*, 90, 377-384

Duke, S.O. (1990). Overview of herbicide mechanisms of action. *Environ Health Persp*, 87, 263-271

Edwards, W.M.; Triplett, G.B. & Kramer, R.M. (1980). A watershed study of glyphosate transport in runoff. *J Environ Qual*, 9, 661-665

Eisler, R. (1994). Acrolein hazard to fish, wildlife and invertebrates: A synoptic review. In: *National Biological Survey. Technical Report Series. Biological Report 23*. Opler, P.A. (Ed.). pp 1-29. U.S. Fish and Wildlife Service, Washington, DC, USA

Environment Canada. (1987). Guidelines for Canadian Drinking Water Quality. Supporting Documents: Glyphosate.

Fairchild, J.F. (2011). Structural and Functional Effects of Herbicides on Non-Target Organisms in Aquatic Ecosystems with an Emphasis on Atrazine. In: *Herbicides and the Environment*, Kortekamp, A. (Ed.), In Tech Open, Rijeka, Croatia, pp. 383-404

FAO (Food and Agriculture Organization). (1999). Specifications and evaluations for plant protection products: Metazachlor. 17 pp. (www.fao.org/ag/AGP/AGPP/Pesticid/Specs/docs/Pdf)

Ferrando, M.D. & Andreu-Moliner, E. (1991). Acute lethal toxicity of some pesticides to *Brachionus calyciflorus* and *Brachionus plicatilis*. *Bull Environ Contam Toxicol*, 47, 479-484

Ferrando, M.D.; Janssen, C.R.; Andreu-Moliner, E. & Persoone, G. (1993). Ecotoxicological studies with the freshwater rotifer *Brachionus calyciflorus*. II. An assessment of the chronic toxicity of lindane and 3,4-dichloroaniline using life tables. *Hydrobiologia*, 255/256, 33-40

Ferrando, M.D.; Sancho, E.; Villarroel, M.J.; Sánchez, M. & Andreu-Moliner, E. (1999). Comparative toxicity of two herbicides, molinate and thiobencarb, to *Brachionus calyciflorus*. *J Environ Sci Health, Part B*, 34, 569-586

Fleeger, J.W.; Carman, K.R. & Nisbet, R.M. (2003). Indirect effects of contaminants in aquatic ecosystems. *Sci Total Environ*, 317, 207-233

Fitzmayer, K.M.; Geiger, J.G. & Van Den Avyle, M.J. (1982). Acute toxicity effects of simazine on *Daphnia pulex* and larva striped bass. *Proc Annu Conf Southest. Assoc Fish and Wildl Agencies*, 36, 146-156

Folmar, L.C.; Sanders, J.O. & Julin, A.M. (1979). Toxicity of the herbicide Glyphosate and several of its formulations to fish and aquatic invertebrates. *Arch Environ Contam Toxicol*, 8, 269-278

Gagneten, A.M. (2002). Efectos del herbicida Paraquat sobre el zooplancton. *Iheringia, Sér Zoo*, 92, 47-56

Golombieski, J.I.; Marchesan, E.; Baumart, J.S.; Reimche, G.B.; Júnior, Ch.R.; Storck, L. & Santos, S. (2008). Cladocers, copepods and rotifers in rice-fish culture handled with metsulfuron-methyl and azimsulfuron herbicides and carbofuran insecticide. *Cienc Rural*, 38, 2097-2102

González-Barreiro, O.; Rioboo, C.; Herrero, C. & Cid, A. (2006). Removal of triazine herbicides from freshwater systems using photosynthetic microorganisms. *Environ Pollut*, 144, 266-271

Gunsolus, J.L. & Curran, W.S. (1998). Herbicide mode of action and injury symptoms. University of Minnesota Extension Service, USA. (www.cof.orst.edu/cof/fs/kpuettmann/FS20533/VegetationManagement)

Hagger, J.A.; Malcolm, M.B.; Leonard, P.; Owen, R. & Galloway, T.S. (2006). Biomarkers and integrated environmental risk assessment: Are there more questions than answers?. *Integr Environ Assess Manage*, 4, 312-329

Hanazato, T. (1998). Growth analysis of *Daphnia* early juvenile stages as an alternative method to test the chronic effect of chemicals. *Chemosphere*, 36, 1903-1909

Hanazato, T. (2001). Pesticide effects on freshwater zooplankton: an ecological perspective. *Environ Pollut*, 112, 1-10

Heckmann, L.H.; Sibly, R.M.; Connon, R.; Hooper, H.L.; Hutchinson, T.H.; Maund, S.J.; Hill, C.J.; Bouetard, A. & Callaghan, A. (2008). Systems biology meets stress ecology: linking molecular and organismal stress responses in *Daphnia magna*. *Genome Biol*, 9, R40 (doi:10.1186/gb-2008-9-2-r40)

Holcombe, G.W.; Phipps, G.L.; Sulaiman, A.H. & Hoffman, A.D. (1987). Simultaneous multiple species testing: acute toxicity of 13 chemicals to 12 diverse freshwater amphibian, fish and invertebrate families. *Arch Environ Contam Toxicol*, 16, 697-710

Horvat, T.; Kalafatic, M.; Kopjar, N. & Kovačević, G. (2005). Toxicity testing of herbicide norflurazon on aquatic bioindicator species the planarian *Polycelis felina* (Daly.). *Aquat Toxicol*, 73, 342-352

Huber, W. (1993). Ecological relevance of atrazine in aquatic systems. *Environ Toxicol Chem*, 12, 1865-1881

Hyne, R.V. & Maher, W.A. (2003). Invertebrate biomarkers: links to toxicosis that predict population decline. *Ecotoxicol Environ Safe*, 54, 366-374

Jenkins, D.G. & Buikema Jr., A.L. (2009). Response of a winter plankton food web to simazine. *Environ Toxicol Chem*, 9, 693-705

Johnson, W.W. & Finley, M.T. (1980). *Handbook of acute toxicity of chemicals to fish and aquatic invertebrates*. U.S. Fish and Wildlife Service. Washington, D.C., USA.

Julli, M. & Krassoi, F.R. (1995). Acute and chronic toxicity of the thiocarbamate herbicide, molinate, to the cladoceran *Moina australiensis* Sars. *Bull Environ Contam Toxicol*, 54, 690-694

Jüttner, I.; Peither, A.; Lay, J.P.; Kettrup, A. & Ormerod, S.J. (1995). An outdoor mesocosm study to assess ecotoxicological effects of atrazine on a natural plankton community. *Arch Environ Contam Toxicol*, 29, 435-441

Kaya, B.; Yanikoglu, A.; Creus, A. & Marcos, R. (2000). Genotoxicity testing of five herbicides in the *Drosophila* wing spot test. *Mutat Res*, 465, 77-84

Kyriakopoulou, K.; Anastasiadou, P. & Machera, K. (2009). Comparative toxicities of fungicide and herbicide formulations on freshwater and marine species. *Bull Environ Contam Toxicol*, 82, 290-295

Leboulanger, C.; Bouvy, M.; Pagano, M; Dufour, R.; Got, P. & Cecchi, P. (2009). Responses of planktonic microorganisms from tropical reservoirs to paraquat and deltamethrin exposure. *Arch Environ Contam Toxicol*, 56, 39-51

Leboulanger, C.; Bouvy, M.; Carré, C.; Cecchi, P.; Amalric, L.; Bouchez, A.; Pagano, M. & Sarazin, G. (2011). Comparison of the effects of two herbicides and an insecticide on tropical freshwater plankton in microcosms. *Arch Environ Contam Toxicol*, DOI 10.1007/s00244-011-9653-3

Lozano, S.J.; O'Halloran, S.L.; Sargent, K.W. & Brazner, J.C. (1992). Effects of esfenvalerate on aquatic organisms in littoral enclosures. *Environ Toxicol Chem*, 11, 35-47

Marcial, H.S. & Hagiwara, A. (2008). Acute toxic concentrations of endocrine disrupting compounds on the copepod, the cladoceran, and the rotifer. *Bull Fac Fish*, 89, 37-44

Martin, T. K.T. & Chu, L.M. (2003). Aquatic toxicity of glyphosate-based formulations: comparison between different organism and the effects of environmental factors. *Chemosphere*, 52, 1189-1197

Meyerhoff, R.D.; Grothe, D.W.; Sauter, S. & Dorulla, G.K. (1985). Chronic toxicity of tebuthiuron to an alga (*Selenastrum capricornutum*), a cladoceran (*Daphnia magna*), and the fathead minnow (*Pimephales promelas*). *Environ Toxicol Chem*, 4, 695-701

Mohr, S.; Feibicke, M.; Berghahn, R.; Schmiediche, R. & Schmidt, R. (2008). Response of plankton communities in freshwater pond and stream mesocosms to the herbicide metazachlor. *Environ Pollut*, 152, 530-542

Moore, M.T.; Pierce, J.R.; Milam, C.D.; Farris, J.L. & Winchester, E.L. (1998). Responses of non-target aquatic organisms to aqueous propanil exposure. *Bull Environ Contam Toxicol*, 61, 169-174

Moretti, M.; Marcarelli, M.; Villarini, M..; Fatigoni, C.; Scassellati-Sforzolini, G. & Pasquini, R. (2002). In vitro testing for genotoxicity of the herbicide terbutryn: cytogenetic and primary DNA damage. *Toxicol In Vitro*, 16, 81-88

Naqvi, S.M.; Leung, T.-S. & Naqvi, N.Z. (1981). Toxicities of paraquat and metribuzin (Sencor®) herbicides to the freshwater copepods, *Eucyclops agilis* and *Diaptomus mississippiensis*. *Environ Pollut Ser. A*, 26, 275-280

Nebeker, A.V. & Schuytema, G.S. (1998). Chronic effects of the herbicide Diuron on freshwater cladocerans, amphipods, midges, minnows, worms, and snails. *Arch Environ Contam Toxicol*, 35, 441-446

Neumann, N.F. & Galvez, F. (2002). DNA microarrays and toxicogenomics: applications for ecotoxicology? *Biotechnol Adv*, 20, 391–419

Newbold, C. (1975). Herbicides in aquatic systems. *Biol Conserv*, 2, 97-118

Nimmo, D.R. & McEwen, L.C. (1994). Pesticides. In: *Handbook of Ecotoxicology*. Calow, P. (Ed.), Vol. 2, Blackwell Scientific Publications, UK.

Nwani, C.D.; Lakra, W.S.; Nagpure, N.S.; Kumar, R.; Kushwaha, B. & Srivastava, S.K. (2010). Toxicity of the herbicide Atrazine: Effects on lipid peroxidation and activities of antioxidant enzymes in the freshwater fish *Channa punctatus* (Bloch). *Int J Environ Res Public Health*, 7, 3298-3312

Pereira, J.L.; Hill, C.J.; Sibly, R.M.; Bolshakov, V.N.; Alvesa, F.G.; Heckmann, L.H. & Callaghan, A. (2010). Gene transcription in *Daphnia magna*: Effects of acute exposure to a carbamate insecticide and an acetanilide herbicide. *Aquat Toxicol*, 97, 268-276

Pérez, G.L.; Vera, M.S. & Miranda, L.A. (2011). Effects of herbicide Glyphosate and Glyphosate-Based formulations on aquatic ecosystems. In: *Herbicides and the Environment*. Kortekamp, A. (Ed.), In Tech Open, Rijeka, Croatia, pp. 343-368

Pérez-Legaspi, I.A.; Quintanar, J.L. & Rico-Martínez, R. (2011). Comparing toxicity endpoints on *Lecane quadridentata* (Rotifera: Monogononta) exposed to two anticholinesterases pesticides. *Environ Toxicol*, DOI 10.1002/tox.20668

Perschbacher, P.W.; Ludwig, G.M. & Slaton, N. (2002). Effects of common aerially applied rice herbicides on the plankton communities of aquaculture ponds. *Aquaculture*, 214, 241-246

Perschbacher, P.W. & Ludwig, G.M. (2004). Effects of diuron and other aerially applied cotton herbicides and defoliants on the plankton communities of aquaculture ponds. *Aquaculture*, 223, 197-203

Persoone, G.; Baudo, R.; Cotman, M.; Blaise, C.; Thompson, K.C.; Moreira-Santos, M.; Vollat, B.; Törökne, A. & Han, T. (2009). Review on the acute *Daphnia magna* toxicity test – Evaluation of the sensitivity and the precision of assays performed with organisms from laboratory cultures or hatched from dormant eggs. *Knowl Manag Aquat Ecosyst*, 393, 1-29

Phyu, Y.L.; Warne, M.S. & Lim, R.P. (2004). Toxicity of atrazine and molinate to the cladoceran *Daphnia carinata* and the effect of river water and bottom sediment on their bioavailability. *Arch Environ Contam Toxicol*, 46, 308-315

Pino, A.; Maura, A. & Grillo, P. (1988). DNA damage in stomach, kidney, liver and lung of rats treated with atrazine. *Mutat Res*, 209, 145-147

Plimmer, J.R.; Bradow, J.M.; Dionigi, C.P.R.; Johnson, R.M. & Wojkowski, S. (2005). Herbicides. In: *Kirk-Othmer Encyclopedia of Chemical Technology*, 5th Edition, Vol. 13, p 281. John Wiley and Sons, Hoboken, NJ, USA

Porter, W.P.; Jaeger, J.W. & Carlson, I.H. (1999). Endocrine, immune, and behavioral effects of aldicarb (carbamate), atrazine (triazine) and nitrate (fertilizer) mixtures at groundwater concentrations. *Toxicol Ind Health*, 15, 133-150

Pratt, J.R. & Barreiro, R. (1998). Influence of trophic status on the toxic effects of an herbicide: a microcosms study. *Arch Environ Contam Toxicol*, 35, 404-441

Prostko, E.P. & Baughman, T.A. (1999). Peanut herbicide injury symptomology guide. Texas Agricultural Extension Service, USA. (www.lubbock.tamu.edu/ipm/peanut/docs/HerbInjurySymptoms.pdf)

Radix, P.; Léonard, M.; Papantoniou, C.; Roman, G.; Saouter, E.; Galloti-Schmitt, S.; Thiébaud, H. & Vasseur, P. (1999). Comparison of *Brachionus calyciflorus* 2-d and Microtox chronic 22-h tests with *Daphnia magna* 21-d for the chronic toxicity assessment of chemicals. *Environ Toxicol Chem*, 18, 2178-2185

Ralston-Hooper, K.; Hardy, J.; Hahn, L.; Ochoa-Acuña, H.; Lee, L.S.; Mollenhauer, R. & Sepúlveda, M.S. (2009). Acute and chronic toxicity of atrazine and its metabolites deethylatrazine and deisopropylatrazine on aquatic organisms. *Ecotoxicology*, 18, 899-905

Relyea, R.A. (2005). The Impact of insecticides and herbicides on the biodiversity and productivity of aquatic communities. *Ecol Appl*, 15, 618-627

Relyea, R.A. & Hoverman, J. (2006). Assessing the ecology in ecotoxicology: a review and synthesis in freshwater systems. *Ecol Lett*, 9, 1157-1171

Relyea, R.A. (2009). A cocktail of contaminants: how mixtures of pesticides at low concentrations affect aquatic communities. *Oecologia*, 159, 363-376

Rider, C.V. & LeBlanc, G.A. (2006). Atrazine stimulated haemoglobin accumulation in *Daphnia magna*: is it hormonal or hypoxic? *Toxicol Sci*, 93, 443-449

Rioboo, C.; Prado, R.; Herrero, C. & Cid, A. (2007). Population growth study of the rotifer *Brachionus* sp. fed with triazine-exposed microalgae. *Aquat Toxicol*, 83, 247-253

Robbens, J.; Van der Ven, K.; Maras, M.; Blust, R. & De Coen, W. (2007). Ecotoxicological risk assessment using DNA chips and cellular reporters. *Trends Biotechnol*, 25, 460-466

Rohm, G. & Haas, M. (1991). Product: STAM Tech 98% DCA herbicide (key 904399-2). Rohm and Haas Company, Philadelphia, PA. pp 10-126

Rohr, J.R. & McCoy, K.A. (2010). A qualitative meta-analysis reveals consistent effects of atrazine on freshwater fish and amphibians. *Environ Health Persp*, 118, 20-32

Sánchez, M.; Andreu-Moliner, E. & Ferrando, M.D. (2004). Laboratory investigation into the development of resistance of *Daphnia magna* to the herbicide molinate. *Ecotoxicol Environ Safe*, 59, 316-323

Sancho, E.; Sánchez, M.; Ferrando, M.D. & Andreu-Moliner, E. (2001). Effects of thiobencarb herbicide to an alga (*Nannochloris oculata*) and the cladoceran (*Daphnia magna*). *J Environ Sci Health, Part B*, 36, 55-65

Sanders, H.O. & Cope, O.B. (1968). The relative toxicities of several pesticides to naiads of three species of stoneflies. *Limnol Oceanogr*, 13, 112-117

Sarma, S.S.S.; Ramírez-Pérez, T.; Nandini, S. & Peñalosa-Castro, I. (2001). Combined effects of food concentration and the herbicide 2,4-Dichlorophenoxyacetic acid on the population dynamics of *Brachionus patulus* (Rotifera). *Ecotoxicol*, 10, 91-99

Sarma, S.S.S. & Nandini, S. (2006): Review of recent ecotoxicological studies on cladocerans. *J Environ Sci Health, Part B*, 41, 1417-1430

Servizi, J.A.; Gordon, R.W. & Martens, D.W. (1987). Acute toxicity of Garlon 4 and Roundup herbicides to salmon, *Daphnia*, and trout. *Bull Environ Contam Toxicol*, 39, 15-22

Snape, J.R.; Maund, S.J.; Pickford, D.B. & Hutchinson, T.H. (2004). Ecotoxicogenomics: the challenge of integrating genomics into aquatic and terrestrial ecotoxicology. *Aquat Toxicol*, 67, 143-154

Snell, T.W.; Moffat, B.D.; Janssen, C. & Persoone, G. (1991). Acute toxicity tests using rotifers. IV. Effects of cyst age, temperature, and salinity on the sensitivity of *Brachionus calyciflorus*. *Ecotoxicol Environ Safe*, 21, 308-317

Snell, T.W. & Moffat, B.D. (1992). A 2-d life cycle test with the rotifer *Brachionus calyciflorus*. *Environ Toxicol Chem*, 11, 1249-1257

Snell, T.W. & Janssen, C.R. (1995). Rotifers in ecotoxicology: a review. *Hydrobiologia*, 313/314, 231-247

Soetaert, A.; Moens, L.N.; Van der Ven, K.; Van Leemput, K.; Naudts, B.; Blust, R. & De Coen, W.M. (2006). Molecular impact of propiconazole on *Daphnia magna* using a reproduction-related cDNA array. *Comp Biochem Physiol C Toxicol Pharmacol*, 142, 66-76

Soetaert, A.; Vandenbroucke, T.; Van der Vem, K.; Maras, M.; Van Remortel, P.; Blust, R. & De Coen, W.M. (2007). Molecular responses during cadmium-induced stress in *Daphnia magna*: integration of differential gene expression with higher-level effects. *Aquat Toxicol*, 83, 212-222

Solomon, K.R.; Baker, D.B.; Richards, R.P.; Dixon, K.R.; Klaine, S.J.; LaPoint, T.W.; Kendall, R.J.; Weisskopf, C.P.; Giddings, J.M.; Giesy, J.P.; Hall, L.W.Jr. & Williams, W.M. (1996). Ecological risk assessment of atrazine in North America surface waters. *Environ Toxicol Chem*, 15, 31-76

Takacs, P.; Martin, P.A. & Struger, J. (2002). Pesticides in Ontario: A critical assessment of potential toxicity of agricultural products to wildlife, with consideration for endocrine disruption. Volume 2: Triazine herbicides, Glyphosate and Metolachlor. Technical Report Series No. 369. Canadian Wildlife Service, Ontario Region, Burlington, Ontario, Canada.

Tennant, A.H.; Peng, B. & Kligerman, A.D. (2001). Genotoxicity studies of three triazine herbicides: *in vivo* studies using the alkaline single cell gel (SCG) assay. *Mutat Res*, 493, 1-10

Thompson, D.G. (2006). The impact of insecticides and herbicides on the biodiversity and productivity of aquatic communities. *Ecol Appl*, 16, 2022-2027

Tomlin, C.D.S. (1997). *The Pesticide Manual. A World Compendium*. British Crop Protection Council, Surrey, UK

Tsui, M.T.K. & Chu, L.M. (2003). Aquatic toxicity of glyphosate-based formulations: Comparison between different organisms and the effects of environmental factors. *Chemosphere*, 52, 1189-1197

Tsui, M.T.K. & Chu, L.M. (2004). Comparative toxicity of glyphosate-based herbicides: Aqueous and sediment porewater exposures. *Arch Environ Contam Toxicol*, 46, 316-323

US EPA. (1986). Pesticide Fact Sheet: Glyphosate EPA 540/FS-88-124. California Department of Food and Agriculture, Medical Toxicology Branch of Toxicology.

USGS (U.S. Geological Survey). (1999). Pesticides in stream sediment and aquatic biota. Current understanding of distribution and major influences. USGS Fact Sheet 092-00. Sacramento, CA. USA

(http://water.usgs.gov/nawqa/pnsp/pubs/fs09200/fs09200.pdf).

Van den Brink, P.J.; van Donk, E.; Gylstra, R.; Crum, S.J.H. & Brock, T.C.M. (1995). Effects of chronic low concentrations of the pesticides chlorpyrifos and atrazine in indoor freshwater microcosms. *Chemosphere*, 3, 3181-3200

Verschueren, K. (1983). *Handbook of environmental data of organic chemicals*. Van Nostrand Reinhold Co. Inc., New York, New York, USA.

Villarroel, M.J.; Sancho, E.; Ferrando, M.D. & Andreu-Moliner, E. (2003). Acute, chronic and sublethal effects of the herbicide propanil on *Daphnia magna*. *Chemosphere*, 53, 857-864

Walker, C.R. (1971). The toxicological effects of herbicides and weed control on fish and other organisms in the aquatic ecosystem. *Proc Eur Weed Res Councyl, 3rd Int. Symp Aquat Weeds*, pp 119-127

Walker, C.H.; Hopkin, S.R.; Sibly, R.M. & Peakall, D.B. (2006). *Principles of Ecotoxicology*. Taylor & Francis. CRC Press. pp 314

Watanabe, H.; Takahashi, E.; Nakamura, Y.; Oda, S.; Tatarazako, N. & Iguchi, T. (2007). Development of a *Daphnia magna* DNA microarray for evaluating the toxicity of environmental chemicals. *Environ Toxicol Chem*, 26, 669-676

Wendt-Rasch, L.; Pirzadeh, P. & Woin, P. (2003). Effects of metsulfuron methyl and cypermethrin exposure on freshwater model ecosystems. *Aquat Toxicol*, 63, 243-256

Wilson, D.C. & Bond, C.E. (1969). The effects of the herbicide diquat and dichlobenil (Casoron) on pond invertebrates. Part 1. Acute toxicity. *Trans Am Fish Soc*, 98, 438-443

10

Herbicide Tolerant Food Legume Crops: Possibilities and Prospects

N.P. Singh[1] and Indu Singh Yadav[1,2]
[1]Indian Institute of Pulses Research, Kanpur
[2]National Research Centre on Plant Biotechnology, New Delhi,
India

1. Introduction

Weeds are one of the major problems in agriculture. Weeds compete with other crops for water and nutrients and, as a result, decrease yields and productivity. Without weed control it is extremely difficult to harvest crops. The advent of mechanization replaced much of the hand labour in the developed world as well as the developing parts of the third world. Mechanical weed control is fraught with high-energy costs, facilitates soil erosion and compaction and has been mostly replaced by chemical weed control using herbicides (Gressel J, 2000). As countries industrialize and develop economically, cheap farm labour becomes unavailable, thus increasing the necessity for cost-effective chemical weed control. In India, weeds cause the highest loss (33%) followed by pathogens (26%), insects (20%), storage pests (7%), rodents (6%) and others (8%). It has been estimated that the potential losses due to weeds in different field crops would be around 180 million tonnes, valued at Rs. 105,0000 millions annually (Anonymous, 2008). Globally, herbicide constitutes 50 percent of the total pesticides sale and in some countries like USA, Germany and Australia; the figure is as high as 60-70 percent. In India, however, the position is different as herbicides form a meager 15 percent of the total pesticide consumption. But still, the consumption has increased rapidly from 4100 metric tons (MT) in 1988-89 to 13,764 MT in 2004 and it is likely to further increase in future (Varshney and Mishra, 2008). Given the harmful economic implications of poor weed management, it is hardly surprising that herbicide production is a main driver of the agrochemical industry. Too often there is no selective chemical that can control a particular weed in a particular crop, as most selectivity between crop and weed are due to catabolic degradation of the herbicide by the crop. Therefore, closely related weeds are to be expected to have similar catabolic pathways as the crop and thus escape the chemical effect. This is one major reason that genetically modified herbicide-resistant crops (GM-HRC) have become so useful, and that biotechnology has been utilized to produce such crops as well as to find new herbicide targets. Selectivity can be enhanced by inserting exogenous resistance genes into the crops or by selecting natural mutations. However, one major concern about transgenic herbicide resistant crops (HRCs) is that the transgene could genetically introgress into related weeds, and make them resistant and therefore, their careful management comes into account.

2. Chemical weed control

The controlling of weeds in the growing crops with weedicides increases their yields and ensures the efficient use of irrigation, fertilizers and plant-protection measures, such as the spraying of insecticides and fungicides. The removal of weeds from the growing crops facilitates easy harvesting and gives a high-quality produce without admixture with weed seeds. Chemical weed control can be adopted quite in time and in situations and under conditions, which make manual or mechanical weeding difficult. A great advantage of this method lies in killing weeds in the crop row or in the immediate vicinity of crop plants. The chemical method is easier, less time-consuming and less costly than weeding by hired laborers. However, there are several disadvantages like environment pollution, human and animal health issues related to its use.

3. Biological weed control

Biological weed control is the action of parasites, predators, or pathogens to maintain another organism's population at a lower average density than would occur in their absence. Biological control is usually thought of as intentional introduction of parasites, predators, or pathogens to achieve control, but it is also a natural phenomenon. Biological control will never be the solution to every weed problem. It is employed as one weed management practice among many. Using tools of biotechnology, it is possible to engineer a more potent parasite, predator or mutant which can be deployed to weed control. The biological weed control can be permanent weed management because once an organism is released, it may be self-perpetuating and control will continue without further human intervention. Besides, there are no chemical environmental residues from biological control other than the organism. Bio control may be the best option for management of invasive species. In ideal cases, initial costs are nonrecurring and usually, once the organism is established, no further inputs are needed. There are some situations where biological control is not appropriate. If a plant is a weed in one place and valued in another place, in the same general geographic region, biological control is inappropriate. Spread of a biological control organism, once introduced, cannot be controlled. Biological control is inherently slow, and results are not guaranteed. Some species are geographically local, minor weeds, and development of a biological control for them would be very expensive and not financially wise because of the small-infested area. Release of a biological control organism can induce competitive suppression or extinction of native biological control organisms and other desirable organisms. Biocontrol, particularly in disturbed cropping situations, will not control as many different weeds as other techniques. It won't eradicate weed problems, but most other techniques won't either.

4. Biochemistry and molecular biology of weed control

The need for developing cost effective chemical weed control systems has led to a vast industrial investment to find and develop selective herbicides and later GM-HRC. Virtually all herbicides marketed are the result of random screening of chemicals. Once success is obtained, further syntheses around the identified chemical are used to find compounds with greater activity and then selectivity. After such compounds have been found and marketed, they become research tools of the physiologists and biochemists, first to find a site of action and then as 'anti-metabolites' to further understand and modulate metabolic pathways. Thus the advent of 2,4-D assisted in understanding auxin action, atrazine and diuron (DCMU) in understanding photosystem II, paraquat for photosystem I, dinitroanilines in tubulin to microtubule assembly, dichlobenil for cellulose biosynthesis, etc. Herbicides are the anti-metabolites of choice in dealing with key enzymes such as glutamine synthase [glufosinate (phosphinothricin)], acetolactate synthase (ALS) (many herbicides), acetyl-CoA carboxylase (ACCase) (many herbicides), dihydropteroate synthase (asulam), enolpyruvate-shikimate phosphate synthase (EPSP) (glyphosate) and phytoene desaturase (many herbicides). The genes for most of these enzymes have been isolated and used in transgenic programs. Such research transcended plant biochemistry and agriculture. For example, it was discovered through comparative genomics that plant and trypanosome β-tubulins were similar to each other and different from mammalian β-tubulin. The dinitro-aniline herbicides then proved to be excellent trypanocides (Chan et al., 1993; Bell, 1998). The repetitive (mis) use of single herbicides in monoculture over many years predictably leads to the evolution of herbicide-resistant weeds (Gressel & Segel, 1978). The advent of triazine resistance was crucial to the understanding of the role of the psbA gene product in the photosystem II binding site, leading to innumerable studies of photosynthesis, biophysics and biochemistry correlated with molecular structure of the gene product. The mutant and natural psbA gene products were crystallized and analyzed, leading to new insights into 'drug' (ligand) binding and design (Michel & Deisenhofer, 1988; Deisenhofer & Michel, 1989). Information from herbicide resistance provided the theoretical underpinning for designing transient drought resistant plants. Harvey and Harper (1982) first promoted the idea that paraquat resistance can be similar to oxidative stress tolerance. This was later extrapolated to being similar to transient drought tolerance (Malan et al., 1990). This has allowed developing quick pre-tests with paraquat to ascertain the level of transient drought tolerance of transgenic plants bearing genes designed to confer oxidative stress resistance. Genes coding for herbicide resistance developed for agriculture became the selectable markers of choice for generating transgenics, supplanting antibiotic resistance, even when there was no plan for registering the herbicide for use in that crop. The huge corporate investment in HRC and Bacillus thuringiensis (Bt) toxin containing crops due to perceived market size resulted in the gain of much of our knowledge on promoters, organelle-specific and transit peptides, as well as more recently in organelle transformation. This corporate investment in basic plant molecular biology was manifold greater than the public sector effort, and the spill-over was great. It is important to understand that the transgenic research is market driven and the market is for weed control.

5. Recombinant DNA technology used to achieve herbicide resistance

The techniques used to achieve herbicide tolerance have been reviewed by Cole (1994). Crops which have been transformed to become herbicide tolerant include are shown in Table 1.

In general, the herbicide tolerance gene is expressed as a determinant which is integrated at a single nuclear locus. Tobacco has often been used as a model crop to study and optimise alien gene performance; this reflects the ease of transformation in this species.

Herbicide	Novel gene product	Gene Function	Gene Source	Transformed agricultural crops
Sulfonylureas	Acetolactate	mts	Higher plant	Chicory, cotton, flax,lettuce, lucerne, melon, sugarbeet, tomato
Imidazolinones	Acetolactate synthase	mts	Higher plant	Tobacco
Glyphosate	Enolpyruvylshikimic acid phosphate synthase	mts	Soil and enteric bacterium, higher plant	Rape, soybean, tomato
	Glyphosate oxidoreductase	detox	Soil bacterium	Maize, rapeseed, soybean
Atrazine	"DI" protein	mts	Higher plant	Soybean
Glufosinate	N-acetyl transferase	detox	Bacterium	Cotton, lucerne
Bromoxynil	Nitrilase	detox	Soil bacterium	Cotton, potato, rape, tomato
2,4-D	Mono-oxygenase	detox	Soil bacterium	Cotton

Table 1. Transformation of crop species for herbicide tolerance.

6. Mechanisms for conferring herbicide tolerance in crops

Tolerance to herbicides can be achieved by various mechanisms and genes:

a.   *bar* gene

Members of the genus *Streptomyces* (Actinobacteria: Actinomycetales) produce hundreds of antibiotics, one of which is bialaphos (also known as bilanafos or PTT). Its chemical structure is given below (Fig 1). Bialaphos is an inhibitor of the key enzyme in the nitrogen assimilation pathway, glutamine synthetase (GS). It becomes active after removal of the alanine residues by intracellullar peptidases. The remaining glufosinate compound inhibits GS and as a result leads to accumulation of toxic levels of ammonia in bacteria and plant cells.

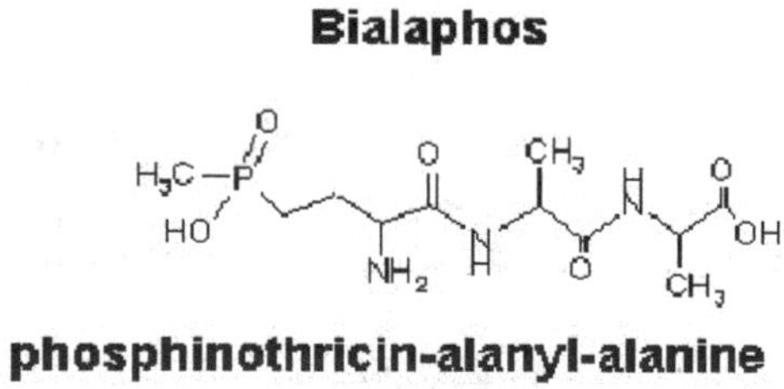

Bialaphos

phosphinothricin-alanyl-alanine

Fig. 1. Structure of bialapos.

Some microorganisms can detoxify glufosinate by producing an enzyme that causes acetylation of the amino group. The gene encoding the acetylating enzyme has been isolated from *Streptomyces hygroscopicus* (Thompson *et al.*, 1987) and from *S. viridochromogenes* (Wohlleben *et al.*, 1988). It has been referred to as *bar* (for bialaphos resistance) and PAT gene, respectively. The *bar* gene encodes a phosphinothricin acetyl transferase (PAT). In the few countries commercial transgenic crops such sugar beet, canola, soybean, rice and maize carrying the *bar* gene has already been released and cultivated commercially.

b.   Detoxifying enzyme coding gene

Continuous search for new herbicides that are highly effective and safe for animals and the environment is the need of the hour. A new class of herbicides that fulfils these needs acts by inhibiting specific amino acid biosynthesis pathways in plants (La Rossa 1984). However, most of these herbicides do not distinguish between weeds and crops. Modifying plants to become resistant to such broad-spectrum herbicides would allow their selective use for crop protection. As a consequence, a major effort has been devoted in several laboratories to engineer herbicide-resistant plants. Two approaches have been followed. In the first, a mutant form of the target enzyme is produced which is still active but less sensitive to the herbicide. In this way, mutant plants producing an altered form of the enzyme acetolactate synthase have been selected which are resistant to the sulfonylurea and imidazolinone herbicides (Shaner and Anderson, 1985). In another example, a mutant form of the bacterial aro A gene was expressed in tobacco and conferred tolerance to the herbicide glyphosate (Comai *et al.*, 1985). The second approach involves overproduction of the target enzyme. It has been demonstrated that overexpression of the plant enzyme 5-enol-pyruvylshikimate- 3 phosphate synthase conferred glyphosate tolerance in transgenic petunia plants (Shah *et al.*, 1986).

Glyphosate was released by Monsanto Chemical Co. in 1971. Its discovery and release were as revolutionary in weed science as the discovery of 2,4-D. The structure of the amino acid glycine is underlined in following Figure.2 Glyphosate, the N-phosphonomethyl derivative of glycine, is a nonselective, foliar herbicide with limited to no soil activity because of rapid and nearly complete adsorption. It controls perennial grasses and has an advantage over paraquat, because glyphosate translocates. It is the only available herbicide that inhibits EPSP synthase. The enzyme is common in the synthetic pathways leading to the aromatic amino acids phenylalanine, tyrosine, and tryptophan. These amino acids are essential in plants as precursors for cell wall formation, defense against pathogens and insects, and production of hormones (Duke, 1990). The enzyme is not found in animals. Low application volume is more effective than high volume, and small plants are more readily controlled than large ones. In contrast, paraquat, a photosynthetic inhibitor, acts quickly (one or two days) on most plants. Glyphosate activity usually cannot be detected as quickly and may take several days to appear after application. One glyphosate formulation is also used as an aquatic herbicide. Transgenic crops resistant to glyphosate have been developed and marketed. Resistant species include Palmer amaranth, common ragweed, hairy fleabane, goosegrass, Italian ryegrass, rigid ryegrass, and buckhorn plantain. Resistance has been found in Australia, Chile, South Africa, Spain, and in 15 US states.

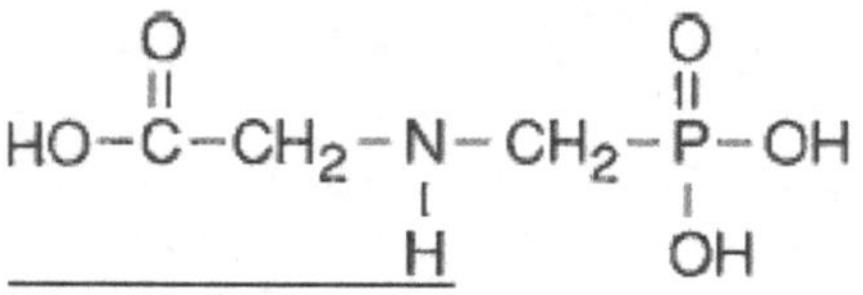

$$HO-\overset{\overset{O}{\|}}{C}-CH_2-\underset{\underset{H}{\|}}{N}-CH_2-\overset{\overset{O}{\|}}{\underset{\underset{OH}{\|}}{P}}-OH$$

Fig. 2. Structure of glyphosate with glycine underlined.

7. Glutamine synthase

Glutamine synthetase (GS) is essential for assimilation of organic nitrogen as ammonia (Duke, 1990). Glufosinate (phosphinothricin) is the only available herbicide that inhibits GS. It is available in the United States for complete weed control in non crop areas and as a directed spray in field- and container-grown nursery stock. It is rapidly degraded in soil with a half-life of seven days. Even though it is not adsorbed tightly, it does not leach because it is degraded quickly. Glufosinate is nearly nonselective. It has been made selective in corn because a gene coding for phosphinothricin acetyl transferase activity was isolated from the soil bacteria, *Streptomyces hygroscopicus*, and cloned into corn. The acetyl transferase enzyme converts glufosinate to its nonphytotoxic acetylated metabolite, enabling crops to achieve resistance by rapidly metabolizing glufosinate.

8. Uses of molecular data in weed control

Not all applications of molecular biology are commercial. There is a necessity to taxonomically classify weeds, as there are differences in selective control among related weeds, as epitomized with weeds of the genus *Amaranthus* (Mayo et al., 1995). Indeed, the classical taxonomy is so complicated that it was found using molecular techniques that many accessions of wild species in the collections were mis-classified (Martin et al., 1997). Molecular taxonomy has been of great assistance and has often provided the decisive data in many cases on whether two similar *Amaranthus* species were actually one, or were separate, or were hybrids, depriving classical taxonomists of their endless battles. Knowing relatedness will be required for assessing the risks of crop gene introgression into weeds, and has been used to trace whether a biotype evolved resistance by introgression, vs. by its own internal mutation (Wetzel et al., 1999b). This is very important to know, as the wild species do not introgress freely with cultivated species and the feral form does exist (Morishima, 1984; Ling-Hwa & Morishina, 1997; Mariam et al., 1996; Majumder et al., 1997; Cohen et al., 1999). Most of the present uses of molecular biology are to find new herbicide targets and to generate HRC.

9. Herbicide tolerance

After using an herbicide continuously to control a weed over time, it may no longer be effective in controlling that weed. In other words, the weed species becomes resistant to the herbicide and is no longer controlled by it. However, the entire population of the weed species may not necessarily behave in the same manner; some members of the same species may still be controlled by the herbicide. There are two basic processes by which herbicide tolerance develops. Most of the time, it develops as a result of selection pressure. In this case, a very small fraction of the population of a particular weed species may possess a slightly different genetic makeup from the rest of the population -- referred to as a biotype -- that makes it tolerate herbicide "X" the first time it is used. Another process by which resistance develops is through mutations. In this case, one or more members of a weed species undergo a change in genetic makeup due to frequent exposure to the herbicide. The modifications usually occur at the site where the herbicide binds at the target site in order for it to be effective.

The mode of action is the mechanism by which a given herbicide travels to a target site within a plant where it exerts activity by inhibiting a growth process vital to the plant. Certain families of herbicides may inhibit the process of photosynthesis. Others may inhibit the synthesis of chlorophyll or amino acids vital to plant's growth, still other groups may cause leaks to plant cells resulting in plant kill. If a weed species develops resistance to two or more herbicides belonging to the same family, the phenomenon is called Cross Resistance. If it develops resistance to two or more herbicides belonging to different families, it is called Multiple Resistance.

There are three physiological mechanisms for natural or induced tolerance or resistance to an herbicide:

1.   Reduced sensitivity at a molecular site of action,
2.   Increased metabolic degradation, and
3.   Avoidance of uptake or sequestration (hiding) after uptake (Duke et al., 1991).

Each of these has potential use in development of resistance in crops. Most of those modified to be resistant to glyphosate and glufosinate are commercially available and grown. Roundup Ready TM soybeans, corn, cotton, and canola have achieved commercial success in the United States and Canada. Other glyphosate-resistant crops are being developed by Monsanto.

The primary cause of herbicide resistance is selection pressure or repeated use of the same herbicide or other herbicides with the same mode of action. Therefore, the most effective step is to use all possible methods of weed control rather than depending upon a single tactic. This helps to avoid the use of the same or similar herbicide repeatedly. An Integrated Pest Management (IPM) method that encompasses cultural, mechanical, chemical, and biological control methods, rotating with different families of herbicides, tank-mixing herbicides having different modes of action, and occasionally using a non selective herbicide to control all weeds are practical methods to reduce resistance buildup. Resistance is real and widely present, but it can be managed. It is well understood that it results from repeated use of the same herbicide or herbicides with the same mode of action in fields. It is not created by the herbicides; it is selected for. The plants that are susceptible are killed. The resistant population survives and comes to dominate. It is a process of evolution by chemical selection. The time for development of resistance has proven to be short. Several species have evolved cross-resistance to more than one herbicide. Since 1982 the number of resistant weeds has more than tripled, and the land area involved has increased 10 times. Multiple resistances have been observed and occur when resistance to several herbicides results from two or more distinct resistance mechanisms occurring in the same species. In general, but not always, there are enough alternative herbicides and other control measures (e.g., rotation, tillage) to manage resistant weeds effectively. Resistance to some of these herbicides has developed in as little as three years. It is equally incorrect to assume that the phenomenon of resistance is the death knell for herbicides. Resistant weeds are not super weeds and are often less fit ecologically than their susceptible relatives. It is important to recognize that resistance is possible and to determine the reasons for it. Management of herbicide resistance will require reducing reliance on herbicides as the primary tool for weed management and developing integrated weed management systems that require the substitution of human intellect and skill for chemical technology (Shaner, 1995).

Most Important herbicide resistant crops are given bellow:

1. Rigid grass   : *Lolium rigidum*
2. Wild oat   : *Avena fatua*
3. Redroot Pigweed   : *Amaranthus retroflexus*
4. Common Lambsquarters   : *Chenopodium album*
5. Green Foxtail   : *Setaria viridis*
6. Bamyardgrass   : *Echinochloa crus-galli*
7. Goosegrass   : *Eleusine indica*
8. Kochia   : *Kochia scoparia*
9. Horseweed   : *Conyza canadensis*
10. Smooth Pigweed   : *Amaranthus hybridus*

10. Herbicide resistance GM crops

Genetically modified crops are the most rapidly adopted technology in agricultural history due to the social and economic benefits these crops may offer. Crops that are genetically altered to be tolerant to herbicide, followed by crops resistant to insects, were the first agricultural biotechnology inventions successfully commercially exploited worldwide. Until the emergence of genetically modified crops, selective herbicides (herbicides that only kill a specific weed) were the answer. The development of selective herbicides is not an easy task and for this reason only a few common weed species could be targeted. Given that each weed requires a different herbicide, herbicide application was frequent, in large volumes and very costly. The advent of herbicide resistant crops caused a major shake-up in the agro-chemical industry. Demand for selective herbicides fell significantly. In certain countries, for the crops that have herbicide resistance, are widely planted and otherwise non-selective (broad spectrum) herbicides are primarily used for weed management. Provided that the field crops are genetically modified to carry gene(s) for herbicide resistance, these broad-spectrum herbicides will not harm the crop. Broad-spectrum herbicides, such as glufosinate and glyphosate, are comparably biodegradable, display low levels of toxicity, and to date, weeds have shown minimal resistance to repeated applications. Resistance to these broad-spectrum herbicides depends upon the genes that have been inserted into the crop plant.

11. Global scenario

The global area planted with transgenic crops is increasing continuously and according to the recent data available (2010); the total global area of transgenic crops is 148 million hectares, a more than 48 fold increase from 1996. More than eight and a half million farmers in 28 odd countries have grown transgenic crops. The majority of the growth occurred in the United States (63%), Argentina (20.5%), Canada (6.5%) and Brazil (4.4%). Almost one third (30%) of the global acreage was grown in developing countries. Total area under transgenic crops in India is around 9.4 million hectare. In 2003, herbicide resistant crops made up 73% of the total genetically modified (GM) crop-growing area, while insect resistant crops constituted 18%. GM crops containing genes for both herbicide resistance and insect resistance comprised 99% of the total GM crop growing area. It is expected that the overall global area of transgenic crops and the number of countries growing transgenic crops will increase in near future. Currently, the agricultural GM market is dominated by a single company, Monsanto. Monsanto produces approximately 90% of genetically engineered crops worldwide. This most likely reflects the ownership by Monsanto of patents on the *bar* gene which confers herbicide resistance as well as patent ownership of various *Bt* toxin genes for insect resistance. Another four companies, Syngenta, Bayer Crop Science, Dow and DuPont produce the remaining 10% of transgenic crops. All major herbicide companies have research programs to incorporate herbicide tolerance through genetic engineering in crops. Success has been achieved with several herbicides. The work has focused on major crops: corn, soybean, wheat, rice, cotton, and tobacco. The technology for agricultural crops was introduced as early as the mid-1980s.

12. Metabolically resistant genetically modified – herbicide resistant crops (GM-HRC)

Many crops bearing transgenes coding for highly specific enzymes that metabolically catabolize herbicides have been generated (Cole & Rodgers, 2000). These include for example, bromoxynil resistance crops bearing a nitrilase (Freyssinet *et al.*, 1996), glufosinate-resistant crops bearing an acetyl-transferase (Vasil, 1996), 2, 4-D resisting crops bearing a highly specific soluble cytochrome P-450 monooxygenase (Streber & Willmitzer, 1989), phenmidipham resisting crops bearing a bacterial gene and dalapon resisting crops bearing a dehalogenase (Buchanon-Wallaston *et al.*, 1992). Of these, only the bromoxynil- and glufosinate resistant crops have reached commercialization. All the herbicide tolerant genes used commercially are of bacterial/ actinomycete origin, despite the fact that plants contain genes for herbicide resistance, which is the basis for most natural metabolic selectivity used for 50 years, yet plant genes conferring metabolic resistance have not been used commercially as yet. There are recent reports using non-prokaryotic genes to confer resistance, but none are yet commercialized, and whether they confer sufficient resistance is not clear. The examples include a rabbit esterase gene conferred resistance to thiazopyr via degradation (Feng *et al.*, 1998), the expression of plant and animal P450 transgenes conferred phenyl urea resistance (Inui *et al.*, 1999; Siminsky *et al.*, 1999). Transgenes encoding maize glutathione transferases increased the level of herbicide resistance (Jepson *et al.*, 1997). The crops generated with metabolic resistance seem to be problem-free, with little metabolic load conferred by generating the small amount of enzyme needed. The toxicology is simplified because the transgene product typically initiates a cascade of events whereby the herbicide eventually disappears. There has been an assumption that one cannot use catabolic enzymes to confer resistance to fast acting herbicides. However, inhibitors of protoporphyrinogen IX oxidase (protox), which actually cause the accumulation of the photodynamically- toxic product induce photodynamic death of plants within 4–6 h in bright sunlight. Beans are immune to some members of this group, e.g. acifluorfen, because they possess a specific homoglutathione transferase and contain enough homoglutathione to stoichiometrically degrade these herbicides before they can damage the crop (Skipsey *et al.*, 1997). Similarly, strains of *Conyza bonariensis* contain a complex of enzymes capable of detoxifying the reactive oxygen species generated by the photosystem I blocker paraquat, and keeping the plants alive until the paraquat is dissipated (Ye *et al.*, 2000; Ye & Gressel, 2000). As almost all herbicides are either degraded in the soil or in some plant species, one should be able to find more genes for catabolic resistance to those herbicides and then be able to rapidly generate herbicide-resistant crops with metabolic resistance than with target site resistance.

13. The success of genetically modified herbicide resistant crops (GM-HRC)

Millions of hectares are being planted with GM-HRC, with insect resistance in second place, with both traits often 'stacked' in the same seeds to enhance their value. The real values of GM-HRC come from instances where there really are no viable weed control methods (e.g. due to evolved herbicide resistances in weeds), and the impact that such GM-HRC could lead to a more sustainable world food production. The easiest way to obtain selectivity among closely related species is to engineer resistance to a general herbicide into the crop. For example, it has already been shown that rice (*Oryza sativa*) is easily controlled by glufosinate. The transgenic rice (Sankula *et al.*, 1997a, b) bearing the *bar* gene confers resistance to this herbicide (Oard *et al.*, 1996). The immediate answer to multiple resistance problems in weeds of wheat in major growing areas is to engineer resistances to inexpensive herbicides (Gressel, 1988). Neither the chemical nor the biotechnological industries have shown particular interest in generating GM-HRC in wheat, rice, millets, pulses or oilseed crops. As too little profit is perceived to come from wheat, rice or other seed or even from generic herbicides, it may be necessary to have wheat, rice, millets and pulses engineered by the public sector. Glufosinate resistance has been engineered into wheat, more as a marker gene than for agronomic utility (Weeks *et al.*, 1993). GM-HR wheat, rice and food legume crops may be an answer to the major problems of these crops. Inserting a gene into wheat or rice conferring resistance to a broad spectrum herbicide can control weeds that evolved resistance in wheat and even closely-related grasses, including red, weedy, and other wild rices (Gressel, 1999a, b, c). The transgenes will allow problems of resistance that have evolved to be overcome especially, the problems of cross-resistances (where one evolutionary step confers resistance to a variety of chemicals) and multiple resistances (where a sequence of evolutionary steps with different selectors, confers resistance to a variety of chemicals). The use of non-plant transgenes may also allow farmers to overcome the natural resistances in weeds closely related to the crop. The problems of interclass cross-resistances and multiple resistances in wheat have necessitated considering the generation of GM-HR oilseed rape, especially in Australia (Gressel, 1999b). Oilseed rape has become an excellent rotational crop alternating with wheat in many places where wheat is grown. There are many agronomic advantages to rotating a dicot with a monocot, especially vis a vis weeds. It should be far easier to eliminate grass weeds in oilseed rape than in wheat, as there are more selective graminicides available for use in dicot crops. There are far too few concrete molecular and biochemical data published about the properties of these crops and thus there are problems in evaluating their properties to allow suggestions for improvements. Thus, some of what will be said below should be considered as speculative. Two types of gene have been used to generate herbicide-resistant crops: (1) where the gene product detoxifies the herbicide; (2) where the herbicide target has been modified such that it no longer binds the herbicide. One could envisage other types such as exclusionary mechanisms, sequestration, etc., but they have yet to be found and thus not utilized. The resulting GM-HRC with each type of resistance is rather different.

14. Herbicide tolerant food legume crops

The food legumes like chickpea, pigeonpea, fieldpea, lentils, urdbean and mungbean are very important for food and nutritional security of poor people in India. These crops suffer to a great extent (33%) due to infestation by weeds. At present no post emergence selective weedicide is available which can be effective to control weeds as these crops are highly sensitive to application of herbicides. Hence, mechanical or manual weeding is considered to be only management options for weed control. In general, food legumes are highly sensitive to available post emergence weedicides. It is, therefore, required to develop resistance/ tolerance in these crops against post emergence weedicides. Development of GM-HRC can be one of the potential options. However, in mid eighty when priorities in area of plant biotechnology were decided in country for developing GM crops, herbicide tolerant was kept out of priority because very cheap agricultural labour were available for these operations in the country. However, with increasing industrialization there is acute shortage of farm labours in the country. Therefore, the need of GM-HRC is realized these days. In view of this, genetic transformation has been successfully attempted in chickpea, pigeonpea and fieldpea with bar gene (used as selectable marker) and stable transformants have been recovered which show considerable degree of resistance to phosphinothricin (Singh et al., 2009). In azuki bean, genetic transformation was done by introducing binary plasmid (pZHBG) comprising the bar gene coding the enzyme, phosphinothricin acetyltransferase which directly inactivates the herbicides phosphinothricin and confers resistance to the commercial herbicides, bialaphos (Confaloneirri et al., 2000). In Cowpea, stable gene transformation was obtained by using particle gun method by Ikea et al., 2003. However, the level of expression of introduced genes in cowpea cells is very low and this accounted for the high mortality rate of progenies under Basta spray. Transgenic plants of the model legume Lotus japonicus were regenerated by hypocotyl transformation using a bar gene as a selectable marker (Lohar et al., 2001). The production of PPT herbicide-resistant L. japonicus plants has shown significant commercial applications in crop production. Brar et al., 1994 developed transgenic plants of peanut of cultivars Florunner and Florigiant, two of the most widely cultivated peanut cultivars in the USA, using the ACCELL® gene delivery method. Gus and bar genes exhibited predictable segregation ratios in the R_1 and R_2 generations and were genetically linked. Integration of the bar gene conferred resistance to BASTA™, a wide-spectrum herbicide, applied at 500 ppm of active ingredient. This work has paved the way to develop herbicide tolerant transgenics in these crops. However, there are far too few concrete molecular and biochemical data published about the properties of these crops and thus there are problems in evaluating them for improvements. With the increasing use of herbicide tolerant crops,there comes an increasing use of glyphosate based herbicide sprays. In some areas glyphosate resistant weeds have developed causing farmers to switch to other herbicides. Some studies also link widespread glyphosate usage to iron deficiencies in some crops, which is both a crop production and a nutritional quality concern, with potential economic and health implications.

15. Risks and concerns of GM HRC

It is generally incurred that the development of herbicide tolerant crop will encourage heavy use of herbicides. Hence, concern has been expressed about water or food contamination from increased herbicide use. Additional concern centres on use of herbicides in crops that do not metabolize the herbicide. Therefore, the unaltered herbicide could be consumed by people. As herbicide resistant crops develop, it is important to remember that no technology is ever proved to be perfectly safe. Scientists look for evidence of harm, and if none is found, conclude that there is none or that it must be looked for in a different way. Second, this technology, like all technologies (e.g., herbicides, cell phones, computers), has both its good and bad uses. We must be cautious about demonizing the potential but unknown bad effects of legitimate uses by good people and weigh them carefully against illegitimate uses by bad people.

Environmental concern: Environmental concern is related to herbicide use. It is suggested that transgenic crops have the potential to create a more sustainable agricultural system than present chemically based systems but will fail "in enabling a fully sustainable agriculture." As genetic traits that have a higher potential of enabling truly sustainable agricultural systems have not been developed due to, the lack of EPA and regulatory policies that specifically promote sustainable traits.

An agricultural biotechnology industry is dominated by agricultural chemical companies.

Patent law and industry policies prevent farmers from saving transgenic seed and thus tailoring transgenic crops to their local ecological conditions.

Social concern. Social concern is related to the following:

1. Fear that the technology will favour large farms and lead to loss of more small farms and small-scale farmers.
2. Cost of food production and food cost to the consumer will rise.

Weed control concerns. There are three major concerns related to weed control:

1. **Development of herbicide resistance:** Herbicide resistance among weeds may become more widespread because of continued use of an herbicide to which a crop is resistant (Sandermann, 2006).
2. **Resistant gene flow to sexually compatible plants:** This is acknowledged as a potential risk of introducing any genetically engineered (transgenic) crop variety. The risk is transfer of desired herbicide resistance from the crop to a weed where undesirable resistance persists by natural selection. It is worth noting that this has happened when genes from herbicide resistant canola moved to a non-weedy relative in the mustard family and then to wild mustard in a short time. The risk may be especially high where the crop and weed are closely related and can interbreed — for example, red rice and rice or Johnson grass and grain sorghum.

Once such gene(s) is transferred within wild populations, it is suggested that a selective advantage could be conferred on the recipients, so altering their biology and influencing their ecological relationship with native genotypes or other species (Lefol *et al*, 1997; Linder, 1998). It is considered that this could constitute a threat to biodiversity. The possibilities for transfer of any trait from crop to weed will depend on the two occurring together, their synchronous flowering, successful pollen transfer and compatibility of the pollen that would allow successful fertilization and embryo development. Any seed produced would then need to germinate and the trait would need to be exhibited in the resulting plant. To maintain the trait, success as a pollen donor, as a seed producer, or both, would be needed. Where a trait carrier is self fertile or where more than one individual has been produced, the F_2 generation may be produced by hybrid mating, though the more likely scenario is for introgression into the recipient species' genome as a result of backcrossing. Crop plants and some weeds are derived from the same ancestors and retain a number of common characteristics. They may also still grow in close association within the geographical area in which both originated and give rise to crop-weed complexes (van Raamsdonk and van der Maesen, 1996) in which introgression of weed characters into the crops and crop characters into the weeds can occur, and may have done so over an extended period of time. Little Seed Canary Grass (*Phalaris minor*) is a monocot weed in the Poaceae family. In India, this weed first evolved resistance to Group C2/7 herbicides in 1991 and infests wheat. Group C2/7 herbicides are known as Ureas and Amides (Inhibition of photosynthesis at photosystem II). Research has shown that these particular biotypes are resistant to isoproturon and they may be cross-resistant to other Group C2/7 herbicides.

3. **Resistant crop plants becoming hard-to-control volunteer weeds:** The quite legitimate concerns of epistasis and pleiotrophy must also be recognized. Another common critique of herbicide resistant crops is that the technology will promote the use of herbicides, not decrease it, while continuing to develop what many view as an unsustainable, intensive monocultural agriculture. It is also suggested that herbicide-resistant crops will reinforce farmers' dependence on outside, petroleum-based, potentially polluting technology. An associated concern is that there is no technical reason to prevent a company from choosing to develop a crop resistant to a profitable herbicide that has undesirable environmental qualities such as persistence, leachability, harm to nontarget species, and so on. It is undoubtedly true that nature's abhorrence of empty niches will mean that other weeds will move into the niches created by removal of weeds by the herbicide used in the newly resistant crop. In other words, herbicide resistance will solve some but not all weed problems. Weeds that are not susceptible to the herbicide to which the crop is resistant will appear. Development of herbicide-resistant crops is proceeding rapidly, and there are important advantages that provide good reasons for their development. Many argue that the technology will provide lower-cost herbicides and better weed control. These are powerful arguments in favour of the technology because both can lead to lower food costs for the consumer. It is also true that herbicide-resistant crops are providing solutions to intractable weed problems in some crops. Glyphosate resistance has been created in several crops. It is an environmentally favourable herbicide, and therefore, it is better to use it in lieu of other herbicides that are not environmentally favourable. An important argument in favour of the technology is that it has the potential to shift herbicide development away from initial screening for activity and selectivity and later determination of environmental acceptability to the latter occurring first. Resistance to herbicides that are environmentally favourable but lack adequate selectivity in any crops or in a major crop so their development will be profitable could be engineered and the herbicide's usefulness could be expanded greatly. This has important implications for minor crops (e.g., vegetables, fruits) where few

herbicides are available because the market is too small to warrant the cost of development. If resistance to an herbicide already successful in a major crop (e.g., cotton) could be engineered into a minor crop, manufacturers and users would benefit. The public doubts about genetic modification of anything are raised, and it is in this context that these doubts must be addressed. Weed scientists and others involved with GMOs often think if we can just educate the public about our science, the problem will be solved as technology is already widely promoted, accepted and used.

16. Future prospects

Plant genetic engineering and biotechnology is now moving from the initial euphoria to the phase of course correction. Several environmental problems related to plant genetic engineering prevent realization of its full potential. One such common concern is the escape of foreign genes through pollen dispersal from transgenic crop plants engineered for herbicide resistance to their weedy relatives creating "superweeds" or causing gene pollution among other crops. Such dispersal of pollen from transgenic plants to surrounding non-transgenic plants has been well documented. The high rate of such gene flow from crops to wild relatives (as high as 38% in sunflower and 50% strawberries) is certainly a serious environmental concern. Clearly, maternal inheritance of foreign genes is highly desirable in such instances where there is no potential for out-cross (Daniell *et al.*, 1998). Since the transgenic crops have been available for some time, we know what has been done with genetically modified herbicide resistant crops. The technology is so new and changing so rapidly that we do not—perhaps cannot—know what might be done. That is, the direction of research is clear, but the final destination is not. We cannot be sure what new possibilities will be discovered as the technology of herbicide resistance continues to develop rapidly. Adoption of molecular-based methods in weed science research will bring a new dimension to the science and can have "far reaching benefits in agriculture and biotechnology" (Marshall, 2001). One potential benefit of genomics research is the discovery of new targets for herbicide action (Hess *et al.*, 2001). Other benefits may include identification and use of genes that contribute to a crop's competitive ability (e.g., early shoot emergence, rapid early growth, fast canopy closure, production of allelochemicals). Genomics may also discover genes that contribute to weediness, a plant's perennial growth habit, seed dormancy, and allelopathy (Weller *et al.*, 2001).

17. References

Annual Report. (2008). National Research Centre on Weeds. Jabalpur. (M.P.).

Bell, A. (1998). Microtubule inhibitors as potential antimalarial agents. *Parasitol. Today*, 14: 234–240.

Buchanan-Wollaston, V., Snape, A. and Cannon, F. (1992). A plant selectable marker gene based on the detoxification of the herbicide dalapon. *Plant Cell Rep.*, 11: 627–631.

Chan, M. M. Y., Grogl, M., Chen, C. C., Bienen, E. J. and Fong, D. (1993). Herbicides to curb human parasitic infections-*in vitro* and *in vivo* effects of trifluralin on the trypanosomatid protozoans. *Proc. Natl. Acad. Sci.*, 90: 5657–5661.

Cohen, M. B., Jackson, M. T., Lu, B. R., Morin, S. R., Mortimer, A. M., Pham, J. L. and Wade, L. J. (1999). Predicting the environmental impact of transgene outcrossing in wild and weedy rices in Asia. In: *Gene flow in agriculture: relevance for transgenic crops.* (pp. 151–157).

Comai, L., Facciotti, D., Niatt, W. R., Thompson, G., Rose, R. E. and Stalker, D. M. (1985). Expression in plants of a mutant *aroA* gene from *Salmonella typhinurium* confers tolerance to glyphosate. *Nature,* 317: 741–744.

Daniell, H., Datta, R., Varma, S., Gray, S. and Lee, S. B. (1998). Containment of herbicide resistance through genetic engineering of the chloroplast genome. *Nat Biotechnol.,* 16: 345–348.

Deisenhofer, J. and Michel, H. (1989). The photosynthetic reaction center from the purple bacterium (*Rhodopseudomonas viridis*). *Science,* 245: 1463–1473.

Duke, S.O. (1990). Overview of herbicide mechanisms of action. *Env. Health Perspectives,* 87:263–271.

Duke, S.O., J. Lydon, J.M. Becerril, T.D. Sherman, L.P. Lettnen, Jr., and H. Matsumoto. (1991).Protoporphyrinogen oxidase-inhibiting herbicides. *Weed Sci.,* 39:465–473.

Feng, P. C. C., Ruff, T. G., Rangwala, S. H. and Rao, S. R. (1998). Engineering plant resistance in thiazopyr herbicide *via* expression of a novel esterase deactivation enzyme. *Pest Biochem Physiol.,* 59: 89–103.

Freyssinet, G., Pelissier, B., Freyssinet, M. and Delon, R. (1996). Crops resistant to oxynils: from the laboratory to the market. *Field Crops Res.,* 45: 125–133.

Gressel, J. and Segel, L. A. (1978). The paucity of genetic adaptive resistance of plants to herbicides: possible biological reasons and implications. *J. Theor. Biol.,* 75: 349–371.

Gressel, J. (1988). Multiple resistances to wheat selective herbicides: New challenges to molecular biology. *Oxford Surveys of Plant Mol. Cell Biol.,* 5: 195–203.

Gressel, J. (1999a). Tandem constructs; preventing the rise of superweeds. *Trends Biotechnol.,* 17: 361–366.

Gressel, J. (1999b). Needed: New paradigms for weed control. In: *Weed Management in the 21st Century: Do we Know Where we are Going?* (pp. 462–486) *Proc. of the 12th Australian Weeds Conference,* Hobart, Tasmania.

Gressel, J. (1999c). Herbicide resistant tropical maize and rice: Needs and biosafety considerations. *Brighton Crop Protection Conference–Weeds* pp. 637–645.

Harvey, B. M. R. and Harper, D. B (1982). Tolerance to bipyridylium herbicides. In: LeBaron H and Gressel J (eds) *Herbicide Resistance in Plants* (pp. 215–234) Wiley, New York.

Hess, F.D., R.J. Anderson, and J.D. Reagan. (2001). High throughput synthesis and screening:The partner of genomics for discovery of new chemicals for agriculutre. *Weed Sci.,* 49:249–256.

Inui, H., Ueyama, Y., Shiota, N., Ohkawa, Y. and Ohkawa, H. (1999). Herbicide metabolism and cross-tolerance in transgenic potato plants expressing human CYP1A1. *Pestic Biochem Physiol.,* 64: 33–46.

Jepson, I., Holt, D. C., Roussel, V., Wright, S. Y. and Greenland, A. J. (1997). Transgenic plant analysis as a tool for the study of maize glutathione transferases. In: Hatzios KK (ed.), *Regulation of Enzymatic Systems Detoxifying Xenobiotics in Plants.* pp. 313–323.

LaRossa, R. A. and Schloss, J. V. (1984). The sulfonylurea herbicide sulfometuron methyl is an extremely potent and selective inhibitor of acetolactate synthase in *Salmonella typhimurium. J. Biol. Chem.,* 25: 8753-8757.

Lefol, E., Séguin-Swartz, G. and Downey, R. K.(1997). Sexual hybridisation in crosses of cultivated *Brassica* species with the crucifers *Erucastrum gallicum* and *Raphanus raphanistrum*: Potential for gene introgression. *Euphytica,* 95: 127-139.

Linder, C. R. (1998). Potential persistence of transgenes: seed performance of transgenic canola and wild x canola hybrids. *Ecological Applications,* 8: 1180-1195.

Ling-Hwa, T. and Morishima, H. (1997). Genetic characterization of weedy rices and the inference on their origins. *Breeding Sci.,* 47: 153-160.

Majumder, N. D., Ram, T. and Sharma, A. C. (1997). Cytological and morphological variation in hybrid swarms and introgressed population of interspecific hybrids (*Oryza rufipogon* Griff _ *Oryza sativa* L) and its impact on evolution of intermediate types. *Euphytica,* 94: 295-302.

Mariam, A. L., Zakri, A. H., Mahani, M. C. and Normah,M. N. (1996). Interspecific hybridization of cultivated rice, *Oryza sativa* L with the wild rice, *O minuta* Presl. *Theor. Appl. Genet.,* 93: 664-671

Martin, C., Juliano, A., Newbury, H. J., Lu, B. R., Jackson, M. T. and Ford Lloyd, B. V. (1997). The use of RAPD markers to facilitate the identification of *Oryza* species within a germplasm collection. *Genet. Resour.& Crop Evol.,* 44: 175-183.

Marshall, G. (2001). A perspective on molecular-based research: Integration and utility in weed science. *Weed Sci.,* 49:273-275.

Malan,C., Greyling, M. M. and Gressel, J. (1990). Correlation between antioxidant enzymes, CuZn SOD and glutathione reductase, and environmental and xenobiotic stress tolerance in maize inbreds. *Plant Sci.,* 69: 157-166.

Mayo, C. M., Horak, M. J., Peterson, D. E. and Boyer, J. E. (1995). Differential control of four *Amaranthus* species by six post emergence herbicides in soybean (*Glycine max*). *Weed Tech.,* 9: 141-147.

Michel, H. and Deisenhofer, J. (1988). Relevance of the photosynthetic reaction center from purple bacteria to the structure of photosystem II. *Biochemistry,* 27: 1-7.

Morishima, H. (1984). Species relationships and the search for ancestors In: Tsunoda S and Takahashi N (eds) *Biology of Rice,* pp. 3-30.

Oard, J. H., Linscombe, S. D., Bravermann, M. P., Jodari, F., Blouin, D. C. and Leech, M. (1996). Development, field evaluation, and agronomic performance of transgenic herbicide resistant rice. *Mol. Breeding,* 2: 359-368.

Sankula, S., Braverman, M. P., Jodari, F., Linscombe, S. D. and Oard, J. H. (1997a). Evaluation of glufosinate on rice (*Oryza sativa*) transformed with the *bar*-gene and red rice (*Oryza sativa*). *Weed Tech.,* 11: 70-75.

Sankula, S., Braverman, M. P. and Linscombe, S. D. (1997b). Response of *bar*-transformed rice (*Oryza sativa*) transformed with the *bar* gene and red rice to glufosinate application timing. *Weed Tech.*,11: 303–307.

Shah, D. M., Horsch, R. B., Klee, H. J., Kishore, G. M., Winter, J. A., Tumer, N. E., Hironaka, C. M., Sanders, P. R., Gasser, C. S., Aykent, S., Siegel, N. R., Rogers, S. G. and Fraley, R. T. (1986).Engineering herbicide tolerance in transgenic plants. *Science,* 233: 478-481.

Shaner, D. L.,Anderson, P. C. and Stidham, M. A. (1984).Potent inhibitors of acetohydroxyacid synthase. *Plant Physiol.*, 76: 545-546.

Siminsky, B., Corbin, F. T., Ward, E. R., Fleishmann, T. J. and Dewey, R.(1999). Expression of a soybean cytochrome P450 monooxygenase cDNA in yeast and tobacco enhances the metabolism of phenylurea herbicides. *Proc. Natl. Acad. Sci.*, 96: 1750–1755.

Skipsey, M., Andrews, C. J., Townson, J. K., Jepson, I. and Edwards, R.(1997). Substrate and thiol specificity of a stress-inducible glutathione transferase from soybean. *FEBS Lett.*, 409: 370–374.

Streber, W. R. and Willmitzer. L. (1989). Transgenic tobacco plants expressing a bacterial detoxifying enzyme are resistant to 2,4-D. *BioTechnology,* 7: 811–816.

Thompson, C. J., Movva, N. R., Tizard, R., Crameri, R., Davies, J. E., Lauwereys, M. and Botterman, J. (1987). Characterization of the herbicide-resistance gene *bar* from *Streptomyces hygroscopicus*. *EMBO J.*, 6: 2519-2523.

Van Raamsdonk, L. W. D. and van der Maesen, L. J. G.(1996). Crop-weed complexes: the complex relationship between crop plants and their wild relatives. *Acta Botanica Neerlandica*, 45: 135-155.

Varshney, J.G. and Mishra , J.S. (2008). Role of herbicides in weed management and food security. *Crop Care* , 34: 67-73.

Vasil, I. K. (1996). Phosphinothricin resistant crops. In: Duke SO (ed.) *Herbicide resistant crops: Agricultural, environmental, economic, regulatory, and technical aspects* pp. 85–92.

Weeks, J. T., Anderson, O. D. and Blechl, A. E. (1993). Rapid production of multiple independent lines of fertile transgenic wheat (*Triticum aestivum*). *Plant Physiol.*, 102: 1077-1084.

Weller, S.C., R.A. Bressan, P.B. Goldsbrough, T.B. Fredenburg, and P.M. Hasegawa. (2001). The effect of genomics on weed management in the 21st century. *Weed Sci.* 49:282–289.

Wetzel, D. K., Horak, M. J., Skinner, D. Z. and Kulakow, P. A. (1999b). Transferal of herbicide resistance from *Amaranthus palmeri* to *Amaranthus rudis*. *Weed Sci.*, 47: 538–543.

Wohlleben, W., Arnold, W., Broer, I., Hillemann, D., Strauch, E. and Pu¨ hler, A. (1988). Nucleotide sequence of the phosphinothricin N-acetyltransferase gene from *Streptomyces viridochromogenes* Tu¨ 494 and its expression in *Nicotiana tabacum*. *Gene,* 70: 25–37.

Ye, B., Faltin, H., Ben-Hayyim, G., Eshdat, Y. and Gressel, J. (2000). Correlation of
    glutathione peroxidase to paraquat/oxidative stress resistance in *Conyza*:
    determined by direct fluorometric assay. *Pestic. Biochem. Physiol.*, 66: 182–
    194.

11

Vegetative Response to Weed Control in Forest Restoration

John-Pascal Berrill and Christa M. Dagley
Humboldt State University
California,
USA

1. Introduction

Longleaf pine (*Pinus palustris* Mill.) stands once occupied an estimated 24 million ha in the southeastern USA (Stout & Marion, 1993). Fire suppression, timber harvest, and land conversion reduced its extent to around one million ha (Outcalt & Sheffield, 1996). In recent times, widespread interest in restoring longleaf pine ecosystems or planting the species for timber production has motivated private landowners, industrial forest owners, and public agencies to establish more longleaf pine forest. Over 33 million longleaf pine seedlings were produced for the 2005-2006 planting season in the southeastern United States (McNabb & Enebak, 2008), and 54 million produced in 2008-2009 (Pohl & Kelly, 2011).

Longleaf pine ecosystems are fire-adapted and support a diverse understory plant community when ground fires are frequent (Peet & Allard, 1993). Longleaf pine seedlings germinate and develop into a grass-like clump, and later transition from this "grass stage" to become woody saplings. Seedlings in the grass stage resist fire, but become vulnerable to fire upon emergence from the grass stage until height growth elevates their terminal bud beyond reach of fire and their bark thickens (Boyer, 1990). Early fire resistance is thought to be an adaptation to frequent fire. During the grass stage, seedlings invest energy in root development in preparation for rapid shoot extension upon emergence. This strategy for re-occupying disturbed sites gives the slower-growing longleaf pine a competitive advantage over less fire-hardy pines and hardwood competitors (Outcalt, 2000). However, in the absence of fire, longleaf pine seedlings are quickly overtopped by competing vegetation. Therefore rapid restoration of longleaf pine forests will necessarily involve some disturbance of competing vegetation. Hardwood regeneration is usually prolific following disturbances such as removal of forest cover. A suite of hardwood species regenerate as stump sprouts and root suckers, developing quickly from established root systems. Grasses and vines also develop quickly after disturbance in the warm humid climate of southeastern USA. Various forms of above- and belowground competition impact on survival and growth of planted longleaf pine seedlings (Harrington et al., 2003; Pecot et al, 2007) and other pine species (e.g., Richardson et al., 1996b; Amishev & Fox, 2006).

Tools available for control of competing vegetation in longleaf pine forest restoration include prescribed fire, mechanical methods, and chemical weed control with herbicides. Prescribed fire most closely mimics the natural disturbance regime in longleaf pine forests, but it may not carry in areas with insufficient quantity or quality of fuels, and it may not be appropriate or acceptable on some ownerships. Mechanical weed control methods include portable saws and machine-mounted mowers or masticators. These methods are more expensive than prescribed fire treatments, but can have similar effects: competing vegetation is disturbed above ground but not always killed; much of it re-sprouts. Herbicides can provide effective and economical control of competing vegetation, but their use may not be appropriate or acceptable in some areas amid concerns over effects on non-target organisms, movement and drift, and persistence in the environment. Fire or broadcast herbicide treatments can eliminate live vegetation cover, exposing soil to erosive forces and temporarily reducing biodiversity. Applying herbicide in spots as opposed to broadcast applications has the advantage of reducing chemical usage while maintaining some continuity of vegetation cover and preserving biodiversity between treated spots (Richardson et al., 1996a).

Research into longleaf pine forest establishment and weed control has focused on the Coastal Plain region of the southeastern USA. Field research on the Coastal Plain indicated that mechanical weed control treatments were inferior to chemical weed control in terms of enhancing longleaf pine seedling survival and growth (Knapp et al., 2006). Chemical weed control with herbicide has proven effective in several longleaf pine restoration studies on the Coastal Plain (Brockway & Outcalt, 2000; Ramsay et al., 2003; Knapp et al., 2006; Haywood, 2007; Freeman & Shibu, 2009; Shibu et al., 2010). Longleaf pine is native to the Coastal Plain, but also occurs naturally in the mountainous regions further inland, and across the Piedmont Region. The Piedmont is a physiographic region extending from the State of New Jersey down to central Alabama, spanning over 200,000 km² of rolling foothills between the Appalachian Mountains and the Coastal Plain (Anon, 2000). Little has been reported on longleaf pine restoration in the Piedmont, but restoration experiments have been established (Berrill & Dagley, 2009).

Data from a replicated field experiment established on degraded Piedmont forest sites are presented here. To our knowledge no other experiment simultaneously addresses questions of repeat herbicide applications versus single treatments each of varying spot sizes, and compares all these weed control treatments to non-herbicide management options. We established non-contiguous single-tree plots in a randomized complete block design with multiple treatment levels nested in a split-plot arrangement within contiguous fixed-area treatment plots. Our objective was to determine the influence of frequency and extent of chemical weed control on planted trees and competing vegetation using commonly-used, widely-available herbicides, and to compare herbicide treatments with mechanical weed control and a no-treatment control. Specifically, we sought to answer the following four questions:

i.   How does planted seedling survival and growth differ between various herbicide treatments and two alternative experimental treatments: mechanical weed control, and zero weed control?

ii.   Is one herbicide treatment sufficient for control of vegetation competing with tree seedlings planted for restoration? Or, will a second 'repeat application' treatment be required?

iii. How large of an area needs to be treated with herbicides around each planted seedling (when making a single herbicide treatment, and/or when making a repeat application)?

More specifically, what is the trade-off between size of treated area (termed 'spot size') and tree seedling response in terms of both survival and growth?

iv. What is the response of weeds to the various treatments? How quickly did each type of weed (grasses, vines, woody vegetation) develop after treatment?

2. Study sites

The restoration experiment was established at four disturbed sites on the 1,900 ha Hitchiti Experimental Forest (N 33⁰ 02' W 83⁰ 42') in Jones County, Georgia, USA. Southern pine beetles (*Dendroctonus frontalis* Zimmerman) had killed patches of even-aged conifer plantation throughout the forest in 2007. The kill areas totalled 10% of the forest area. Salvage harvesting in 2007 was followed by broadcast burning that consumed most of the scattered woody debris and residual hardwoods. Fire failed to carry through some areas due to lack of fuels. Containerized 1-0 'mountain variety' longleaf pine seedlings were hand planted in late March 2008 at a spacing of approximately 3.65 x 3.65 m (740 stems/ha).

Vegetation naturally regenerating throughout the study sites consisted primarily of hardwood stump sprouts and root suckers, vines, forbs, and various grasses. Natural regeneration of 22 tree species was recorded, including an abundance of dogwood (*Cornus florida* L.), loblolly pine (*P. taeda* L.), persimmon (*Diospyros virginiana* L.), sweetgum (*Liquidambar styraciflua* L.), and water oak (*Quercus nigra* L.). Five shrub species, 49 forb species, and eight vine species were recorded. The most common forb was American burnweed (*Erechtites hieracifolia* (L.) Raf.). Throughout the four sites selected as experimental replicates for the restoration study, muscadine grapevines (*Vitis rotundifolia* Michx.) were abundant and expanding laterally to occupy the disturbed sites.

Elevation of the four study sites ranged from 120-150 m above sea level. Soils were classified as a mixture of Davidson and Vance soil series with remnants of loamy surface layer over clay subsoil. The rolling hills were incised by a series of narrow, shallow gullies (Brender 1952). Before the beetle attack in 2007, the four study sites were forested with planted stands of loblolly pine 24-100 years old. Site index ranged from 24.4 m to 27.4 m at base age 25 years for loblolly pine (Clutter & Lenhart, 1968).

Climate at the study site is humid and warm in summer months, and cool in winter. Monthly average low temperatures range from -1°C in January to 19°C in July, and monthly highs range from 13°C in January to 32°C in July. Extreme temperatures were the record high of 40°C in July 1986 and the record low of -20°C in January 1985. The average annual rainfall of 1180 mm is distributed throughout the year; March being the wettest month with 140 mm, and October the driest with 70 mm average monthly rainfall (www.weather.com).

2.1 Experimental design

One experimental replicate block was established in each of four beetle-killed areas at different locations across the forest. Within each replicate block (study site), four treatment plots were established. The 25 x 25 m square treatment plots were surrounded by 4 m wide buffers. Treatments applied to each plot were either mechanical weed control, chemical weed control (repeated in two plots), and control (i.e., no weed control). In a split-plot arrangement, each chemical weed control treatment measurement plot (considered the experimental unit for main treatments) was divided into approximately 12 replicates of three single-tree plots where a single longleaf pine seedling became the experimental unit. Within a split-plot replicate of three adjacent longleaf pine seedlings, each of the three seedlings was randomly assigned a different 'spot size' spray area treatment: a small, medium, or large circular herbicide spot sprayed around the planted seedling.

2.1.1 Weed control treatments

Chemical and mechanical treatments were applied approximately three months after planting, in late June 2008. The objective of the chemical weed control treatment was to reduce above- and belowground competition in the vicinity of longleaf pine seedlings. Glyphosate in the form of isopropylamine salt of N-(phosphonomethyl) glycine was delivered using a backpack sprayer with 2% active ingredient in water at a rate of 6.9 liters active ingredient in 360 liters of solution per ha (D'Anieri et al., 1990). Longleaf pine seedlings were covered with large paper cups prior to spraying. One week after glyphosate application, competing vegetation was mowed close to ground level, and cut stumps of woody species within each randomly-assigned spot treatment area immediately treated with an 8% triclopyr water-based solution of triethylamine salt (5.74% triclopyr acid equivalent). Triclopyr was only used when woody vegetation was present within the treatment spots. Therefore the volume of triclopyr applied differed between small, medium, and large spots, and due to variations in density of woody vegetation within and between study sites. Across all sites, the sum of all spot areas in herbicide plots (0.133 ha) and surrounding buffers (0.060 ha) was 0.193 ha. A total of 0.132 liters of triclopyr active ingredient was applied in these spots, giving an average application rate of 0.69 liters per hectare. These application rates would equate to the volume applied per hectare if the entire area was treated. We applied much less volume to our herbicide treatment plots (total area 0.89 ha at four sites) because it was only applied in spots. The chemical weed control treatment applied in circular spots around each longleaf pine seedling resulted in very different volumes of active ingredient being applied in small, medium, and large spots. We calculated that if, for example, three land managers each prescribed one of the spot size treatments we tested, then the prescription with medium size spots would require approximately four times more active ingredient per hectare than the small spots we tested, and four times less herbicide than if the large spots were prescribed (Table 1). Therefore, even with a second 'repeat' application of herbicide in the same spot size, total chemical usage in small spots sprayed a second time would be half the volume used in a single application in medium size spots, and so forth. Implementing the largest spot size across an area would result in 74% of the ground area being treated if 740 stems/ha were planted (Table 1).

Spot size	Small	Medium	Large
Spot radius (m)	0.455	0.892	1.784
Spot area (m^2)	0.650	2.500	10.000
Treated area (ha)	0.048	0.185	0.740
Glyphosate usage (liters ai/ha)	0.332	1.280	5.110

Table 1. Herbicide spot treatment sizes, and comparison of anticipated chemical usage assuming each spot size treatment was applied to 740 seedlings planted on one hectare i.e., treated area is the combined area of 740 spots, and glyphosate usage is the total volume of active ingredient (ai) needed to implement 740 small, medium, or large herbicide spots.

Mowing in the chemical weed control treatment plots extended beyond the circular spots to cover the entire plot area to uniformly reduce aboveground competition. Vegetation in the mechanical treatment plot was also mowed close to ground level manually using motorized brush saws.

Prior to treatment in late June 2008, the following data were collected in all 25 x 25 m measurement plots: longleaf pine seedling status (live/dead), health and physical condition (brown spot infection, sparseness of live foliage, damaged/covered), and total height (if emerging from grass stage). Within 30 cm of each longleaf seedling (0.3 m² sample area), herbaceous ground cover percent was estimated occularly and maximum height of herbaceous cover was measured. Within approximately 50 cm of each longleaf seedling (1 m² sample area), vine cover percent and woody vegetation cover percent were recorded, and the maximum height of woody vegetation measured. Survival was also assessed at the end of the first growing season, in October 2008. This did not include assessment of competing vegetation due to seasonal discrepancies in cover caused by loss of leaf area among annual plants and deciduous perennials (Fig. 1).

The vegetation assessments were repeated in early June 2009, 11 months after the first assessment and the first set of weed control treatments were applied. All competing vegetation within 1 m² quadrats centred on each longleaf pine seedling was assessed. Immediately after the year-two assessment, chemical weed control was re-applied in one of the two chemical treatment blocks at each study site. This repeat herbicide application treatment was named treatment "H2". No treatments were applied in year two to the other chemical weed control plot at each study site. This 'single application' herbicide treatment was named treatment "H1". The mechanical weed control treatment (named "M") was repeated at each study site in year two, reducing aboveground competition from herbaceous vegetation, vines, and woody perennials in the measurement plot and surrounding buffer. Mowing was also applied in the H2 treatment in year two, completing reduction of above- and belowground competition. No treatments were applied to control plots (named "C"). We returned annually thereafter to monitor the development of planted longleaf pine seedlings and competing vegetation, assessing longleaf pine seedling survival, emergence from the grass stage, height of emerged longleaf pine seedlings, and competing vegetation height and cover percent.

Seedling survival and growth data were subjected to monthly growth adjustment assuming an 8-month growing season from April to November. This procedure gave seasonally-adjusted age estimates for seedlings at each assessment event i.e., data for assessments in the first growing season were assigned age 0.5 years (end of June) and 0.875 years (October), with subsequent assessments at age 1.375 years in June of the second growing season, age 2.25 years in May of the third season, and age 3.5 years in July of the fourth growing season. Seedlings were assigned age 0 years at the time of planting in the winter month of March 2008.

3. Survival of planted seedlings

Survival of longleaf pine seedlings was assessed post-treatment, twice in the first growing season, and annually thereafter. Survival over the year immediately following the first treatment (herbicide and mechanical) was highest following chemical control of competing

Fig. 1. Study sites at Hitchiti Experimental Forest in the first spring after broadcast burning and planting of longleaf pine seedlings (A), close-up of newly-planted seedling (B), middle of first growing season, before treatment (C) and immediately following mechanical treatment (D), herbicide spots at end of first growing season (E), and no-treatment control at end of first growing season. *Photo credit: J-P. Berrill (A-D) & Rex Dagley (E, F).*

vegetation (treatments H1 & H2), intermediate following mowing of competing vegetation (M), and lowest in the no-treatment control (C). The repeat application of herbicide to competing vegetation in the second growing season (H2) enhanced survival, whereas survival declined in the mechanically-treated areas where competing vegetation was rapidly recovering from mowing (Fig. 2).

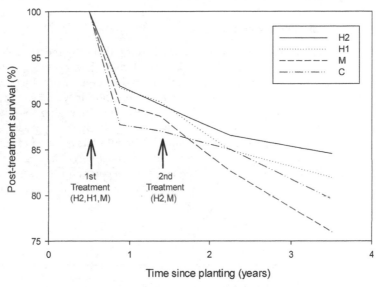

Fig. 2. Survival of planted longleaf pine seedlings from the time of application of the first series of weed control treatments: herbicide applied once in year 1 (H1), herbicide applied twice (H2), mechanical weed control (M), and no-treatment control (C). Sample size: n=627 (H2: n=157, H1: n=145, M: n=159, C: n=166).

4. Growth of planted seedlings

Rapid restoration of longleaf pine forest requires that seedlings emerge from the grass stage and sustain a higher rate of height growth than adjacent competing vegetation (Fig. 3).

Fig. 3. Longleaf pine seedlings in grass stage (left) and emerged from grass stage (right). *Photo credit: David Combs, USDA Forest Service Southern Research Station, Athens, GA.*

4.1 Emergence from grass stage

The number of seedlings emerging from the grass stage was compared between mechanical and chemical weed control treatments, and between different herbicide spot sizes.

4.1.1 Mechanical vs. chemical weed control

Longleaf pine seedlings treated with herbicide were more likely to emerge from the grass stage sooner than seedlings receiving mechanical weed control or no weed control. Over 60% of seedlings receiving a single herbicide treatment had emerged from the grass stage by the fourth growing season. The repeat application of herbicide in year two resulted in a modest enhancement in emergence with 75% of seedlings emerging by the time of assessment midway through the fourth growing season (Fig. 4). By this time, across the four study sites, the number of emerged seedlings in measurement plots equated to 468, 368, 284, and 236 stems/ha in the H2, H1, M, and C treatments, respectively. The highest frequency of emergence among seedlings occurred sometime between consecutive assessments of the experiment in the months of June in the second growing season and May in the third growing season.

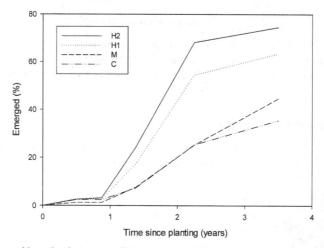

Fig. 4. Proportion of longleaf pine seedlings emerged from grass stage in each weed control treatment: herbicide applied once in year 1 (H1), herbicide applied twice (H2), mechanical weed control (M), and no-treatment control (C). Sample size: n=627 (H2: n=157, H1: n=145, M: n=159, C: n=166).

4.1.2 Herbicide spot size

The number of seedlings emerging from the grass stage in the year after the initial herbicide treatment ranged from 11-23% and was not significantly affected by size of herbicide spot. The repeat application of herbicide appeared to promote a modest 'wave' of emergence from the grass stage, but without any apparent relation to herbicide spot size (Fig. 5).

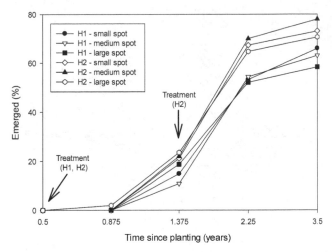

Fig. 5. Proportion of longleaf pine seedlings emerged from grass stage in each herbicide spot size weed control treatment: herbicide applied once in year 1 (H1) and herbicide applied twice (H2), in small (0.65 m²), medium (2.5 m²), and large (10 m²) spots around each planted seedling. Sample size: n=302 (H1-S: n=49, H1-M: n=47, H1-L: n=49, H2-S: n=53, H2-M: n=52, H2-L: n=52).

4.2 Planted seedling height growth

The height development of longleaf pine seedlings that had emerged from the grass stage was compared between mechanical and chemical weed control treatments, and between different herbicide spot sizes.

4.2.1 Mechanical vs. chemical weed control

Height growth of individual seedlings was variable within and between treatments (Fig. 6). Among seedlings that emerged from the grass stage within a year of the first treatments being applied, average height development was most rapid after repeat application of herbicide. Height growth was similar in plots receiving either mechanical treatment or a single herbicide treatment, and slowest in the un-treated control (Fig. 7).

Seedlings emerging at different times caused the average height to rise and fall; the average height of seedlings emerging early increased over time, while later emergence introduced new, shorter seedlings to the calculation of average height. This presented challenges for analysis and testing for differences between treatments. Isolating height data for seedlings that emerged between two consecutive re-measurements somewhat mitigated the problem, and allowed us to test for differences in periodic height increment (rate of growth over a specified period) among seedlings that emerged within the same time period. The periodic average height increment between the third and fourth growing seasons was significantly greater after repeat application of herbicide (78 cm/yr; p = 0.03). Periodic height growth was similar in plots receiving either mechanical treatment (64 cm/yr) or a single herbicide treatment (63 cm/yr), and slowest on average in the un-treated control (48 cm/yr).

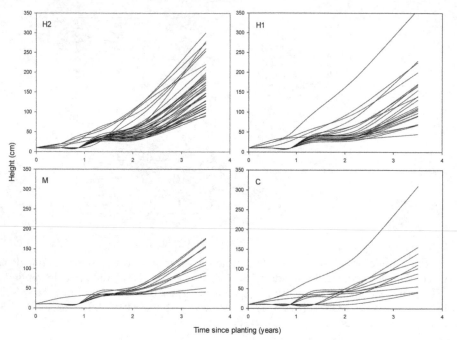

Fig. 6. Height development of individual longleaf pine seedlings that had emerged from grass stage between the time of planting and the middle of the second growing season in each weed control treatment: herbicide applied once in year 1 (H1), herbicide applied twice (H2), mechanical weed control (M), and no-treatment control (C). Sample size: n=83 (H2: n=37, H1: n=24, M: n=11, C: n=11).

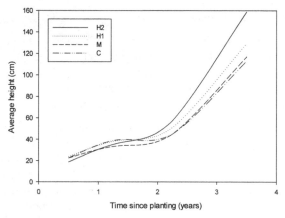

Fig. 7. Average height of longleaf pine seedlings that had emerged from grass stage between the time of planting and the middle of the second growing season in each weed control treatment: herbicide applied once in year 1 (H1), herbicide applied twice (H2), mechanical weed control (M), and no-treatment control (C). Sample size: n=83 (H2: n=37, H1: n=24, M: n=11, C: n=11).

4.2.2 Herbicide spot size and height growth

Average height development of longleaf pine seedlings that emerged within the year following application of herbicide treatments in year one was enhanced by the repeat application of herbicide. Among spot sizes tested, average height was greatest within large spots and lowest in medium-sized spots (Fig. 8). Part of these differences between treatments was likely caused by a random variable that we were not able to control for: variations in timing of emergence from the grass stage and initiation of height growth. This problem was mitigated by examining the rate of longleaf pine seedling height growth between the third and fourth growing seasons. This 'periodic' height increment was greater on average among seedlings receiving a repeat application of herbicide (Fig. 9). However, differences in height growth between the repeat herbicide applications in small, medium, and large spots were not significant (p = 0.43). These repeat treatments resulted in significantly greater seedling height growth than among seedlings treated once with the smallest size of herbicide spot (p = 0.03). The statistical significance of differences between spot size treatments was likely understated because: (i) our sample sizes decreased when we restricted the analysis to seedlings emerging within one year of the first herbicide, and (ii) due to variability in periodic height growth data among young longleaf pines in each treatment.

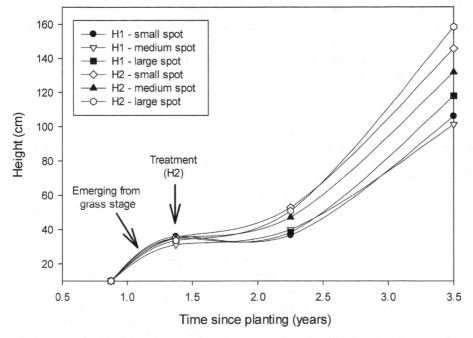

Fig. 8. Average height of longleaf pine seedlings receiving herbicide weed control treatment once (H1) and twice (H2) in small (0.65 m²), medium (2.5 m²), and large (10 m²) spots around each planted seedling. Height data represent average height of seedlings that emerged from grass stage within one year of the first herbicide application. Sample size: n=51 (H1-S: n=7, H1-M: n=4, H1-L: n=8, H2-S: n=11, H2-M: n=10, H2-L: n=11).

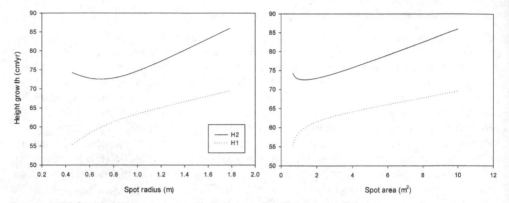

Fig. 9. Relationship between herbicide spot size, number of herbicide applications, and height growth of longleaf pine seedlings emerging from grass stage within one year of the first herbicide application. Height growth calculated as the periodic height increment between the third and fourth growing seasons. Sample size: n=51 (H1-S: n=7, H1-M: n=4, H1-L: n=8, H2-S: n=11, H2-M: n=10, H2-L: n=11).

5. Control of competing vegetation

The extent of competing vegetation cover and its composition were monitored over consecutive growing seasons. Assessment of $1m^2$ quadrats centred on each longleaf pine seedling gave estimates of the percent cover and type of vegetation adjacent to, and presumably competing with, the planted seedlings.

5.1 Weed coverage and composition

Competing vegetation developed quickly in the first growing season. Approximately half of the bare ground around planted seedlings was covered by grasses and forbs, vines, and woody vegetation by the time of the first treatments, three months after planting longleaf pine. The herbicide treatment removed competing vegetation cover in the vicinity of planted seedlings, but only temporarily. Competing vegetation re-occupied herbicide-treated spots at a slower rate than before treatment. Total vegetation cover at the end of the first growing season was only 20% after herbicide treatment, whereas it had attained over 60% cover in the absence of any treatment and following mechanical treatment. In the second growing season, competing vegetation expanded to cover approximately 90% of ground area surrounding planted seedlings in the no-treatment control area and after mechanical treatment. It only covered approximately 50% of ground area in plots receiving a single herbicide treatment by the end of year two, and approximately 25% of ground area in plots receiving a repeat herbicide application in the second growing season. Grasses increased in relative abundance following mechanical treatment. Vine cover increased at the same rate in the control and mechanical treatment areas. Woody vegetation increased in relative abundance, at the expense of grass cover, in the no-treatment control areas. Herbicide treatments had a lasting impact on the development of woody vegetation cover, especially after herbicide was re-applied in the second growing season (Fig. 10).

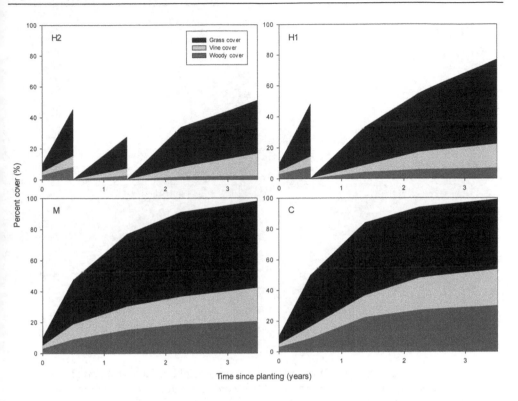

Fig. 10. Weed coverage of ground around planted longleaf pine seedlings. Cover percent is the average cover of each type of competing vegetation in each treatment: herbicide applied once in year 1 (H1), herbicide applied twice (H2), mechanical weed control (M), and no-treatment control (C). Sample size: n=627 (H2: n=157, H1: n=145, M: n=159, C: n=166).

5.2 Weed height development

Calculating the average of height data for the tallest competing vegetation adjacent to each longleaf pine seedling gave an approximation of the 'top height' or 'dominant height' of the vegetation canopy. The dominant height and percent cover of competing vegetation recovered from each treatment at similar rates, with one exception: mechanical treatment appeared to stimulate height growth of competing vegetation (Fig. 11). Calculating dominant height for different components of the competing vegetation gave separate estimates for woody vegetation and for herbaceous vegetation (grasses and forbs). The height of woody vegetation increased steadily, whereas the height of herbaceous vegetation appeared to attain its maximum within two years of treatment. The time taken for vegetation cover or height to return to pre-treatment levels – referred to as 'treatment persistence' – was shorter (rapid recovery; low treatment persistence) for herbaceous vegetation height than for woody vegetation height or total vegetation cover. The repeat application of herbicide doubled herbicide treatment persistence in terms of vegetation cover, and checked hardwood height development by approximately three years (Fig. 11).

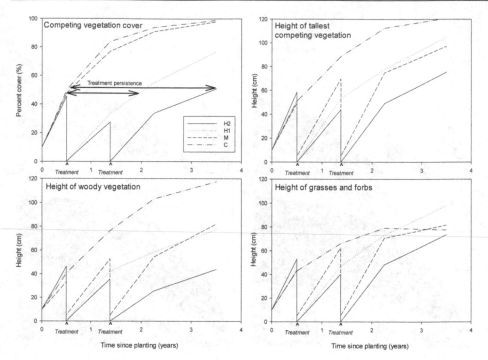

Fig. 11. Development of competing vegetation cover and height in each treatment: herbicide applied once in year 1 (H1), herbicide applied twice (H2), mechanical weed control (M), and no-treatment control (C). Height of competing vegetation represented by average height of the tallest individual competitor (herbaceous or woody vegetation) in 1 m² quadrat centred on each longleaf pine seedling. Sample size: n=627 (H2: n=157, H1: n=145, M: n=159, C: n=166).

6. Comparing growth of crop trees and woody competitors

The most vigorous individuals in any cohort of planted trees are of notable importance in forest restoration. The expectation is that these trees will dominate and form the main forest canopy. Woody vegetation could represent an ongoing threat to successful restoration of longleaf pine because, unlike herbaceous vegetation, it can sustain height growth and compete with the longleaf pines for light and growing space over the longer term. Longleaf pines that outsize their competitors by several meters should be able to maintain long live crowns, remain vigorous, and retain dominance over competing vegetation. We compared height growth of the tallest longleaf pine seedlings, in terms of average height of the tallest 200 stems/ha, with height growth of their major competitor: naturally-regenerating woody vegetation. The repeat application of herbicide in year two was the only treatment that allowed longleaf pine seedlings to gain a substantial height advantage over adjacent woody vegetation by the fourth growing season. The average height of the tallest 200 stems/ha of longleaf pine in the H2 treatment was 115 cm greater than the average height of competing woody vegetation. By the fourth growing season, the tallest 200 longleaf pine seedlings/ha in no-treatment control plots were an average of 45 cm shorter than the average height of competing woody vegetation in the absence of mechanical or herbicide treatment (Fig. 12).

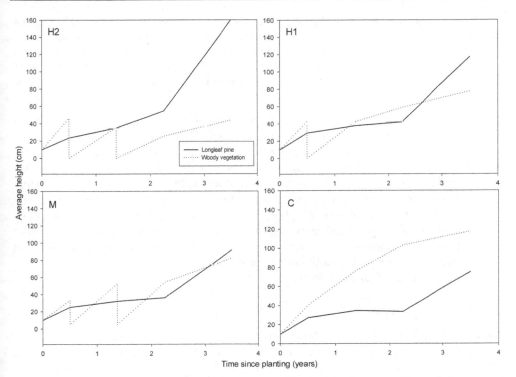

Fig. 12. Height development of the tallest 200 stems/ha longleaf pine seedlings and competing woody vegetation in each treatment: herbicide applied once in year 1 (H1), herbicide applied twice (H2), mechanical weed control (M), and no-treatment control (C). Height of competing vegetation represented by average height of the tallest woody vegetation in 1 m² quadrat centred on each longleaf pine seedling. Sample size: n=200 longleaf pine seedlings (n=50 per treatment, representing 200 stems/ha), and n=454 quadrats containing woody vegetation (H2: n=113, H1: n=105, M: n=120, C: n=116).

7. Conclusion

Mechanical control of competing vegetation provided an early enhancement in survival and emergence of longleaf pine seedlings planted in beetle-killed areas, but the beneficial effects were short lived. Herbaceous vegetation exhibited the most aggressive early response to mechanical treatment. The mechanical treatment also appeared to stimulate height development of woody vegetation, resulting in low treatment persistence. Our data suggest that mechanical treatments may need to be repeated regularly if sufficient numbers of longleaf pine are to overtop the competing vegetation. Repeat application of herbicide provided lasting control of competing vegetation, enhanced survival and emergence from the grass stage, and promoted rapid height growth of longleaf pine seedlings planted on the four sites in the central Georgia Piedmont region.

Seedlings emerging from the grass stage began their height growth at different times, providing challenges for summary and analysis of treatment effects on height growth. The problem was not completely mitigated by examining a subset of data for seedlings that emerged during a single time period between consecutive re-measurements of the experiment; sample size was reduced and differences in timing of emergence still introduced variability in height growth estimates. More frequent re-measurements should overcome this problem by allowing for the study of subsets of seedlings emerging from the grass stage at similar times.

We found no evidence that treating larger areas around planted seedlings with herbicide would promote earlier emergence from the grass stage. Once emerged, the seedlings grew marginally more rapidly, on average, in larger spots. Height growth was significantly more rapid following the repeat application of herbicide in the second growing season than among seedlings receiving only one herbicide treatment in the smallest spot size. Therefore if only one treatment will be applied in future restoration projects, we recommend a larger size of herbicide spot treatment. However, total chemical usage is lower when implementing smaller spots, and more vegetation cover is maintained between the smaller spots. If repeat herbicide treatments are planned, then our results suggest that smaller spot sizes applied twice will provide adequate enhancement of survival, emergence, and growth among planted longleaf pine seedlings.

8. Acknowledgement

We are indebted to the late John Moore – manager of the Hitchiti Experimental Forest - for valuable advice and support for our research project, and for preparing the beetle-killed sites and planting longleaf pine seedlings. We gratefully acknowledge the support of the USDA Forest Service Southern Research Station at Athens, Georgia, and appreciate being involved in this collaboration between the Georgia Forestry Commission, Mac Callaham and his team at the Center for Forest Disturbance Science at the Southern Research Station, and Humboldt State University. David Combs of the Southern Research Station provided technical support and has spent considerable time in the field since the project was initiated in 2008. Rex Dagley has tirelessly volunteered field assistance since 2008. We also thank the Humboldt State University Sponsored Programs Foundation for supporting dissemination of original research results by covering page charges for this open-access publication.

9. References

Amishev, D. Y. & Fox, T. R. (2006). The effect of weed control and fertilization on survival and growth of four pine species in the Virginia Piedmont. *Forest Ecology and Management* 236: 93-101.

Anon. (2000). *Piedmont.* The Columbia Gazetteer of North America. University of Columbia Press. New York, NY.

Berrill, J-P.; Dagley, C.M. (2010). Assessing longleaf pine (*Pinus palustris*) restoration after southern pine beetle kill using a compact experimental design. *Scandinavian Journal of Forest Research* 25(8): 75-85.

Boyer, W. D. (1990). Longleaf pine *Pinus palustris* Mill. In R. M. Burns, & B. H. Honkala (Eds.), *Silvics of North America. Agriculture Handbook 654, Vol. 1: Conifers* (pp. 405-412). USDA Forest Service. Washington, D.C.

Brender, E. V. (1952). *A guide to the Hitchiti Forest Research Center*. (USDA Forest Service Station Paper No.19). USDA Forest Service. Southeastern Forest Experiment Station, Asheville, North Carolina.

Brockway, D. G. & Outcalt, K. W. (2000). Restoring longleaf pine wiregrass ecosystems: Hexazinone application enhances effects of prescribed fire. *Forest Ecology and Management* 137: 121-138.

Clutter, J. L. & Lenhart, J. D. (1968). *Site index curves for old-field loblolly pine plantations in the Georgia Piedmont* (Rep. No. 22, Series No. 1). Georgia Forest Research Council. Macon, Georgia.

D'Anieri, P., Zedaker, S. M., Seiler, J. R. & Kreh, R. E. (1990). Glyphosate translocation and efficacy relationships in red maple, sweetgum, and loblolly pine seedlings. *Forest Science* 36(2): 438-447.

Freeman, J.E., & Shibu, J. (2009). The role of herbicide in savanna restoration: effects of shrub reduction treatments on the understory and overstory of a longleaf pine flatwoods. *Forest Ecology and Management* 257: 978-986.

Harrington, T. B., Dagley, C.M. & Edwards, M. B. (2003). Above- and belowground competition from longleaf pine plantations limits performance of reintroduced herbaceous species. *Forest Science* 49(5): 681-695.

Haywood, J. D. (2007). Influence of herbicides and felling, fertilization, and prescribed fire on longleaf pine establishment and growth through six growing seasons. *New Forests* 33: 257-279.

Knapp, B. O., Wang, G. G., Walker, J. L. & Cohen, S. (2006). Effects of site preparation treatments on early growth and survival of planted longleaf pine (*Pinus palustris* Mill.) seedlings in North Carolina. *Forest Ecology and Management* 226: 122-128.

Littell, R. C., Milliken, G. A., Stroup, W. W., Wolfinger, R. D. & Schabenberger, O. (2006). *SAS for Mixed Models* (2nd ed.). SAS Institute Inc. Cary, North Carolina.

McNabb, K. & Enebak, S. (2008). Forest tree seedling production in the southern United States: the 2005-2006 planting season. *Tree Planters' Notes* 53(1): 47-56.

Outcalt, K. W. (2000). The longleaf pine ecosystem of the South. *Native Plants Journal* 1(1): 43-51.

Outcalt, K. W. & Sheffield, R. M. (1996). *The longleaf pine forest: trends and current conditions*. (USDA Forest Service Station Resource Bulletin SRS-9). USDA Forest Service. Southern Research Station, Asheville, North Carolina.

Pecot, S.D., Mitchell, R.J., Palik, B.J. Moser, E.B. & Heirs, J.K. (2007). Competitive responses of seedlings and understory plants in longleaf pine woodlands: separating canopy influences above and below ground. *Canadian Journal of Forest Research* 37: 634-648.

Peet, R. K. & Allard, D. J. (1993). Longleaf pine vegetation of the southern Atlantic and eastern Gulf Coast regions: a preliminary classification. In S. M. Herman (Ed.), *Proceedings of the Tall Timbers Fire Ecology Conference, No. 18, The longleaf pine ecosystem: ecology, restoration, and management* (pp. 45-81). Tall Timbers Research Station. Tallahassee, Florida.

Pohl, R., Kelly, S. 2011. Growing trees in Georgia. *Tree Planters' Notes* 54(1): 4-9.

Ramsay, C. L., Jose, S., Brecke, B. J. & Merritt, S. (2003). Growth response of longleaf pine (*Pinus palustris* Mill.) seedlings to fertilization and herbaceous weed control in an old field in southern USA. *Forest Ecology and Management* 172: 281-289.

Richardson, B., Davenhill, N., Coker, G., Ray, J., Vanner, A. & Kimberley, M. (1996a). Optimising spot weed control: first approximation of the most cost-effective spot size. *New Zealand Journal of Forestry Science* 26(1/2): 265-275.

Richardson, B., Vanner, A., Ray, J., Davenhill, N. & Coker, G. (1996b). Mechanisms of *Pinus radiata* growth suppression by some common forest weed species. *New Zealand Journal of Forestry Science* 26(3): 421-437.

SAS Institute Inc. (2004). *SAS/STAT 9.1 user's guide, Vols. 1-7* (1st ed.). SAS Institute Inc. Cary, North Carolina.

Shibu, J., Ranasinghe, S. & Ramsey, C.L. (2010). Longleaf pine (*Pinus palustris* P. Mill.) restoration using herbicides: overstory and understory vegetation responses on a coastal plain flatwoods site in Florida, U.S.A. *Restoration Ecology* 18(2): 244-251.

Stout, I. J. & Marion, W. R. (1993). Pine flatwoods and xeric pine forest of the southern (lower) coastal plain. In W. H. Martin, S. G. Boyce & A. C. Echternacht (Eds.), *Biodiversity of the southeastern United States, lowland terrestrial communities* (pp. 373-446). John Wiley and Sons. New York, NY.

Thatcher, R. C., Searcy, J. L., Coster, J. E. & Hertel, G. D. (Eds.). (1980). *The Southern Pine Beetle*. (USDA Technical Bulletin 1631). Expanded Southern Pine Beetle Research and Application Program, Science and Education Administration, USDA Forest Service. U.S. Government Printing Office. Pineville, Louisiana.

12

Sugar Beet Weeds in Tadla Region (Morocco): Species Encountered, Interference and Chemical Control

Y. Baye[1], A. Taleb[2] and M. Bouhache[3]
[1]*Centre Régional de la Recherche Agronomique de Tadla, Beni Mellal,*
[2,3]*Institut Agronomique et Vétérinaire Hassan ll, Rabat,*
Morocco

1. Introduction

Sugar beet occupies each year about 65.000 hectares in Morocco which allows a production that approaches or exceeds three million tons of roots, with an average yield of 46 tonnes per ha (54% of national needs sugar consumption). Since its introduction in Morocco in 1962-1963, sugar beet yield increased significantly in quantity and quality. In Morocco, the sugar beet is a very important crop because of its products and by-products, mainly:

- Production of sugar for sugar consuming population.
- Producing leaves, beet tops and pulp wet and dry food that are essential for cattle sheep that is either intended for milk production or to that of meat. It is important to note that major investments such as installation of various agro-industrial units were made.

In Morocco, sugar beet is planted from September through June - July. Yield obtained by farmers, averaging 46T/ha, is significantly below the request potential that would be 90 to 100 T/ha. Many factors contribute to low sugar beet production. Poor stand establishment, inadequate weed control, inadequate insect control and inadequate nitrogen fertilization are the main causes of low tonnage and poor quality sugar beet in Morocco.

The sugar beet is an important strategic crop in the irrigated perimeter of Tadla. During these 5 last years, an annual surface of 12000 ha is emblaved by this crop representing 23% / of the national area. The average yield obtained in the region is approximately 45 to 50 T/ha, which is very low compared to the potential yield.

Sanitary problems particularly weed management is a great constraint to sugar beet production and weeds may cause high yield losses (Rzozi et al., 1990). This paper presents the main results of investigations and experiments conducted in Tadla region to improve the weed management program by identifying mains weed species encountered in sugar beet field, studying the effect of weeds on sugar beet growth and estimating yield losses and determining the critical period of weed control and evaluating herbicide treatments.

2. Sugar beet weeds

2.1 Introduction

The sugar beet is an important strategic crop in the irrigated perimeter of Tadla. During these 5 last years, an annual surface of 12000 ha is emblaved by this crop. The average yield obtained in the region is approximately 45 to 50 T/ha, which is very low compared to the potential yield which would be of 100 T/ha. Several constraints of technical order are at the origin of this low production, among which the weak control of sanitary problems particularly weed management. In order to achieve a good control of weeds, these last must be well identified. Tanji and Boulet (1986) drew up a general floristic and biological inventory of these weeds in Tadla area (All crops included). The objective of this work was to study thoroughly this inventory in sugar beet.

2.2 Material and methods

2.2.1 Presentation of the study area

The plain of Tadla is located at the foot of the Middle Atlas Mountain (Center of Morocco) (Figure 1). This plain has an area of about 360,000 hectares. The altitude varies between 250 m and 500 m and an average of 400 m. According to Emberger climagram , the plain of Tadla has an arid climate with mild winter for the area north of the Oued Oum Er Rabia; winter to charge for the south as well as some of Beni Amir.

In general, natural vegetation is limited to the most degraded soils, the shallower and less suitable for agriculture are sheltered pastures. The average rainfall varies between 556 mm in Beni Mellal as maximum and 327 mm in Dar Ould Zidouh and is averaging 346.6 mm.These datas are decreasing because of climate change. Average monthly temperatures range from 10.2 ° C in January to 28 ° C in August. Minimum monthly temperatures range from 3.23 ° C in January and 18.5 ° C in August and the average maximum temperatures range from 17.8° C in January and 37.5°Cin August.

Fig. 1. Localisation of the studied region in Morocco map (12).

2.2.2 Prospecting and sampling

A total of 126 sugar beet fields were explored. Only fields not chemically treated and weedy full kept by farmers were prospected. A stratified sampling according to Gounot (1969) was established taking account of some factors mainly type of soil, rainfall and temperatures. Meanwhile, farmers were questioned about cultural practices and soil samples were taken in order to characterize soil texture and total calcium content.

The method of the "tower field" has been adopted to identify the weed species present (Maillet, 1981), for which an Abundance-Dominance Index (ADI) (+, 1, 2, 3, 4, 5) according to the scale of Montegut (Not dated) modified by Boulet et al. (1989) has been assigned. This index is as follows:

+: Very rare species (1 to 5 feet), virtually no recovery.
1: scarce species, recovery very low, irregular distribution
2: averagely abundant species, low recovery, irregular distribution
3: abundant species, covering less than 50%, regular distribution
4: abundant species, recovery of 50 to 75%, regular distribution
5: very abundant species, recovery from 75 to 100%, regular distribution

The agronomic importance of each species is judged based on its relative frequency and covering. The estimation of the average abundance of species during the reading was conducted assuming equivalences between the ADI and its average covering in percentage (Boulet et al, 1984). The methodology was as follows:

ADI	Covering	Average covering
5	75 – 100	87,5
4	50 – 75	65,5
3	25 - 50	37, 5
2	5 – 25	17, 5
1	1 – 5	5
+	< 1	1

These values allow calculating the average covering R of each species at reading time. The combination of this index, of the absolute frequency of species and their ethological type, allowed attribution of a "Partial Nuisibility Index" (PNI) to species (Bouhache et al, 1984).

PNI = (Sum of coverage/number of reading) x 100. The perennial species are underlined and only species with a frequency higher than 20% are taken in consideration.

Species encountered were identified by using some documents such as Flora Europea (Tutin et al., 1964- 1984), Catalogue des Plantes du Maroc (Jahandiez and Maire, 1931-34) and Mauvaises herbes des regions arides et semi arides du Maroc occidental (Tanji et al., 1988). The ethological type for each species was determined according to classification elaborated by Raunkiaer (1905). The biogeographical origin of weed species was derived from Quezel and Santa (1962-63) and Negre (1961-62) on flora investigations.

2.3 Results and discussion

2.3.1 Systematic aspect

A total of 144 weed species including volunteer wheat belonging to 30 botanical families (Table 1) were inventoried in the 162 sugar beet fields prospected. This number correspond respectively to 43,6% and 17,2% of the total weed flora of Tadla region (Tanji and Boulet, 1986) and of Central West Morocco (Boulet et al., 1989) and is relatively low compared to that observed in the Gharb region (Tanji et al., 1984), more important than that showed in Doukkala region (Bouhache and Ezzahiri, 1993) and similar to that found in Moulouya region (Taleb and Rzozi, 1993).

Dicotyledonous species are prevalent (118 species) and correspond to 81,9% of total encountered. Similar results are shown in other regions where sugar beet is grown. Six families dominated particularly the weed flora (Table 1): Asteraceae, poaceae fabaceae, brassicaceae apiaceae and caryophyllaceae. They provide 51.8% of the total. Representing 81 species , these six families are also dominant in sugar beet in Gharb region (Tanji et al.,1984), in cereals (Taleb and Maillet, 1994) and generally for the national flora (Bouhache and bouleT, 1984 and Ibn Tatou and Fennane, 1989). The most dominant family is the asteraceae, that is represented by 19 species, representing 13.2% of the weed flora found. The Asteraceae is also the richest family in species by about 20.000 species worldwide (Taleb, 1995)

2.3.2 Ethological aspect

According to RAUNKIAER classification, the 144 species surveyed belong to five ethological types (Table 2). The ethological spectrum is dominated by annuals (therophytes) with 119 species or 82,6% of the total. This data is similar to that obtained by the main regional botanical and floristic studies of sugar beet weed flora (Tanji et al., 1984; Bouhache and Ezzahiri, 1993; Taleb and Rzozi, 1993. The Geophytes follow with 17 species (11.8%), bisannuals (hemicryptophytes) with 5 species (3.5%) and the chamaephytes and others with 2 species (2.1%). The geophytes encountered are mainly monocotyledonous species with rhizomes, bulbs and tubers. The most important geophytes species inventoried are Convolvulus arvensis L., Solanum elaeagnifolium Cav. And Cynodon dactylon (L.) Pers. They cause serious problems to the crop.

2.3.3 Biogeographical distribution of species

The Mediterranean weed species (broadly defined) dominate the flora inventoried with 56.2%. This high rate of Mediterranean species confirms those of other authors (Bouhache and Boulet, 1984; Loudyi, 1985; Tanji and Boulet, 1986; Careme, 1990; Taleb, 1995; Wahbi, 1994; Bensellam, 1994) or for the entire Moroccan flora (about 2 / 3 according to Braun-Blanquet and Maire, 1924). European and eurasiatic species represent 5,5 and 4,9 % of the total. Cosmopolitan and sub- cosmopolitan are well represented (8,3%). This seems to be high comparatively to that reported by Bouhache and al., 1993. Concerning endemic species to north west of Africa, they are represented only by Diplotaxis tenuissiliqua Del., also reported by Tanji and Boulet (1986).

Famillies	Number of species	Contribution (%)	Ranking
Asteraceae	19	13,2	1
Poaceae	19	13,2	1
Fabaceae	18	12,5	3
Brassicaceae	9	6,2	4
Apiaceae	8	5,5	5
Caryophyllaceae	8	5,5	5
Amaranthaceae	6	4,2	7
Chenopodiaceae	5	3,5	8
Euphorbiaceae	4	2,8	9
Liliaceae	4	2,8	9
Papaveraceae	4	2,8	9
Plantaginaceae	4	2,8	9
Polygonaceae	4	2,8	9
Rubiaceae	4	2,8	9
Convolvulaceae	3	2,1	15
Malvaceae	3	2,1	15
Solanaceae	3	2,1	15
Lamiaceae	3	2,1	15
Boraginaceae	2	1,4	19
Geraniaceae	2	1,4	19
Ranunculaceae	2	1,4	19
Scrophulariaceae	2	1,4	19
Araceae	1	0,7	30
Cyperaceae	1	0,7	30
Iridaceae	1	0,7	30
Portulacaceaa	1	0,7	30
Primulaceae	1	0,7	30
Rhamnaceae	1	0,7	30
Urticaceae	1	0,7	30
Verbenaceae	1	0,7	30

Table 1. Specific contribution of botanical families encountered.

Biological type	%
Therophytes (Annuals)	82.6
Geophytes (Perennials)	11.9
Hemicryptophytes (Bisannuals)	3.4
Chamaephytes and nanophanerophytes	2.1

Table 2. Ethological aspect of sugar beet weed flora in Tadla.

2.3.4 Agronomic aspect

The number of weed species per Sugar beet field varied from 9 to 26 and averaged 17,5. It is relatively low compared to that reported at Doukkala region. The weed survey allowed

Species	PNI
Group 1: species with IPN>1000	
Lolium rigidum Gaudin.	1919
Phalaris brachystachys Link.	1530
Triticum aestivum L.	1209
Triticum durum L.	1112
Avena sterilis L.	1059
Convolvulus arvensis L.	1024
Group 2: species with 500<IPN<1000	
Lolium multiflorum Lam.	910
Cichorium endivia L.	787
Anagallis foemina Miller	768
Papaver rhoeas L.	700
Ridolfia segetum L.	672
Medicago polymorpha L.	651
Melilotus sulcata Desf.	642
Phalaris minor Retz.	640
Galium tricornitum Dandy	638
Chenopodium murale L.	635
Chenopodium album L.	590
Sonchus oleraceus L.	572
Lamium ampexicaule L.	570
Sinapis arvensis L.	528
Solanum elaeagnifolium Cav.	521
Malva parviflora L.	501
Fumaria parviflora Lam.	501
Group 3: species with 250<IPN<500	
Emex spinosa (L.) Campd.	401
Rumex pulcher L.	381
Chrysanthemum coronarium L.	325
Bromus rigidus L.	315
Calendula Arvensis L.	301
Vicia sativa L.	270
Chrysanthemum segetum L.	250
Group 4: species with IPN<250	
Polygonum aviculare L.	237
Phalaris paradoxa L.	220
Antirrhinum orontium L.	201
Reseda alba L.	187
Plantago afra L.	150
Scorpiurus vermiculatus L.	132
Vaccaria hispanica Med.	120
Lathyrus ochrus (L.) DG.	104
Cynodon dactylon (L.) Pers.	92

Table 3. Partial Nuisibility Index (PNI) of the most frequent weed species in sugar beet.

identifying 39 major weed species including volunteer wheat that are relatively frequent and cause serious problems and yield loss for the crop (table 3). These species were divided into four groups on the basis of their PNI.

Weeds belonging to group 1 are mainly monocotyledonous species such as *Lolium rigidum* Gaudin., *Phalaris brachystachys* Link., *Avena sterilis* L. And volunteer wheat (*Triticum aestivum* L. And *Triticum durum* L.). This later generally precede sugar beet in the plot. These species competes highly with sugar beet because of their relatively high covering and early emergence in the season. The perennial rhizomatous weed *Convolvulus arvensis* L. is also a dangerous species and it is very difficult to control because of its important vegetative multiplication.

Group 2 contain many species with PNI between 500 and 1000 that also could be noxious for the crop regarding their covering. These weeds are mainly dicotyledonous species such as *Anagallis foemina* Miller, *Papaver rhoeas* L., *Medicago polymorpha* L., *Chenopodium album* L., *Sinapis arvensis* L., *Galium tricornitum* Dandy. *Solanum elaeagnifolium* Cav. is deep rooted weed and a very troublesome species in all Tadla region

Other species with relatively low covering (Groupe 3 and 4) are often encountered in sugar beet field but they are less competitive compared to those belonging to group 1 and 2: *Rumex pulcher* L., *Chrysanthemum coronarium* L., *Bromus rigidus* L., *Calendula Arvensis* L., *Vicia sativa* L., *Chrysanthemum segetum* L., *Reseda alba* L., *Plantago afra* L., *Scorpiurus vermiculatus* L., *Vaccaria hispanica* Med.

2.4 Conclusion

The sugar beet weed flora in Tadla region is much diversified. Effectively, 144 species belonging to 30 botanical families were encountered in the 126 field prospected. The most represented families are asteraceae, poaceae, fabaceae, bracassicaceae, apiaceae and caryophyllaceae. Therophytes (annuals) and dicotyledonous species dominate with 82,6% and 81,9 respectively. The floristic diversity vary from 9 to 26 species per field and it average 17, 5. The weed survey allowed identifying 39 major weed species including volunteer wheat that are relatively frequent and cause serious problems and significant yield losses for the crop.

3. Weed interference and critical period

3.1 Introduction

Weeds compete with crop plants for water, light nutrients and space and cause considerable yield losses. Integrate weed management (IWM) involves a combination of cultural, mechanical, biological, genetic and chemical methods for effective and economical weed control (Swanton and Weise, 1991). The principles of IWM should provide the foundation for developing optimum weed control systems and efficient use of herbicides. The critical period for weed control (CPWC) is a key component of an IWM program. Weeds are limiting factors in sugar beet production (Cooke and Scott, 1993). Integrated weed control management is necessary for minimizing weeds interference and maximizing the crop yield (Schweizer, 1983; Cooke and Scott, 1993).

The critical period of weed interference refers to the period during which a crop must be kept free of weeds in order to prevent yield loss. It represents the time interval falling between two separate components: (a) the minimum length of time after seeding that a crop must be kept weed-free so that later-emerging weeds do not reduce yield, and (b) the maximum length of time that weeds which emerge with the crop can remain before they become large enough to compete for growth resources (Radosevich and Holt, 1984; Zimdahl, 1988; Weaver *et al.*, 1992; Baziramakenga and Leroux, 1994; Ghadiri, 1996).

Sugar beet can tolerate weeds until 2-8 weeks after emergence, depending on the weed species, planting date, the time of weed emergence relative to crop and environmental conditions (Cooke and Scott, 1993). The presence of weeds can decrease sugar beet yield by 90%. For example, a single presence of barnyardgrass *Echinochloa crus-galli* (L.) *Beauv.* plant per 1.5 m2 resulted in yield reduction of 5 to 15 % (Norris, 1996). The earliest date at which weeding could cease in sugar beet without significant yield loss has been shown to be between 4 and 12 weeks, depending on sowing date, rainfall and weed infestation (Link and Koch, 1984; Scott et aI., 1979; Singh et aI., 1996). Studies on the competitive effect of weeds in sugar beet have been numerous under temperate climates (Dawson, 1965; Farahbakhsh and Murphy, 1986; Schweizer and Dexter, 1987; Scott et aI., 1979; Zimdahl and Fertig, 1967). Continuous post-planting hand-weeding for 17 weeks and 15 weeks in 1990, and for 15 weeks and 12.5 weeks in 1991 were required to limit sugar beet root yield loss to 5% and 10%, respectively In Gharb region (Alaoui et al., 2003). Based on 10% loss of yield, the beginning of the critical period of weed control (CPWC) was 25 and 5 days after planting for the first year and the second year, respectively. On this basis, the end of the critical period of weed control was 78 days for the first year and 88 after planting for the second year (Salehi et al., 2006).

This research was conducted to study (i) the effect of weed competition on sugar beet growth parameters and (ii) determine the minimum period sugar beet should be kept weed-free after planting (CPWC) in the Tadla region to limit yield loss from late emerging weeds

3.2 Material and methods

3.2.1 Experimental site localization and characterization

Field experiment was conducted during two growth seasons 2003- 2004 and 2004-2005 at Afourer experimental station of the National Institute of Agricultural Research in Tadla region. The soil characteristic are as follows: 2.72 % organic matter, 11% sand, 37.2% silt, 51.6 % clay, and pH 8.1. Plots were plowed, disked three times and harrowed for seedbed preparation. Sugar beet cv. 'lydia', a mono germ variety, was seeded manually in a 2 cm deep in 70-cm wide rows with a spacing of 10 cm between seeds (population of 83,000 plants/ha) on October 15 in 2003 and November 25 in 2004.

Fertilization, irrigation and diseases and predators control were achieved in experimental plots according to those recommended by the sugar regional comity.

3.2.2 Competition duration

To determine the critical period of weed control in sugar beet, an experiment was conducted and consisting of 16 treatments. Weed free treatments included the removal of weeds at 4, 7,

9, 11, 13, 17 and 21 weeks after emergence (WAE) of sugar beet. In weed infested treatments, weeds were allowed to interfere with sugar beet crop 4, 7, 9, 11, 13, 17 and 21 weeks after emergence sugar beet crop. Two control treatments (full-season control of weeds and full-season interference of weeds) were also included. Individual plots consisted of 10 rows, each 10 m long.

3.2.3 Experimental design and statistical analysis

The experiment was a randomized complete block design with four replicates. Data on weeds and on sugar beet growth parameters and yield components were subjected to an analysis of variance using statistical STATITCF software. The means were compared using Fisher's protected LSD ($\alpha = 0.05$).

3.2.4 Measurements

Weed density is not as reliable as biomass to assess weed interference in a crop (Scott et al., 1979; Tomer et al., 1991; Wilson and Peters, 1982), especially for species which have a high capacity to compensate for low densities through tillering and branching. Therefore, the impact of weed-free and weedy duration on weed growth and on crop growth and crop yield was assessed through weed dry weight. Weed dry weight were measured during the entire growing season for all individual plots. Four 0.5 m x 0.5m quadrates per plot were placed randomly over the plot. Weeds within the sampling area were removed by hand, taken to laboratory and dried at 60 C for 48 h to determine total weed dry weight. Sugar beet growth was assessed at the same time as weed sampling. Six sugar beet plants without root were taken randomly in plot but not on central rows that served for estimating yield. The number of leaf per plant, leaf area and dry matter was determined. Because of unavailability of an electronic leaf area meter, a graduated table was used for measuring leaf area. Sucrose percentage and the concentration of impurities (sodium, potassium, amino-N) were measured at the regional sugar factory.

3.3 Results and discussion

3.3.1 Effect of Weed free and weedy periods on weed dry matter

The dominant weeds observed in 2003 were volunteer wheat (ADI = 4), *Phalaris brachystachys* Link.(3), *Avena sterilis* L. (2), *Cichorium endivia* L. (4), *Papaver rhoeas* L. (3), *Ridolfia segetum* L. (3), *Sinapis arvensis* L. (3), and *Galium tricornitum* Dandy (2). With the exception of field bindweed (*Convolvulus arvensis* L.) (3), the same weed species were dominant in 2004. Weed free periods resulted in lower weed dry matter and weedy periods resulted in high weed dry matter (Figure 2). Maximum total weed dry weight generally decreased as weed-free duration was increased. The statistical analysis showed a highly significant difference (Not shown).

These findings are similar to those observed by Salehi et al. (2006), Rzozi (1993) and Alaoui et al. (2003). Weed growth was reduced drastically after a weed free duration greater than 17 WAE in both years. Same results were obtained for all the two years 2003 an 2004. For the later, weed dry matter was relatively lower because the later date of sowing results generally in low weeds density.

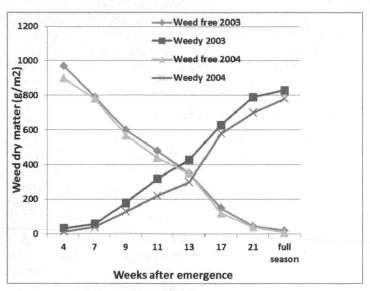

Fig. 2. Effect of competition duration on weed dry matter.

3.3.2 Effect of weed free and weedy periods on sugar beet growth parameters

All sugar beet growth parameters were affected by the presence of weeds. Effectively, the sugar beet leaf number decreased as weedy periods increased and in contrast it increased as weed free periods increased (Figure 3). Also, the leaf area decreased as weedy periods increased. This parameter was highly significantly reduced because of the important competitive effect of weeds. (Figure 4). The crop leaf dry matter was also significantly reduced by the weed competitive effect. The longer the weedy period the lower sugar beet dry matter. The later increased as the weed free period increased (Figure 5). These results confirm those of Alaoui et al. (2003)   reporting that the leaf area and the other growth parameters are vigorously decreased by the competitive effect of weeds.

3.3.3 Effect of weed free and weedy periods on sugar beet yield, on sugar yield and sugar content

Weed infestation reduced root yield in all treatments. The presence of weeds during the entire growing season decreased root yield by 97.6 % and 68.9 % in 2003 and 2004, respectively, as compared to full season weed free check. Although sugar content did not show any significant difference between various treatments in both years, weed infestation decreased sugar yield, their corresponding yields decreased considerably in infested treatments. For example, season-long weed infestation decreased sugar yield by 89.8% and 81.1 % in 2003 and 2004, respectively, as compared to weed free check (data not shown). The concentration of sugar beet impurities such as potassium, sodium and amino nitrogen were not affected by weed competition (data not shown).

In most years in Morocco, weeds can cause more than 75% yield reduction (Rzozi, unpublished data; Rzozi et al., 1990). Such reductions indicate complete crop failure because small sugar beet roots produced under severe weed competition cannot be processed. In other countries, weeds also seriously suppress sugar beet yield (Schweizer and Dexter, 1987).

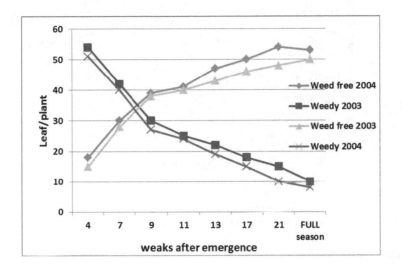

Fig. 3. Effect of weeds on sugar beet leaf number.

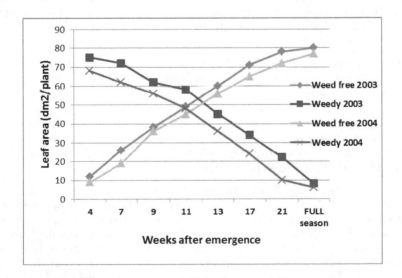

Fig. 4. Effect of weeds on sugar beet leaf surface.

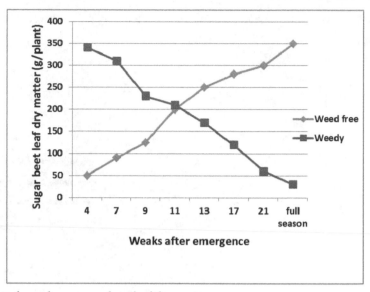

Fig. 5. Effect of weeds on sugar beet leaf dry matter.

3.3.4 Critical period of weed control

Weed interference caused a sharp decline in sugar beet root yield in both years (Figure 6 and 7). Based on 10 % permissible decrease in root yield, weeding should start from 4 WAE and 7 WAE in 2003 and 2004, respectively (Figure 6 and 7). For the given 10% root yield reduction, weed control should be continued until 15 WAE and 12 WAE in 2003 and 2004, respectively (Figure 5 and 6). Weed interference caused a sharp decline in sugar yield (data not shown). Based on 10 % permissible decrease in root yield, weeding should start from 3.5 WAE and 7 WAE and must be continued until 15 WAE and 11 WAE in 2003 and 2004 respectively.

The results show that the critical period begins earlier in 2003 and its duration is longer comparatively to that observed in 2004 which is shorter and begins relatively later. This may be due to date of sowing. Effectively, in 2003, sugar beet was sown October 15 and this allows to many weed species, particularly gramineous including volunteer wheat, to germinate and emerge in great number and vigorously at the same time of the crop germination and emergence. In 2004, sugar beet was sown 25 November. At this time, a great number of weed species (mainly gramineous) has germinated and emerged from soil and destructed during the seedbed preparation.

Emergence time of weeds influences the critical period of weed control (Zimdahl, 1987; Weaver et al., 1992; Mesbah et al., 1994; Ghadiri, 1996). In Shahrekord, sugar beet is planted in May and June; this delay in seedbed preparation and planting may lead to earlier germination of weeds over the sugar beet crop. Therefore, critical period of weed control starts earlier and its duration is longer. At early growth stages, sugar beet has a low competitive ability against weeds; as a result critical period would start sooner. In 2003, presence of weeds for the entire growing season reduced root yield by 97.6% relative to weed free control. In 2004, the reduction was 68.6 %. A similar 71% root yield reduction was also observed by Shahbazi and Rashed Mohassel (2000). Dawson (1977) showed that annual weeds that germinate during a 2-week period after planting or a 4-week period after two-leaf stage in sugar beet reduce root yield by 26 to 100%. Therefore, effective control of weeds at early stages seems to be more important than that of later developed stages. The closure of crop canopy at later growth stages suppresses the late-emerging weeds. The increased period of weed competition reduces the photosynthesis and crop growth

(Zimdahl, 1987; Ghadiri, 1996). Longer presence of weeds caused more use of environmental resources (light, water, and nutrients) and more accumulation of dry matter in weeds, making the critical period longer and, therefore reducing root and white sugar yield of the sugar beet crop.

3.4 Conclusion

A field experiment was conducted during two growing seasons 2003/2004 and 2004/2005 to assess the effect of weeds on sugar beet growth parameters and sugar beet yield and to determine the critical period of weed control (CPWC). Weed free treatments and weed infested treatments included the removal (or not) of weeds at 4, 7, 9, 11, 13, 17 and 21 weeks after emergence of sugar beet. Dry matter of weed, sugar beet leaves/plant, sugar beet leaf area and sugar beet dry weight was measured during all growing season. Weed free periods resulted in lower weed dry matter and weedy periods resulted in high weed dry matter. Maximum total weed dry weight generally decreased as weed-free duration was increased. The presence of weeds during the entire growing season decreased root yield by 97.6 % and 68.9 % in 2003 and 2004, respectively. All crop growth parameters were significantly reduced by weed infestation.

The critical period of weed control began at 4 and 7 weeks after sugar beet emergence (WAE) and continued until 15 and 12 WAE in 2003/2004 and 2004/2005 respectively depending on sowing period. It was concluded that the CPWC is longer in 2003/2004 than in 2004/2005.

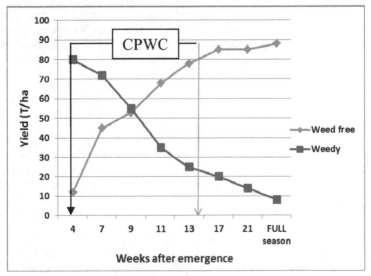

Fig. 6. Critical period of weed control (2003/2004).

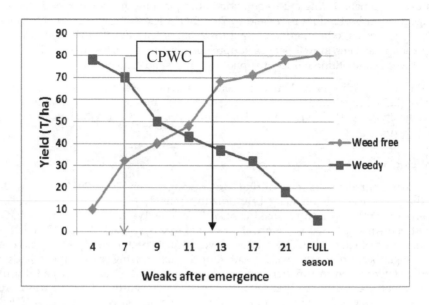

Fig. 7. Critical period of weed control (2004/2005).

4. Chemical control of sugar beet weeds

4.1 Introduction

In the area of Tadla, sugar beet is regarded as an important crop. Weeds constitute a great constraint to crop production improvement and cause important yield losses (Rzozi and al., 1990). The farmers do not use herbicides efficiently. Generally, only one herbicide is applied and the results are not satisfactory (Baye et al, 2004). This work aims to develop a chemical weed control program by evaluating the effectiveness of some herbicide treatments.

4.2 Material and methods

4.2.1 Field experiment localization

A field experiment was conducted during the two sugar beet growing seasons 2003/2004 and 2004/2005 in three location Fqih Ben Salah, Afourer and Deroua to assess the efficacy of some herbicide treatments. These locations were chosen in order to have diversified weed flora and then have maximum information about herbicide activity spectrum.

4.2.2 Herbicides and herbicide treatments studied

The main and important herbicides homologated on sugar beet and registered in Morocco such as ethofumesat, desmedipham, phenmedipham, metamitron, triflusulfuron methyl and lenacil were experimented (Table 4). Thes active ingredients were tested either alone or in mixture (Table 5). A hand weeding taked place for all treatments when it was necessary.

4.2.3 Observations on weeds

The importance of weeds encountered in field experiments was estimated according to the Abundance – Dominance-Index (ADI).

4.2.4 Evaluation of herbicide efficacy

Weed dry weight were measured during at 60 days after treatments (DAT) for all individual plots. Four 0.5 m x 0.5m quadrates per plot were placed randomly over the plot. Weeds within the sampling area were removed by hand, taken to laboratory and dried at 60° C for 48 h to determine total weed dry weight. The efficacy in percentage (%) for each treatment is calculated comparing its dry matter to that of the check.

4.2.5 Observations on the crop

The sugar beet yield was estimated on the two central rows at harvest. Sucrose percentage and the concentration of impurities (sodium, potassium, amino-N) were measured at the regional sugar factory.

4.2.6 Experimental design and statistical analysis

The experiment was a randomized complete block design with four replicates. Individual plots were 4m x 8m size. Data on efficacy (%) were first transformed to Arc Sin% if necessary. Sugar beet yield and efficacy data were subjected to an analysis of variance using statistical STATITCF software. The means were compared using Fisher's protected LSD (α = 0.05).

Commercial product	Active ingredient
Tramat Combi	30 % ethofumesat + 12 % lenacil
Betanal Progress	16 g/l desmedipham + 62 g/l phenmedipham + 128 g/l ethofumesat
Goltix	70 % metamitron
Safari	70 % triflusulfuron methyl
Venzar	80 % lenacil
Fusilad Super	125 g/l Fluazifop p-butyl

Table 4. herbicides tested.

4.3 Results and discussion

4.3.1 Importance of weed flora

In Fqih Ben Salah location, weed flora is dominated by gramineous mainly volunteer wheat. Some dicotyledonous species such as *Malva parviflora, Medicago polymorpha, Emex spinosa* and *fumaria parviflora* are important (Table 6). In Deroua location, infestation by gramineous was low and *cichorium endivia, Sinapis arvensis* and *convolvulus arvensis* were dominant in 2003/2004 and *Rumex pulcher, Papaver rhoas* and *Ridolfia segetum* dominated the weed flora in 2004/2005. Concerning Afourer, *Cichorium endivia, Sinapis arvensis, Polygonum aviculare, Lamium amplexicaule and Ridolfia segetum* were the most important species in both two growing season.

4.3.2 Efficacy of the herbicide treatments

Generally, fluazifop- p- butyl (Fusilade Super) achieved a good gramineous control (data not showed). However, it is important to mention that some ray grass (*Lolium* spp) population had recently developed resistance to this herbicide.

Treatments	Herbicide treatments tested
T1	Tramat Combi (3,5l/ha) in post sowing preemergence
T2	Goltix (5kg/ha) applied in post sowing preemergence
T3	Goltix (5kg/ha) in 2 applications (2,5 + 2,5) kg/ha post emergence (2 true leaves stage)
T4	Safari (60g/ha) in 2 applications (30+30) g/ha) post emergence (2 true leaves stage)
T5	Betanal Progress (5l/ha) in 2 applications (3 + 2) l/ha g/ha) post emergence (2 true leaves stage)
T6	(Safari (30g/ha) + Venzar (200g/ha) applied twice post emergence (2 true leaves stage)
T7	(Betanal Progress(1,25l/ha) + Safari (30g/ha)) applied twice post emergence (2 true leaves stage)
T8	(Betanal Progress (1,25l/ha) + Goltix (1kg/ha)) applied twice post emergence (2 true leaves stage)
T9	(Goltix (1kg/ha) + Safari (30g/ha) applied twice post emergence (2 leaf stage)
T10	(Betanal Progress (1L/ha) + Goltix (300g/ha) + Venzar (100g/ha)) applied twice post emergence (2 true leaves stage)
T11	(Betanal Progress (0,8L/ha) + Safari (30g/ha) + Goltix (300g/ha) + Venzar (100g/ha)) applied twice post emergence (2 true leaves stage)
T12	Hand weeding (Three times in the season)
T13	Check (Not treated)

Table 5. Herbicides treatments experimented.

In order to control gramineous species, all first post emergence application are mixed with Fusilade Super (1l/ha); an oil concentrate adjuvant (Seppic at 1/ha) is adjusted to the two application to obtain satisfactory activity. The second application is made 10 days after the first.

Concerning post sowing preemergence application treatments, Tramat combi (T1) provided good efficacy (90 % and more) and protected then the crop for a long period more than 2 months (Table 7). This allowed to sugar beet to grow vigorously. The treatment controlled both dicotyledonous and monocotyledonous species except *Emex spinosa* that showed some tolerance to this herbicide. The other treatment applied preemergence (T2) showed not satisfactory with efficacy lower than 68 %. This herbicide did not control monocotyledonous (volunteer wheat included) and many other dicotyledonous species such as *Medicago polymorpha and Melilotus sulcata*.

For post emergence applications, it was noted that when treatments were applied alone (not mixed), the efficacy was not satisfactory. Effectively, efficacy was generally below 70 % except T5 in 2004/2005 at Deroua (Table 7). In this case, the percent control is above 80 %.

This difference in efficacy is explained mainly by the herbicide activity of each one. Safari (T4) provides low control against *Papaver rhoeas, Chenopodium album, Anagallis foemina, stellaria media, Cichorium endivia and Fumaria parviflora*. In contrast, it achieves good control against many other important species particularly malvaceae, *Malva parviflora*, apiaceae such as *Ridolfia segetum* and *Ammi majus* and brassicaceae maily *Sinapis arvensis*. Goltix (T3) did not control apiaceae, malvaceae and other species; however, it provides good control of polygonaceae such as *Rumex pulcher* and *Emex spinosa*. Betanal Progress presented the most large herbicide activity spectrum and controlled great number of species even applied alone in some times. This is the case of Deroua in 2003/2004. The efficacy obtained is 82%.

Generally, treatments achieved good efficacy when applied in tank mixtures than when applied individually alone because of their complementarily in eliminating maximum weed species. So, this must be taken in consideration in a weed chemical management program.

Species	Fqih Ben Salah		Afourer		Deroua	
	2003/04	2004/05	2003/04	2004/05	2003/04	2004/05
Volunteer wheat	4	3	1	2	2	2
Phalaris brachystachys	1	3	3	3	2	2
Lolium rigidum	3	2	2	1	1	1
Avena sterilis	2	2	1	2	2	2
Bromus rigidus	+	+	+	+	+	1
Malava parviflora	4	3	+	+	+	1
Emex spinosa	3	1	2	+	+	2
Rumex pulcher	+	+	1	1	2	4
Anagallis foemina	1	4	3	3	+	2
Chenopodium album	2	3	3	1	1	2
Fumaria parviflora	3	3	1	1	+	1
Cichorium endivia	+	2	4	4	4	2
Convolvulus arvensis	+	1	2	3	3	3
Sinapis arvensis	1	1	4	4	1	2
Sonchus oleraceus	1	1	2	2	1	2
Polygonum aviculare	1	+	4	3	+	1
Lamium amplexicaule	1	+	4	3	1	2
Medicago polymorpha	4	+	3	2	1	2
Melilotus sulcata	2	1	2	+	1	1
Papaver rhoeas	2	+	3	2	+	4
Ridolfia segetum	+	+	2	4	1	3
Ammi majus	-	+	+	+	+	+
Stellaria media	-	+	+	+	+	3
Veronica polita	-	+	+	1	-	1
Torilis nodosa	-	-	+	+	-	-
Euphorbia exigua	-	-	+	+	-	-
Galium aparine	-	-	1	1	-	1
Capsella bursa-pastoris	-	-	+	+	2	3

Table 6. Weed species encountered in field experiments.

4.3.3 Effect of herbicide treatments on sugar beet yield

Weed presence in sugar beet during all season caused yield losses between 86 and 93% following the nature of weed flora and the location. Herbicide treatments did not affect the sugar content percentage (Data not showed). Sugar beet yield was significantly affected by the herbicide treatments (Table 8). The post sowing preemergence treatment (Tramat Combi) achieved a satisfactory yield averaging 75 T/ha. This is due to its good weed control achievement during a long period. When used in tank mixtures (particularly 3 and 4 products), herbicide treatments provide high yields (Table 8). It is important to mention that weed chemical treatment alone is generally not sufficient to provide good root sugar beet production and it must be followed by other weed control methods such as mechanical, cultivation and hand weeding.

	Fqih Ben Salah		Afourer		Deroua	
	2003/04	2004/05	2003/04	2004/05	2003/04	2004/05
Treatments						
T1	89.5a	86a	87.3a	86.9a	90.1a	92.6a
T2	62c	60.3c	65c	65.9b	64bc	68.4b
T3	69.3b	60.7c	65.2c	66.8b	69.5b	65.2bc
T4	65.4bc	69b	63.5c	65.2b	61.4c	50.1d
T5	72b	79ab	75b	70b	82a	62.9c
T6	75b	72.2b	65c	69b	72.6b	62c
T7	84a	86.4a	87.8a	85.7a	86.9a	66bc
T8	75.1b	79b	76b	62b	80.1a	74.1b
T9	72b	75.6b	69.3c	67b	65b	72.b
T10	75b	77b	62c	60b	69b	75b
T11	88.2a	86.7a	88.1a	86.3a	88.6a	80.6a
T12	70b	73b	74b	69b	73b	65bc

Means within columns followed by different letters are significantly different at α = 0.05.

Table 7. Efficacy of herbicide treatments (%) at 60 DAT.

Treatments	Fqih Ben Salah		Afourer		Deroua	
	2003/04	2004/05	2003/04	2004/05	2003/04	2004/05
T1	74a	72.9a	75a	74.6a	78a	80.2a
T2	52.3c	53.6c	51c	49c	53.3bc	51.6c
T3	54.3bc	51c	50.9c	52.3c	53.4bc	50.3c
T4	52c	53.2b	52.6c	51c	50c	46c
T5	60b	62b	60.3b	59.2b	68a	51c
T6	62.1b	61.4b	54bc	58bc	60.6b	50.6c
T7	69a	70.5a	71.2a	72a	71.9a	73.2a
T8	61.6b	63b	64.5b	65.1b	69a	61b
T9	53c	60b	51c	50c	49c	52.3c
T10	60.8b	62.6b	57bc	55.4c	59b	61b
T11	70.2a	71.3a	72.6a	71.9a	73.3a	72.8a
T12	49.8c	50.9c	51c	52c	50.2c	51.9c
T13	7.2d	5d	3.9d	8d	4.8d	6.3d

Means within columns followed by different letters are significantly different at $\alpha = 0.05$.

Table 8. Effect of herbicide treatments on sugar beet yield.

Many studies relative to sugar beet weed chemical control were achieved in Morocco and other counties. Bensellam et al. (1993) reported that phenmediham + pyrazone achieved good control of weeds in sugar beet. Rzozi et al. (1990) found that nor metamitrone followed by phenmedipham neither chloridazone applied preemergence gave good efficacy. El Antri (2002) reported that triflusulfuron methyl + lenacil + clopyralid achieved good control of weeds in sugar beet. El Ghrasli and Allali (2002) estimated that farmers in Gharb region could use Safari, Goltix, Betanal and Venzar to control weeds in sugar beet. The pre sowing and preemergence herbicides: Tramat Combi and Goltix and the post emergence safari, Goltix, Betanal Progress and Venzar are widely used in France (Anonymous, 1999) and in USA (Stachler, 2011).

4.4 Conclusion

A field experiment was conducted during two growing seasons 2003/2004 and 2004/2005 in three locations in Tadla region to evaluate the effectiveness of some herbicides treatments. The main and important herbicides homologated on sugar beet and registered in Morocco such as ethofumesat, desmedipham, phenmedipham, metamitron, triflusulfuron methyl and lenacil were experimented individualy alone or in tank mixtures.

Tramat combi (Ethofumesate + lenacil) applied post sowing preemergence provided good efficacy (90 % and more) and protected then the crop for a long period more than 2 months.

Generally when applied post emergence, herbicides ethofumesate, metamitron, triflusulfuron methyl, phenmedipham, desmedioham and lenacil achieved good efficacy in tank mixtures than applied individually alone because they are complementarily in eliminating maximum weed species. So, this must be taken in consideration in a weed chemical management program. These herbicide treatments allow to crop to grow without weed competitiveness nearly until the end of the critical period and are often followed by a mechanical cultivation or a hand hoeing.

5. General conclusion

In Morocco, sugar beet is an important strategic crop. It is planted from September through June - July. Yield obtained by farmers, averaging 50 T/ha, is significantly below the request potential that would be 90 to 100 T/ha. Many factors contribute to low sugar beet production. Poor stand establishment, inadequate weed control, inadequate insect control and inadequate nitrogen fertilization are the main causes of low tonnage and poor quality sugar beet in Morocco.

This paper presents the main results of investigations and experiments conducted in Tadla region to improve the weed management program by identifying mains weed species encountered in sugar beet field, studiying the effect of weeds on sugar beet growth and estimating yield losses and determining the critical period of weed control and evaluating herbicide treatments.

One hundred twenty six (126) fields of sugar beet were surveyed by stratified sampling in Tadla region (Center of Morocco). In total, 144 weed species belonging to 30 botanical families were recorded. Six among them asteraceae, poaceae, fabaceae, brassicaceae, apiaceae and caryophyllaceae account 81 species (56,1% of total species). Dicotyledonous (81,9%), annuals (82,6%) and the Mediterranean floristic element (56,2%) were predominant and characterized the weed flora. The agronomic study made it possible to distinguish 24 species and volunteer wheat causing appreciable problems to the crop. Statistical analysis using soil-climatic factors allowed distinguishing four ecologic groups.

To determine the critical period of weed control in sugar beet, an experiment was conducted and consisting of 16 treatments. Weed free treatments included the removal of weeds at 4, 7, 9, 11, 13, 17 and 21 weeks after emergence (WAE) of sugar beet. In weed infested treatments, weeds were allowed to interfere with sugar beet crop 4, 7, 9, 11, 13, 17 and 21 weeks after emergence sugar beet crop. Weed infestation reduced root yield in all treatments. The presence of weeds during the entire growing season decreased root yield by 97.6 % and 68.9 % in 2003 and 2004, respectively. Based on 10 % permissible decrease in root yield, weeding should start from 4 WAE and 7 WAE in 2003 and 2004, respectively. For the given 10% root yield reduction, weed control should be continued until 15 WAE and 12 WAE in 2003 and 2004. The results show that the critical period begins earlier in 2003 and its duration is longer (77 days) comparatively to that observed in 2004 which is shorter (35 days) and begins relatively later.

A field experiment was conducted during two sugar beet growing seasons 2003/2004 and 2004/2005 in three locations to assess the efficacy of some herbicide treatments. These locations were chosen in order to have diversified weed flora and then have maximum information about herbicide activity spectrum. Concerning post sowing preemergence application treatments, Tramat combi (T1) provided good efficacy (90 % and more) and protected then the crop for a long period more than 2 months. This allowed to sugar beet to grow vigorously. The treatment controlled both dicotyledonous and monocotyledonous species. For post emergence applications, it was noted that when treatments were applied alone (not mixed), the efficacy was not satisfactory. Generally, herbicides (ethofumesate, metamitron, triflusulfuron methyl, phenmedipham, desmedioham and lenacil) achieved good efficacy when applied in tank mixtures than when applied individually alone because they are complementarily in eliminating maximum weed species. So, this must be taken in consideration in a weed chemical management program. These herbicide treatments allow to crop to grow within weed competitiveness nearly until the end of the critical period and are often followed by a mechanical cultivation or a hand hoeing.

6. Acknowledgments

The authors thank M. Ait El Alia and O. Bennig for helping with collecting the research data.

7. References

Alaoui, S. B., Donald L. Wyse and A. G. Dexter. 2003 Minimum Weed-free Period for Sugarbeet (Beta vulgaris L.) in the Gharb Region of Morocco. Journal of Sugar Beet Research Vol 40 No 4: 251 -272

Anonymous. 1999. Désherbage des betteraves pour 1999. La Technique betteravière 1999; Protection de la culture, p: 45- 48.

Baziramakenga, R. and Leroux, G. D. 1994. Critical period of quackgrass (Elytrigia repens) removal in potatoes (Solanum tuberosum). Weed Science 42: 528-533.

Baye, Y., M. El Antri, M. Bouhache and A. Taleb. 2004. Situation actuelle du désherbage de la betterave à sucre au Tadla. 5 ème congrés de l'AMPP, 30-31 Mars, Rabat, p : 183-194.

Baziramakenga, R. and Leroux, G. D. 1994. Critical period of quackgrass (Elytrigia repens) removal in potatoes (Solanum tuberosum). Weed Science 42: 528-533.

Bensellam E.H., M.Bouhache and S.B.Rzozi. 1993. Effet du peuplement et de la stratégie du désherbage sur le rendement et la qualité technologique de la betterave à sucre dans le Gharb. Sucrerie Maghrébine, n°50-5: 13-24.

Bensellam, E.H. 1994. Etude floristique des adventices des vergers d'agrume dans le Gharb. Rapport de fin de stage de titularisation. Institut National de Recherche Agronomique, Département de Recherche de Phytiatrie du Gharb. 45 p.

Bouhache, M. and C. Boulet. 1984 . Etude floristique des adventices de la tomate dans le Souss. Homme, terre et eau; 14(57), 37-49.

Bouhache, M. C. Boulet and A. Chougrani. 1993. Aspects floristico agronorniques des mauvaises herbes de la région de Loukkos (Maroc). Weed Research, 24, 19-126.

Bouhache, M. and B. Ezzahiri. 1993. Caractérisation de la flore adventice associée à la betterave à sucre dans les Doukkala. Sucrerie Maghrébine, 56/57, 32-38.

Boulet, C., A. Tanji and A. Taleb . 1989. Index synonyrnique des taxons présents dans les milieux cultivés ou artificialisés du Maroc Occidental et Central. Actes Inst.gro Vél.(Maroc), 9(3&4),65-99.

Cooke, D. A. and Scott, K. 1993. The sugar beet crop. Translated by: A. Koucheki, and A. Soltani. Jahad Daneshgahi of Mashhad Pub. Mashhad.

Dawson, J.H. 1965. Competition between irrigated sugar beets and annual weeds. Weeds, 13:245-249.

Dawson, J. H. 1977. Competition of late-emerging weeds with sugar beets. *Weed Science* 25:168–170.

El Antri, M. 2002. Intérêt de l'utilisation du clopyralid dans des programmes de désherbage de la betterave à sucre au Maroc. Compte rendu de la section méditerranéenne de l'IIRB. 23 -25 Octobre, p : 234-239.

El Ghrasli, D. and M.Allali. 2002. Désherbage de la betterave à sucre au Gharb (Maroc). Compte rendu de la section méditerranéenne de l'IIRB. 23 -25 Octobre, p : 224-233.

Farahbakhsh, A. and K.J. Murphy. 1986. Comparative studies of weed competition in sugarbeet. Aspects of App. Biology. 13: 11-17.

Ghadiri, H. 1996. Concept and application of critical period of weed control. *Collections of full papers of 4th Iranian crop Production & Breeding Congress,* Isfahan, 257-265.

Ghersa, Jahandiez, E. and R. Maire. 1931-193. Catalogue des plantes du Maroc, 3 tomes. ed Lechevallier, Paris, 913 p.

Gounot, T, M. 1969. Méthodes d'études quantitatives de la végétation, Paris, France. Lechevalier. 3 tomes.

Ibn Tatou, M. and M. Fenanne. 1989. Aperçu historique et état actuel des connaissances sur la flore vasculaire du Maroc. Bull. ln. Sci. Rabat. (13), 85-94.

Link, R. and W. Koch. 1984. Analyse des effets de ['envahissement par les mauvaises herbes aux differents stades de vegetation sur Ie rendement des betteraves sucrieres. Proc. EWRS 3rd. Symp. On weed problems in Mediterranean area. 121-128.

Loudyi, M.C. 1985. Etude botanique et écologique de la végétation spontanée du plateau de Meknès (Maroc). Thèse 3ème cycle. V.S.T.L., Montpellier, 147 p.

Maillet, J. 1981. Evolution de la flore adventice dans le Montpellierais sous la pression des techniques culturales. Thèse Docteur-Ingénieur, Montpellier, France, U.STL. 200 P.

Maillet, J. 1992. Contribution et dynamique des communautés de mauvaises herbes des vignes de France et des rizières de Carnargne. Thèse de Doctorat d'Etat ES Science. Université Montpellier II, 179 p.

Mesbah, A., S.D. Miller, K.J. Fornstrom and D.E. Legg. 1994. Sugar beet-weed interactions. University of Wyoming. Agricultural Experiment Station. B-998.

Negre R. 1961-1962. Petite flore des régions arides et semi-arides du Maroc occidental CNRS, Paris, 2 tomes, 279 p.

Norris, R. F. 1996. Sugar beet integrated weed management. Publication UC IPM Pest Management Guidelines: Sugar beet, Veg Crops/Weed Science, UC Davis UC DANR Publication 3339.

Quezel, P. and S. Santa . 1962-1963. Nouvelle Flore de l'Algérie et des régions désertiques, méridionales. CNRS, Paris. 2 tomes. 1 170 p

Radosevich, S. R. and Holt, J. S. 1984. Weed ecology: Implications for vegetation management. John Wiley and sons. New York. pp 265.

Raunkiaer, C.1905. Types biologiques pour la géographie botanique:.(Kgl. *Donskevidcnskabcrne sclskabs Forhandt,* 5, *347-437.).* Bull.Acad. R Sc Danemark, 5, 347-437.

Rzozi, S.B., R. El Hafid and M. El Antri. 1990. Résultats préliminaires sur le désherbage de la betterave betterave à sucre dans le perimetre de Tadla. Actes Inst.Agr. Vet., vol. 10, (2): 49-59.

Rzozi, S.B. 1993. Weed Interference and Control in sugarbeet *(Beta vulgaris* L.) in the Gharb region (Morocco). Ph.D. Thesis, University of Minnesota.

Salehi, F., H. Esfandiari1 and H. Rahimian Mashhadi. 2006. Critical Period of Weed Control in Sugar beet in Shahrekord Region. Iranian Journal of Weed Science, Vol 2, No. 2, 1-12.

Schweizer, E. E. 1983. Common lambsquarters *(Chenopodium album)* interference in sugar beets *(Beta vulgaris). Weed Science* 31:5-8.

Schweizer, E.E. and A.G. Dexter. 1987. Weed control in sugarbeets *(Beta vulgaris* L.) in North America. *In.* Reviews of Weed Science, W.S.S.A., Vol. 3: 113-133. 272 Journal of Sugar Beet Research Vol40No4.

Scott, R.K, S.J. Wilcockson and FR. Moisey. 1979. The effects of time of weed removal on growth and yield of sugarbeet. Agric. Sci. Camb.93:693-708.

Stachler, F. 2011. Timing of Herbicide Applications is Critical for Effective Weed Control in Sugarbeet. Weed Science (May 26, 2011) (From Internet).

Singh, M., Saxena, M. C., Abu- Irmaileh, B. E., A1–Thahabi, S.A. and Haddad.,N.I. 1996. Estimation of critical period of weed control. *Weed Science* 44: 273-283.

Swanton, C.J. and S.F. Weise. 1991. Integrated weed management: the rational and approach. Weed Technology. 5:648- 656

Taleb, A. and S.B. Rzozi. 1993. Etude botanique et agronomique des adventices de la betterave à sucre dans la basse Moulouya. Sucrerie Maghrébine. 56/57 : 25 – 31.

Taleb,A. and J. Maillet. 1994. Mauvaises herbes des cereals de la Chaouia. Aspect floristique. Weed Research. 34 : 345- 352.

Taleb, A. 1995. Flore illustrée des mauvaises herbes des cultures du Gharb. Doctorat d'Etat Es Sciences Agronomiques. Institut Agronomique et Vétérinaire Hassan II, Maroc, 2 volumes, 303p.

Tanji, A., C. Boulet and M. Hammoumi. 1984. Inventaire phytoécologique des adventices de la betterave à sucre dans le Gharb (Maroc). Weed Research, Vol.24., 391 – 399.

Tanji, A. and C. Boulet. 1986. Diversité biologique et biologique des adventices de la region de Tadla (Maroc); Weed Research. Vol.26 : 159 – 166.

Tanji, A., C. Boulet and Regehr. 1988. Mauvaises herbes des régions arides et semi- arides du Maroc occidental. 397 p.

Tutin, T.G., V.H.Heywood, N.A. Burgnes, D.M. Moore, D.H. Valentine and D.A. webb. 1964 –1984. Flora Europea. Cambridge University Press, 5 volumes.

Wahbi, M. 1994. Etude de la flore adventice des agrumes et des cultures maraichères de la region de Souss: Aspect botanique, agronomique et écologique; Mémoire de "ème Cycle, Phytiatrie. Inst.Agron. Vet ; Hassan II, Complexe d'Agadir. 86 p.

Weaver, S. E., Kropff, M. J. and Groeneveld, R. M. W. 1992. Use of ecophysiological models for crop-weed interference: The critical period of weed interference. *Weed Science* 40:302-307.

Zimdahl, R.L. and S.N. Fertig. 1967. Influence of weed competition on sugarbeets. Weeds 15:336-339.

Zimdahl, R. L. 1987. The concept and application of the critical weed-free period. In: *Weed management in agroecosystems: Ecological Approaches.* (Eds. M. A. Altieri and M. Liebman). CRC Press, Inc. Florida, USA. pp 145-155.

13

Herbicides in Winter Wheat of Early Growth Stages Enhance Crop Productivity

Vytautas Pilipavičius
Aleksandras Stulginskis University
Lithuania

1. Introduction

Herbicides are chemical substances destroying undesirable plants (weeds) or suppressing their growth.

Wheat (*Triticum* spp.) is a cereal that is cultivated worldwide. It is the most important human food grain (Hanee M. Al-Dmoor, 2008). Traditionally, herbicides in winter wheat are applied from two leaf stage till the end of tillering (Triasulphuron - *Logran*) and in spring at tillering (Propoxycarbazone-sodium - *Atribut*, Sulphosulphuron - *Monitor*, Iodosulphuron-methyl-sodium - *Husar*) and from tillering till booting (Florasulam + 2.4-D 2-ethylhexyl ester - *Mustang*). *Atribut, Monitor and Husar* best fit for control of annual monocotyledonous weeds as *Apera spica-venti, Avena fatua, Poa pratensis* etc. and some annual dicotyledonous weeds as *Galium aparine, Tripleurospermum perforatum, Viola spp., Lamium spp.* etc. *Logran* best fits for control of annual dicotyledonous weeds as *Sinapis spp., Capsella bursa-pastoris, Thlaspi arvense* and etc. while *Mustang* is designed for control of dicotyledonous weeds as *Chenopodium album, Centaurea cyanus, Myosotis arvensis, Sonchus arvensis, Cirsium arvense* and others (Rimavičienė, 2005). Appropriate selected and in time applied herbicides destroy spreading weeds in crop or suppress weed growth and new seed production. Crop weediness is considerably reduced when soil is adequately cultivated, herbicides are applied and crop rotations are practiced (Barberi et al., 1997). It was determined that the field crops of cultured plants are plant associations or so called agrophytocenoses, and that the total biomass of a crop stand (crop and weed biomass) is more or less constant and that the crop yield is inversely proportional to the weed biomass (Lazauskas, 1990, 1993). Effectiveness of chemical weed control is determined by three main specifications: selection of an adequate herbicide, its optimal norm and duration of application. In the process of weed control it should be remembered that wet climate, cold spring weather, long autumn are the factors that help them grow and develop. Another important factor is the ratio of weed biological groups. The prevailing weeds in Lithuania are short-lived annual dicotyledons that comprise 70-90% of all the weeds (Pilipavičius, 2005). Many annual weeds successfully survive till spring because of climate warming; however, they were naturally frozen during winter time just 10-15 years ago. Global warming is the increase in the average temperature of the Earth's near-surface air and the oceans since the mid-twentieth century and its projected continuation. Including uncertainties in future greenhouse gas concentrations and climate sensitivity, the IPCC, scientific intergovernmental body set up by the World Meteorological Organization (WMO) and by the

United Nations Environment Programme (UNEP), anticipates a warming of 1.1°C to 6.4°C by the end of the 21st century, relative to 1980–1999 (Summary for policymakers, Climate change, 2007). Conventionally herbicides are used in spring for weed control in winter cereals, therefore, perennial and winter annual weeds have favourable conditions to grow and compete with cereals when vegetation in spring is renewing. Winter wheat *Triticum aestivum* L. is sensitive to weed competition in early stages of growth and development. Therefore, intensive agricultural systems seek to destroy all growing weeds in crops and avoid of weed seed bank replenishment with new matured weed seeds that can survive in soil for decades (Koch & Hurle, 1978; Niemann, 1981). Weed seed bank in the soil changes in two directions: regularly cleans from seeds and is replenished by them. Balance between these processes decides seed bank change dynamics in the soil (Pilipavičius, 2004, 2007b). Many researchers (House, 1989; Faravani & Khaghani, 2004; Sikkemaa et al., 2007; Stasinskis, 2009) have investigated the effect of herbicide application on field weediness of wheat crops. However, there are reported data on conventional standard time of use of pre-emergence herbicides in autumn or post-emergence herbicides in spring. The potential use of herbicides at early stages of development of winter wheat in autumn has not been clearly considered, and published research data are still insufficient. Intensive use of herbicides following the traditional crop growing technologies, however, does not entirely solve the problem of weediness.

The work hypothesis: application of herbicides in autumn will control weeds that survive during winter time and winter wheat will not be damaged and better conditions for crop competition in spring after renewing of vegetation would be created.

The aim of this work was to evaluate various herbicide active substance applications in autumn at early stages of winter wheat *Triticum aestivum* L. development, its influence on crop weediness and productivity.

2. Chemical weed control development in winter wheat

The research was carried out in Kaunas county, Prienai district, Ašminta region, Strielčiai village. Winter wheat fore-crop was black fallow. Experimental field was ploughed in autumn using semi-helical plough Overum 4 to the depth of 24 cm, cultivated using a cultivator with comb harrow KPŠ-15 to the depth of 10 cm and the surface of soil was levelled off with a roller PP-7. The field was fertilized in autumn (10 September, 2005) by amofos 100 kg ha-1 and potassium chloride 200 kg ha-1. Winter wheat cv. Ada was sown on 12 September, 2005. The sowing-machine SPU – 6 (inter-beds 12.5 cm) was used. The amount of seeds comprised 240 kg ha-1 and they were sown in the depth of 4-5 cm. Winter wheat cv. Ada develops well on all soils growing it by the conventional technologies. Cultivar Ada has high overwintering qualities evaluated by 8-9 points from 9. It is resistant to frost, when at tillering node temperature subsides till minus 140C died just 6% of plants. Average productivity 6.36 t ha-1; stem medium high 90-94 cm (Characteristics of wheat varieties ..., 2011; Lithuanian national list of plant varieties, 2011). The investigated active substances of herbicides according to the scheme of research were applied in autumn at BBCH 14-15 of winter wheat (7 October, 2005).

Experimental design:

1.   Control treatment, not sprayed with herbicides in autumn*
2.   Monitor 75% g. (Sulphosulphuron 750 g kg-1), 26.7 g ha-1

3. Atribut 70% w.s.g. (Propoxycarbazone-sodium 700 g kg^{-1}), 0.120 g ha^{-1}
4. Mustang 458.75 g L^{-1} c.s. (Florasulam + 2.4-D 2-ethylhexyl ester 6.25 + 452.5 g L^{-1}), 0.5 L ha^{-1}
5. Logran 20% w.s.g. (Triasulphuron 200 g kg^{-1}), 0.03 g ha^{-1}
6. Husar 5% w.s.g. (Iodosulphuron-methyl-sodium 50 g kg^{-1}), 0.200 g ha^{-1}.

Note: g – granules; w.s.g. – water-soluble granules; g L^{-1} c.s. – grams in a litre of concentrated suspension.

The experiment was carried out in four replications. The experimental data were evaluated using analysis of variance by *Selekcija* (Tarakanovas, 1997) and correlation-regression analysis by *SigmaPlot 8.0* (SPSS Sciences, 2000).

For the first fertilization in spring (12 April, 2006) 250 kg ha^{-1} of ammonium nitrate was used and for the second fertilization (8 May, 2006) 200 kg ha^{-1} of ammonium nitrate was applied. A *composite of herbicides Sekator 300 g ha^{-1} and MCPA 1 L ha^{-1} with a growth regulator Cycocel 1.5 L ha^{-1} was used for conventional spraying at BBCH 22-23 (2 May, 2006, sprayed all experimental field, standard technology). At the beginning of winter wheat stem elongation (25 May, 2006) the composite of insecticide Fastak 100 g ha^{-1}, fungicide Folikur 0.75 L ha^{-1} and complex fertilizer Wuxal 5 L ha^{-1} were sprayed.

2.1 Meteorological conditions

Lithuanian territory occupies intermediate geographical position between west Europe oceanic climate and Eurasian continental climate. Climate of the Lithuania territory forms in different radiation and circulation conditions. Differences in these conditions hardly cross the boundaries of microclimatic differences; therefore, Lithuania belongs to western region of the Atlantic Ocean continental climatic area (Basalykas et al., 1958). During 2005-2006 meteorological conditions were favourable for winter wheat crop establishment. Autumn of 2005 was warm and rainy, i.e. suitable for crop emergence and early growth. The beginning of winter delivered well balanced conditions for wintering, however, during January – March 2006 the temperatures were rather low with insufficient snow cover on the soil. April – June 2006 was cool with high variation of rainfall which principally did not exceed the long-term average. Significant increase of rainfall in August resulted in complicated conditions for winter wheat maturing and harvesting (Pilipavičius et al., 2010b). Meteorological conditions during vegetation of winter wheat crop experiment are summarised in figure 1. September of 2005 was enough warm, average air temperature during the second ten day period was higher by 2^{0}C comparing with long-term average. However, the first ten day period of September was very rainy with rainfall of 68.3 mm while during the second ten day period it compounded 21.5 mm and the third ten day period pasted without rainfall. The first ten day period of October was warm but already at the second and the third ten day periods average air temperatures dropped to 8^{0}C which were by 1.1^{0}C and 1.4 ^{0}C warmer comparing with long-term average, accordingly. Rainfall during the first ten day period in October reached 31.3 mm while during the second and the third ten day periods decreased till 6.0 mm and 10.4 mm, accordingly. Hence, warm and humid first ten day period in October formed favourable conditions for winter wheat tillering. During November average air temperatures decreased till 1.6^{0}C, 5.6^{0}C and 6.1^{0}C and amount of rainfall consisted of 6.1 mm, 12.2 mm and 6.7 mm. Gradual decrease of temperatures and rainfall was adequately fitting biological needs of winter wheat. During December average air temperatures fell down below 0, i.e. till -1.7^{0}C while long-term average was -2.2^{0}C. Precipitation consisted 6.0 mm at the first ten day period of December,

17.7 mm at the second and 22.4 mm at the third. It was lower than long-term average and formed proper conditions for winter wheat cv. Ada wintering.

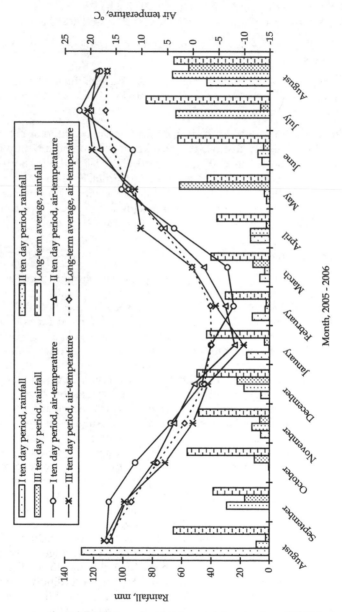

Note. Long-term average of rainfall (600.4 mm) and average air temperature (6.8°C) by ten day periods during 1974-2004, Kaunas Hidrometeorological station

Fig. 1. Meteorological conditions, rainfall and average air temperature by ten day periods during vegetation of winter wheat crop, Kaunas county, Prienai district, Lithuania.

Meteorological conditions during January, February and March were very unfavourable for winter wheat overwintering, because there was few precipitation formed insufficient snow cover with rather low average air temperatures of -7.2⁰C, -6.3⁰C and -2.7⁰C accordingly. During January and February precipitation was by 54% and 42% lower comparing with long-term average. April weather was cool, just at the second ten day period it started to warm till +10.2⁰C with rainfall of 29.3 mm while long-term average was 36.4 mm. During May average air temperatures reached 12.5⁰C that was slightly lower than long-term average of 12.6⁰C. The third ten day period of May was very rainy reaching 61.9 mm that consisted 83% of the whole month standard comparing with long-term norm average. The first ten day period in June was cool 11.7⁰C comparing with long-term average 15.5⁰C while during the second and the third ten day periods average air temperatures reached 18.1⁰C and 19.7⁰C accordingly. June was very dry with 18 mm rainfall when long-term average is 63 mm. It was by 71% lower than long-term average and it was very inappropriate for winter wheat growth and development (Fig. 1). As a consequence, average winter wheat grain yield 2.46 t ha⁻¹ in Lithuania in 2006 was the lowest comparing with 2005-2010 (Statistics Lithuania, 2011). July was warm with average air temperature of 20.9⁰C and 64.3 mm of rainfall that was by 20.3 mm lower than long-term average. The rainiest month during winter wheat vegetation was August with 165.6 mm rainfall exceeding long-term average by 99.9 mm and created very unfavourable conditions for winter wheat grain maturing and wet soil aggravated grain harvesting.

2.2 Soil weed seed bank

2.2.1 Soil agrochemical characteristics

Soil samples for agrochemical analysis and establishment of weed seed bank, seed varietal composition and quantity, were taken from 0-20 cm soil layer at the end of September from 10 sites of all treatments and their replications, making combined samples. Soil agrochemical characteristics were established in the Centre of Agrochemical Research, Lithuanian Agricultural Institute and weed seed bank composition was established at the Lithuanian University of Agriculture. Experimental field soil was *Gleyic Cambisols CMg*. Topsoil layer was alkaline, average in humus, rich in phosphorus and average rich in potassium (Table 1).

Agrochemical soil properties			
pH	Humus %	P_2O_5 mg kg⁻¹	K_2O mg kg⁻¹
7.1	2.17	152	146

Table 1. Experimental field soil agrochemical characteristics of 0–20 cm soil layer.

2.2.2 Weed seed bank

Weed seeds from soil samples were washed through 0.25 mm sieve and separated by saturated solution of high specific mass of NaCl (Rabotnov, 1958; Warwick, 1984, Pilipavičius, 2004). Seeds of 12 weed species (10 annual and 2 perennial) were identified in the soil weed seed bank. Annual weed seeds dominated with 88.0%-95.7% of weed seed bank, from them winter annual weed seeds (*Viola arvensis, Tripleurospermum perforatum,*

Thlaspi arvense etc.) comprised 21%-60% (36%-54% from the whole seed bank). Seeds of the perennial weeds (*Cirsium arvense, Rumex crispus*) were in the minority with 4.3%-12.0% from the whole soil weed seed bank (Table 2). In winter wheat crop perennial weeds were in the minority as well (Fig. 3 & 4). From separated weed seed species, the seeds of *Chenopodium album* prevailed in the soil seed bank. They comprised 34%-48% of the whole soil weed seed bank. However, *Chenopodium album* was recessive weed in the crop as it is summer annual weed and consequently is freezing during winter time (see subchapter 2.3). *Viola arvensis* was the other dominant weed in the soil seed bank that covered 16%-30% of seed bank (Table 2). The main change in the number of weed species was influenced by appearance and disappearance of weed seeds that were low in number. It was either actual for the crop.

Weeds	Treatment					
	Control	Monitor	Atribut	Mustang	Logran	Husar
	Weed seeds					
Chenopodium album L.	3.0	2.0	4.25	2.75	4.0	4.75
Cirsium arvense (L.) Scop.	0.0	0.0	0.25	0.0	0.0	0.0
Fallopia convolvulus L.	0.0	0.0	0.0	0.0	0.5	0.0
Galeopsis tetrahit L.	0.0	0.0	0.25	0.0	0.0	0.0
Myosotis arvensis (L.) Hill.	0.0	0.25	0.25	1.0	0.25	1.75
Persicaria lapathifolia L.	0.0	0.0	0.0	0.0	0.0	0.25
Rumex crispus L.	0.75	0.25	1.0	0.75	0.75	1.5
Sinapis arvensis L.	0.25	1.0	0.0	0.25	0.25	0.25
Stellaria media (L.) Vill.	0.75	0.5	1.75	0.0	0.5	1.75
Thlaspi arvense L.	0.0	0.0	0.25	0.5	0.25	0.25
Tripleurospermum perforatum (Merat) M. Lainz	0.50	0.5	0.75	0.0	0.0	0.0
Viola arvensis Murray	1.0	1.25	3.75	2.0	2.5	2.75
All weed seeds	6.25	5.75	12.5*	7.25	9.0	13.25*
LSD$_{05}$	4.34					

Note. * - essential differences at 95% level of probability, compared to control treatment

Table 2. Weed seed bank in winter wheat crop soil of 0–20 cm layer, weed seeds in 100 g of air-dry soil (Pilipavičius et al., 2010a).

Direct chemical soil weed seed bank control is rather indeterminable, therefore ecological and cultural weed control methods should be applied for weed seed control. One of possibilities to control weed seed bank is harvesting cereal at earlier stage of maturity for the whole plant silage. It is essentially important factor, decreasing the amount of coming new weed seeds to the soil terminating weed seed rain while only a small part of weed seeds pours at milk or milk–dough stages of cereal maturity (Pilipavičius & Lazauskas, 2000, Pilipavičius, 2002, 2006) and delivering more fodder for animals from the same plot area (Pilipavičius, 2007b, 2012). When the amount of new weed seeds getting into the soil decreases, soil is cleaning quicker from them (Pilipavičius, 2004). Another important factor is weed seed position in the soil. The more weed propagation rudiments are decreased in the top soil layer, the less is weediness of the crop – the number and the mass of weeds (Pilipavičius, 2007a, Pilipavičius et al., 2009). Existing weed seed bank and vegetative weed parts in soil can be managed in non-chemical way. According to theoretical preconditions and data of the experiments, it is proved that total turnover of the arable soil layer in organic agriculture is a very important means of weed control decreasing weediness of the crop and increasing harvest. Comparing technological ploughing processes and ploughs, it was concluded that two layer ploughs can help to carry out this process the most effectively in organic agriculture (Lazauskas & Pilipavičius, 2004) and can be successfully applied in conventional agriculture.

2.3 Crop weediness

Weediness of winter wheat crop was established by a quantitative-weight method. Four samples with wire rim 50 x 50 cm (0.25 m²) were taken from each experimental plot to establish weed density and mass (Pilipavičius, 2005) in autumn and spring during the winter wheat tillering before and after spraying with herbicides. Collected weeds were air-dried and distributed into species. Nomenclature of Latin plant names was based on the Institute of Botany's edition *Vascular plants of Lithuania* (Gudžinskas, 1999).

Twenty weed species in experimental field were established in autumn before spraying with herbicides. Seventeen of them were annual and three perennial ones. After autumn spraying with herbicides number of established weed species increased by one annual and one perennial, however, weed biomass essentially decreased (Fig. 2).

After overwintering the number of weed species in the crop principally did not change while after conventional spring spraying with herbicides it decreased by one annual weed species. The main change in the number of weed species was influenced by appearance and disappearance of weeds as *Chenopodium album* L., *Erysimum cheiranthoides* L., *Sinapis arvensis* L., *Myosotis arvensis* (L.) Hill, *Veronica arvensis* L., *Cerastium arvense* L., *Fumaria officinalis* L., *Viola arvensis* Murray, *Galeopsis tetrahit* L., *Polygonum aviculare* L. and some other weeds that were low in number. It was confirmed that more weed species were established in the crop (Fig. 3 & 4) than in soil weed seed bank (Table 2). Before autumn application of herbicides there were no considerable differences in weediness of winter wheat but after herbicide spraying the number of weeds decreased by 15.4 – 28.4% and their air-dry biomass lessened even up to 56.8% (Fig. 2). Monitor (Sulphosulphuron) was the most effective herbicide in destroying weeds in winter wheat crop in autumn compared to the control treatment with no herbicide application in autumn and other herbicides used. Crop weediness decreased by 18 weeds in m⁻² and by 9.5 g m⁻² of air-dry mass; the reduction comprised 28.4% and 56.8% respectively. Assessing the effectiveness of different herbicide active substances, it was established that after autumn spraying the number of weeds decreased from 32.4% to 91.7% compared to not sprayed by herbicides in autumn control. Assessing winter wheat crop in spring, it was determined that autumn herbicide application resulted in reduced weediness also after crop wintering. It was established that herbicide application in autumn at early stages of winter wheat development significantly decreased crop weediness as well in spring vegetation after over-wintering (Fig. 2). Later, Latvian researchers received analogous results while winter wheat crop in plots where herbicides were applied in autumn was more even, denser and better developed than in plots just with spring herbicide application. In the spring-treated plots the crops became thin and in open places weed plants that were not controlled by the herbicides could regrow and develop well during the growing season up to harvest time (Vanaga et al., 2010).

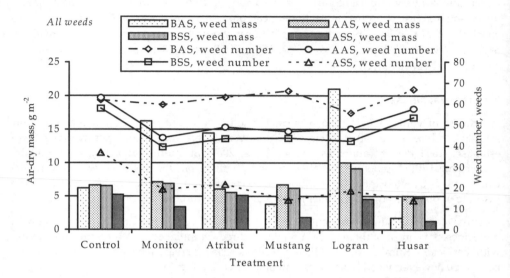

Note. LSD$_{05}$=24.18 for air-dry weed mass g m^{-2} before autumn application of herbicides; LSD$_{05}$=7.34 for air-dry weed mass g m^{-2} after autumn application of herbicides; LSD$_{05}$=22.85 for weed number before autumn application of herbicides; LSD$_{05}$=23.28 for weed number after autumn application of herbicides; BAS – before autumn spraying, AAS – after autumn spraying, BSS – before spring spraying, ASS – after spring spraying

Fig. 2. Winter wheat crop weediness before and after autumn and spring application of herbicides.

Annual weeds such as *Centaurea cyanus* (Fig. 6), *Raphanus raphanistrum* (Fig. 8), *Thlaspi arvense* (Fig. 11) and *Tripleurospermum perforatum* (Fig. 12) prevailed in winter wheat crop whereas among perennial weeds only of *Sonchus arvensis* and *Plantago major* and a few plants of *Antennaria dioica* and *Poa trivialis* emerged in the crop (Fig. 4). Short-lived annual weeds in the crop of winter wheat in autumn and after application of the intended herbicides comprised 96%-100% and 93%-100% respectively, whereas in spring before and after application of background spring herbicides they comprised 89%-100% and 99%-100% respectively (Pilipavičius et al., 2010a) (Fig. 2 & 3).

This means that perennial weeds are better adapted to wintering than the short-lived ones because the increase of air-dry biomass up to 11% of perennial weeds was established in spring before the application of chemical weed control measures (Fig. 4). Assessing the effectiveness of diverse herbicides, it was established that the number of weeds after spraying in autumn decreased by 32–91% compared to the control treatment plot not sprayed in autumn. Assessing winter wheat crop in spring, it was determined that autumn application of herbicides resulted in lessened crop weediness after its wintering even before spring spraying (Pilipavičius et al., 2010a). The number of weeds decreased by 70–92% compared to the control (Fig. 2) with no autumn herbicide application.

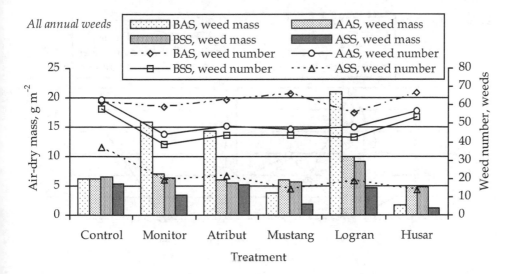

Note: Annual weed species: *Capsella bursa-pastoris* (L.) Medik (Fig. 5), *Centaurea cyanus* L. (Fig. 6), *Galium aparine* L. (Fig. 7), *Raphanus raphanistrum* L. (Fig. 8), *Raphanus sativus* L. (Fig. 9), *Stellaria media* (L.)Vill (Fig. 10), *Thlaspi arvense* L. (Fig. 11), *Tripleurospermum perforatum* (Merat) M.Lainz (Fig. 12), other annual weeds - *Chenopodium album* L., *Erysimum cheiranthoides* L., *Sinapis arvensis* L., *Myosotis arvensis* (L.) Hill, *Veronica arvensis* L., *Cerastium arvense* L., *Fumaria officinalis* L., *Viola arvensis* Murray, *Galeopsis tetrahit* L., *Polygonum aviculare* L.;
BAS – before autumn spraying, AAS – after autumn spraying, BSS – before spring spraying, ASS – after spring spraying

Fig. 3. Annual weeds in winter wheat crop before and after autumn and spring application of herbicides.

Comparing weed over-wintering possibilities as crop weediness change dynamics, is important to pay attention for the development of one average weed plant air-dry mass (g per plant). In our experiment, it was shown that short-lived annual weeds successfully survived winter frosts even increasing its average one weed plant mass (Table 3).

Though, moderated increase of annual weed mass during winter time was in conformity with the research hypothesis that many annual weeds successfully survive winter time as earlier was not usual. Naturally, perennial weeds have the highest tolerance to winter frosts as biologically well adapted to over-wintering. The highest mass of one its over-wintered plant before spring application of herbicides reaches 6.4 gram increasing it from 2.52 gram in autumn. However, perennial weeds were rare in our experimental field (Fig. 4) and even were not present in some experimental plots overall (Table 3). Received analogous cereal crop weediness variations mostly depend on experimental field weediness heterogenity especially in intensive operating fields with low weediness as each weed observation is made randomly (Pilipavičius, 2005). Similar trend of annual and perennial weed populations has been noticed by Geisselbrecht-Taferner et al., 1997; Colbach et al., 2000; Rew et al., 2001 and other researchers. Dominating annual weeds in the winter wheat crop directly influenced all weed average plant mass that remained analogous to average annual weed plant mass (Table 3).

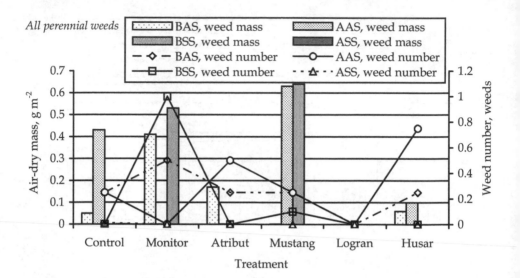

Note. Perennial weed species - *Antennaria dioica* L., *Plantago major* L., *Poa trivialis* L., *Sonchus arvensis* L.; BAS – before autumn spraying, AAS – after autumn spraying, BSS – before spring spraying, ASS – after spring spraying.

Fig. 4. Perennial weeds in winter wheat crop before and after autumn and spring application of herbicides

Spring spraying with composite of herbicides *Sekator* and *MCPA* was low effective as standard technology in control treatment (average weed plant mass increase from 0.10 gram in autumn till 0.14 gram in spring after application). Winter wheat crop spring spraying with herbicides was either low effective as average weed plant mass has tendency to increase comparing weed average mass in autumn after experimental application of herbicides or in spring before application of standard chemical weed control technology with average weed one plant mass after spring application of herbicides (Table 3).

Weed air-dry biomass in the crop before spraying by herbicides in spring regularly depended on left weed air-dry mass in the crop after autumn spraying by herbicides r = 0.608** using chemical weed control (according to the experimental design) in early winter wheat growth and development stages. The reliable linear dependence (1) best described this regularity.

$$y = 4.077 + 0.335\ x;\ P = 0.0016 \tag{1}$$

Weeds left in the crop after autumn spraying by herbicides reliably increased crop weediness in spring before conventional spraying. Weed air-dry biomass of 1 g m^{-2} left in winter wheat crop in autumn increased crop weediness by 0.335 g m^{-2} in spring after renewing of vegetation (Pilipavičius et al., 2010b). Other researchers (Spiridonov et al., 2006) have affirmed high biological and economical efficiency of autumn application of herbicide (*Difezan*) in winter wheat crop in comparison with the conventional spring period of treatment. The successful post emergence control of weeds in the winter cereal crops at the BBCH 11-25 in autumn with herbicide *Atlantis* was reported by Brink and Zollkau 2004.

Treatment	Crop weediness average one weed plant air-dry mass, g per weed			
	Before autumn application of herbicides	After autumn application of herbicides	Before spring application of herbicides	* After spring application of herbicides Sekator 300 g ha⁻¹ and MCPA 1 L ha⁻¹ composite
All annual weeds				
Control	0.10	0.11	0.11	0.14
Monitor	0.27	0.16	0.17	0.18
Atribut	0.23	0.13	0.13	0.24
Mustang	0.06	0.13	0.13	0.13
Logran	0.38	0.21	0.22	0.25
Husar	0.03	0.09	0.09	0.09
All perennial weeds				
Control	0.20	1.72	-	0.10
Monitor	0.82	-	0.53	-
Atribut	0.68	-	-	-
Mustang	-	2.52	6.4	-
Logran	-	-	-	-
Husar	0.24	0.13	-	-
All weeds				
Control	0.10	0.11	0.11	0.14
Monitor	0.27	0.16	0.17	0.18
Atribut	0.23	0.12	0.13	0.24
Mustang	0.06	0.14	0.14	0.13
Logran	0.38	0.21	0.22	0.25
Husar	0.03	0.09	0.09	0.09

Note. See experimental design.

Table 3. Average one weed plant air-dry mass change in winter wheat crop with autumn and spring applications of herbicides.

Separate weed species of winter wheat crop responded adequately to the autumn application of herbicides as the whole crop weed community. Winter annual weed *Capsella bursa-pastoris* (Fig. 5) was moderate in density and mass, however, it had tendency to increase in winter wheat crop without autumn application of herbicides. However, higher initial *Capsella bursa-pastoris* population in plots of *Monitor* and *Logran* treatments was inhibited and decreased with modern autumn application of herbicides (Fig. 5). Winter annual weed *Centaurea cyanus* (Fig. 6) was dominant in winter wheat crop agrophytocenoses before autumn application of herbicides. It comprised 4.4%-8.4% of weed density and 1.6%-17.3% of total weed biomass.

Density and mass of *Centaurea cyanus* had tendency to decrease after herbicide application in autumn. In overwintered crop *Centaurea cyanus* decreased by 5% in density and by 79.6%-19.8% in mass comparing it with the crop before winter time. Decrease of *Centaurea cyanus* density was similar in control treatment without autumn application of herbicides, however its mass decreased just by 3.8%, i.e. the *Centaurea cyanus* mass decrease after overwintering was by 5.2-20.9 times lower than in plots with autumn chemical weed control (Fig. 6). The other winter annual weed *Galium aparine* (Fig. 7) in winter wheat was less numerous and less in biomass comparing to *Centaurea cyanus* (Fig. 6), *Capsella bursa-pastoris* (Fig. 5) or *Raphanus raphanistrum* (Fig. 8). During wintering *Galiun aparine* without autumn herbicide application, density in the crop increased in mass by 56.5% and in density by 16.7%. It formed more competitive weed community against winter wheat in spring. Autumn application of herbicides subserved effectively *Galium aparine* control (Fig. 7). Summer annual weed *Raphanus raphanistrum* (Fig. 8) initial population made 3.2-9.8 weeds per square meter and 1.6-8.2 g m^{-2} of air-dry mass. It comprised 0.9%-2.6% of total weed density and 2.5%-13.1% of total weed mass. After autumn application of herbicides, *Raphanus raphanistrum* density and mass decreased by 21.9%-36.2% and 38.1%-65.5% accordingly.

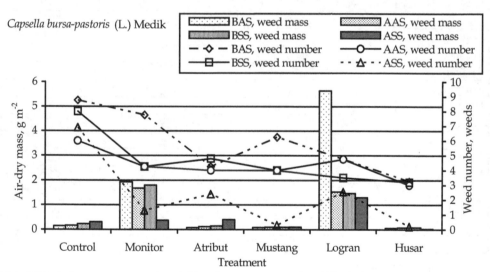

Note. BAS – before autumn spraying, AAS – after autumn spraying, BSS – before spring spraying, ASS – after spring spraying

Fig. 5. *Capsella bursa-pastoris* (L.) Medik in winter wheat crop before and after autumn and spring application of herbicides.

In spring, renewing crop vegetation, *Raphanus raphanistrum* plants made 0.8%-1.9% of total weed density and 2.2%-6.2% of total weed mass. It was visible decrease of this weed population, hence, it showed that biologically summer annual weeds already can successfully survive during winter while conventionally it should not happen (Fig. 8). Other winter wheat crop weed belonging to *Brassicaceae* family, *Raphanus* genus was *Raphanus sativus* (Fig. 9). *Raphanus sativus* was present just in one treatment, was low in number and biomass. *Raphanus sativus* separately had no substantial effect on crop weediness and belonged to temporal weed flora element in the crop.

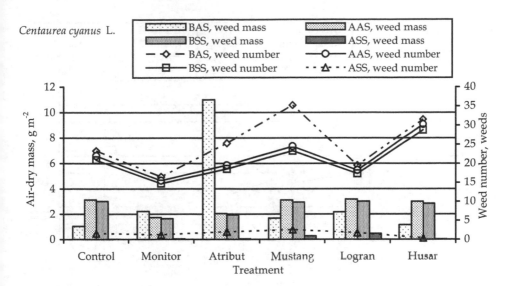

Note. BAS – before autumn spraying, AAS – after autumn spraying, BSS – before spring spraying, ASS – after spring spraying

Fig. 6. *Centaurea cyanus* L. in winter wheat crop before and after autumn and spring application of herbicides.

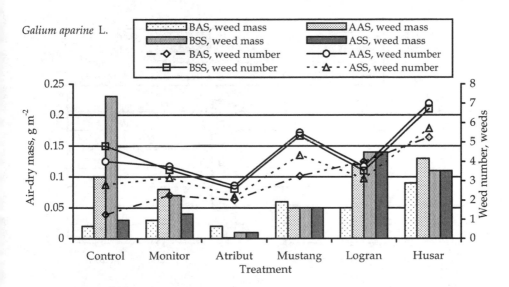

Note. BAS – before autumn spraying, AAS – after autumn spraying, BSS – before spring spraying, ASS – after spring spraying

Fig. 7. *Galium aparine* L. in winter wheat crop before and after autumn and spring application of herbicides.

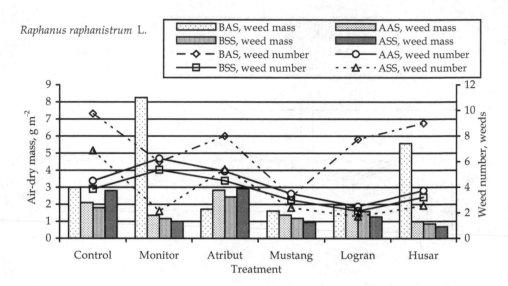

Note. BAS – before autumn spraying, AAS – after autumn spraying, BSS – before spring spraying, ASS – after spring spraying

Fig. 8. *Raphanus raphanistrum* L. in winter wheat crop before and after autumn and spring application of herbicides.

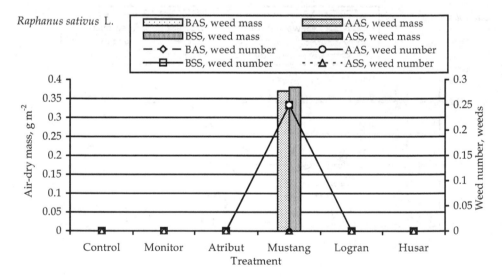

Note. BAS – before autumn spraying, AAS – after autumn spraying, BSS – before spring spraying, ASS – after spring spraying

Fig. 9. *Raphanus sativus* L. in winter wheat crop before and after autumn and spring application of herbicides.

More important winter wheat crop weed was *Stellaria media* (Fig. 10). Biologically belonging to the summer annual ephemeral weeds *Stellaria media* showed ability to survive during winter (either as *Raphanus raphanistrum*) that was not usual for the conventional Lithuanian conditions (Aleksandravičiūtė et al., 1961). *Stellaria media* overwintering was not effected even by the unfavourable wintering conditions (see meteorological conditions, subchapter 2.1). Chemical weed control, especially in autumn was not successful for control of this weed. On the contrary, *Stellaria media* was initiated to growth after herbicide application in autumn (Fig. 10). It could be influenced by the *Stellaria media* biological quality to launch new branches (stems and roots) from each damaged or fresh node. Consequently, it makes *Stellaria media* control and evaluation even more complicated.

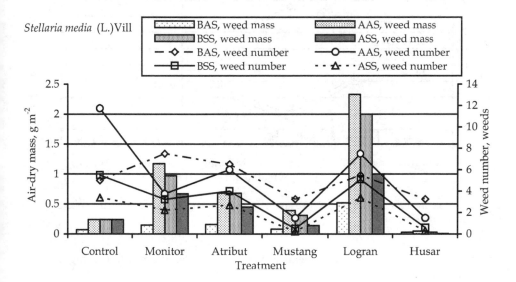

Note. BAS – before autumn spraying, AAS – after autumn spraying, BSS – before spring spraying, ASS – after spring spraying

Fig. 10. *Stellaria media* (L.)Vill in winter wheat crop before and after autumn and spring application of herbicides.

Winter annual weed *Thlaspi arvense* made 1.8%-3.2% of total weed density and 0.23%-13.6% (the highest excess in *Logran* treatment plot) of total crop weed mass before autumn herbicide application (Fig. 11). After autumn chemical weed control applied in winter wheat crop it had tendency to decrease in number by 2.4-1.6 times and till 11 times in mass. Over wintered *Thlaspi arvense* has trivial increase in mass of control and *Husar* treatments, trivial decrease in *Mustang* treatment and sustained autumn level in other treatment plots. Autumn application of herbicides as *Mustang, Logran* and *Husar* in winter wheat crop lead to decease of *Thlaspi arvense* mass after spring spraying while after autumn application of *Monitor* and *Atribut Thlaspi arvense* mass increased 1.6 and 6.6 times after spring spraying accordingly. In mentioned last two autumn treatment cases (*Monitor* and *Atribut*) standard spring application of herbicides was ineffective.

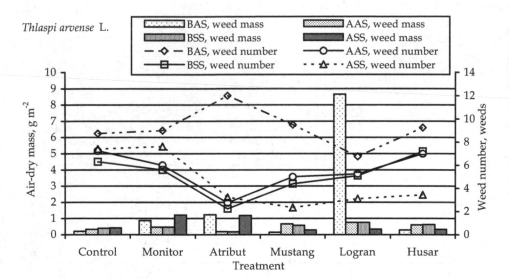

Note. BAS – before autumn spraying, AAS – after autumn spraying, BSS – before spring spraying, ASS – after spring spraying

Fig. 11. *Thlaspi arvense* L. in winter wheat crop before and after autumn and spring application of herbicides.

Winter annual weed *Tripleurospermum perforatum* was spread enough homogenously in the experimental field except two excesses in *Monitor* and *Logran* treatments at early winter wheat crop development stages in autumn before applying chemical weed control means, that was reduced till averagely general *Tripleurospermum perforatum* weediness in the autumn crop after herbicides application (Fig. 12).

In spring *Tripleurospermum perforatum* plant development was limited by standard spring herbicide application (composite of herbicides Sekator 300 g ha-1 and MCPA 1 L ha-1 with a winter wheat growth regulator Cycocel 1.5 L ha-1 at BBCH 22-23) while in control treatment without autumn herbicide application it was not effective. It could be concluded that even without significant crop weediness decrease in autumn growth period of winter wheat weeds are damaged and weaken what is essentially highlighted at conventional spring application of herbicides in winter cereals.

Tripleurospermum perforatum (Merat) M.Lainz

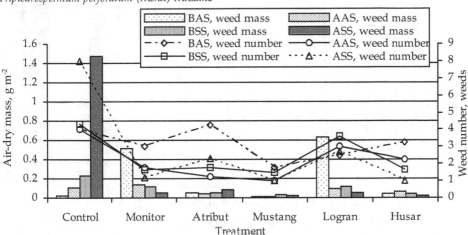

Note. BAS – before autumn spraying, AAS – after autumn spraying, BSS – before spring spraying, ASS – after spring spraying

Fig. 12. *Tripleurospermum perforatum* (Merat) M.Lainz in winter wheat crop before and after autumn and spring application of herbicides.

2.4 Crop productivity

Winter wheat grain yield was expressed in moisture of 14% and absolutely clean mass. Grain moisture was established by drying grains in a thermostat at the temperature of 105 °C until they reached the constant weight. The biggest winter wheat yield 4.55 t ha⁻¹ was got using *Husar* (Iodosulphuron-methyl-sodium) in autumn and the least one 3.68 t ha⁻¹ in control treatment without autumn application of herbicides (Fig. 13). Average winter wheat grain yield in Lithuania in the same year was 2.46 t ha⁻¹ (Statistics Lithuania, 2011). Average winter wheat grain yield in Lithuania in 2006 comparing it with 2005 decreased by 35% and general yield by 43.6% (Market research, 2007). Modern technologies of herbicide application at early stages of winter wheat development in autumn are very promising while comparing winter wheat grain yield in our experiment with average winter wheat yield in Lithuania. During 2005-2010 the highest winter wheat grain average yield was 4.76 t ha⁻¹ in 2008 (Statistics Lithuania, 2011), i.e. just by 4.4% higher than in our best treatment. Though, winter wheat vegetation 2007-2008 meteorological conditions for winter wheat growing were better than during 2005-2006 vegetation.

In our experiments (Pilipavičius et al., 2010b) essential winter wheat grain yield increase by 0.62-0.87 t ha⁻¹ was established after use of *Monitor* (Sulphosulphuron), *Mustang* (Florasulam + 2.4-D 2-ethylhexyl ester), *Logran* (Triasulphuron) and *Husar* (Iodosulphuron-methyl-sodium) compared with in autumn unsprayed control. Winter wheat grain yield increase reached 25%, 20%, 19% and 24% accordingly, and after spraying by *Atribut* (Propoxycarbazone-sodium) grain yield had tendency to increase (Fig. 13).

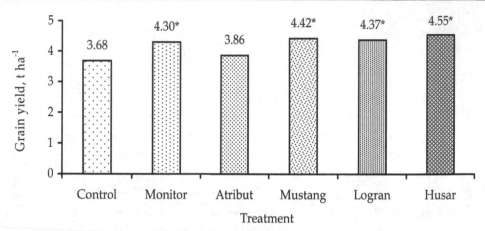

Fig. 13. Winter wheat grain yield of crop with autumn application of herbicides.
(Pilipavičius et al., 2010b)

Evaluating grain chemical composition (Table 4), it was established that independently of
used herbicides, nutritional composition of grain was not radically different. The amount of
crude protein in winter wheat grain changed from 8.9% to 9.9% and there was found from
1.43% to 2.1% of crude fat, from 1.93% to 2.67% of crude fibre and from 1.2% to 1.7% of
crude ash (Pilipavičius et al., 2010b).

Treatment	in grain dry matter							
	Crude protein		Crude fat		Crude fibre		Crude ash	
	%	t ha^{-1}	%	t ha^{-1}	%	t ha^{-1}	%	t ha^{-1}
Control	8.9	0.328	1.53	0.056	2.23	0.082	1.2	0.044
Monitor	9.4	0.404	1.43	0.061	2.7	0.116	1.5	0.064
Atribut	9.1	0.351	1.69	0.065	2.59	0.099	1.2	0.046
Mustang	9.9	0.437	2.1	0.092	2.43	0.107	1.3	0.057
Logran	9.3	0.406	2.00	0.087	1.93	0.084	1.7	0.74
Husar	9.5	0.432	1.82	0.082	2.67	0.120	1.7	0.077

Table 4. Grain chemical composition of winter wheat crop with autumn application of
herbicides (Pilipavičius et al., 2010b).

A statistically reliable reverse linear correlation r = -0.565** (Pilipavičius et al., 2010b) was
established between crop weed number after spring spraying by herbicides and winter
wheat grain yield (2).

$$y = 4.517 - 0.091 \; x; P = 0.004 \tag{2}$$

Evaluating dependence between weed air-dry biomass after herbicide application in spring
and winter wheat grain yield (3), it was established statistically reliable reverse linear
correlation r = -0.438* (Pilipavičius et al., 2010b).

$$y = 4.454 - 0.0128 \; x; P = 0.032 \tag{3}$$

It was in conformity as dependence of winter wheat grain yield on weed density (2) and coincided (3) with the law of crop productivity (Lazauskas, 1990, 1993; Pilipavičius et al., 2009).

3. Conclusion

Winter wheat *Triticum aestivum* L. is sensitive to weed competition in early stages of growth and development. In conventional technologies herbicides in winter cereals are applied in spring. Therefore, perennial and annual (especially winter annual) weeds over-wintered successfully in cereals have favourable conditions to grow and compete with cereals when vegetation in spring is renewing.

Perennial weeds are well adapted to over-wintering biologically while their air-dry mass increases till 11% during wintering. Conventionally, it is opposite to annual weeds. However, it was established that during winter time in winter wheat crop annual weeds, even some summer annual ones, had increase adaptivity of successful surviving winter frosts and accumulated higher one plant average mass by 5-6% during winter time. Moderated increase of annual weed mass during winter time was in conformity with the research hypothesis that many annual weeds successfully survive winter time as earlier was not usual.

Separate weed species of winter wheat crop responded adequately to the autumn application of herbicides as the whole crop weed community. Standard spring spraying as conventional technology with composite of herbicides is insufficient effective while average weed plant mass increase from 0.10 gram in autumn till 0.14 gram in spring after herbicide application, i.e. increase make 40%. Weeds had tendency to spread in winter wheat crop without autumn application of herbicides and formed more competitive weed community against winter wheat in spring. It can be concluded that even without significant crop general weediness decrease in autumn by herbicides weeds are damaged and weaken what is essentially highlighted at conventional spring application of herbicides in winter cereals. Weeds left in the crop after autumn spraying by herbicides reliably by 33.5% increased crop weediness in spring before conventional spraying, as described in regression equation y = 4.077 + 0.335 x.

Winter wheat yield and its agrophytocenoses weed air-dry mass in spring crop was in conformity with the law of crop productivity. Winter wheat grain yield depended on weed air-dry mass and was described by the reverse linear correlation r =-0.438* and regression y = 4.45–0.013x analyses.

Modern technologies of herbicide application at early stages of winter wheat development in autumn are very promising while in our experiment it gives increase in grain yield till 25% and comparing it with average winter wheat yield in Republic of Lithuania during the same period it was got increase in grain yield from 50% to 85%.

For the best weed control and winter wheat yield results herbicides should be applied in autumn, especially when the weather is favourable for prolonged development of weeds even at low density of perennial ones in the crop.

4. Acknowledgment

We thank Mrs. Vilma Pilipavičienė for her help in English paper correction.

We thank Journal of Food, Agriculture and Environment (publishing house - WFL Publisher OY, Finland, www.world-food.net) editorial board for permission to use research material from published author articles in Journal of Food, Agriculture and Environment, volume 8, No1 and No2 of 2010 issue year (Letter of allow from JFAE-Editorial Office Team on 29th August, 2011).

5. References

Aleksandravičiūtė, B.; Apalia, D.; Brundza, K. et. al. (1961). *Flora of Lithuania / Lietuvos TSR flora.* Vol.3., (pp. 272), 662 p. Mokslas, Vilnius, Lithuania

Barberi, P.; Silvestri, N. & Bonani, E. (1997). Weed communities of winter wheat as influenced by input level and rotation. *Weed Research,* Vol.37, No.5, pp. 301-313, ISSN 0043-1737

Brink, A. & Zollkau, A. (2004). Optimisation of grass weed control with Atlantis®WG in winter cereals. *Journal of Plant Diseases and Protection,* Sp.iss.XIX, pp. 637-646, ISSN 0938-9938

Basalykas, A.; Bieliukas, K. & Chomskis, V. (1958). *Lietuvos TSR fizinė geografija / Physical geography of Lithuania.* Vilnius, Vol.1, pp. 501-504

Characteristics of wheat varieties, listed in national list of plant varieties (2011). Kviečių veislių, įrašytų į nacionalinį augalų veislių sąrašą, charakteristikos (2011). 02.08.2011, Available from <http://www.avtc.lt/assets/files/veisliu%20aprasymai/Kvieciu%20veisliu%20ira sytu%20i%20NS%20charakteristikos.doc>

Colbach, N.; Dessaint, F. & Forcella, F. (2000). Evaluating field-scale sampling methods for the estimation of mean plant densities of weeds. *Weed Research,* Vol.40, pp. 411-430, ISSN 0043-1737

Faravani, M. & Khalghani, J. (2004). Synchronized weed chemical control and wheat harvesting. *Journal of Food, Agriculture & Environment,* Vol.2, No.1, pp. 202-204, ISSN 1459-0255

Geisselbrecht-Taferner, L., Geisselbrecht, J. & Mucina, L. (1997). Fine-scale spatial population patterns and mobility of winter-annual herbs in a dry grassland. *Journal of Vegetation Science,* Vol.8, pp. 209-216, ISSN 1100-9233

Gudžinskas, Z. (1999). *Vascular plants of Lithuania,* 212 p., ISBN 9986-662-14-1, Institute of Botany, Vilnius, Lithuania

Hanee, M. Al-Dmoor. (2008). Quality profile of the most commonly grown wheat varieties in Jordan. *Journal of Food, Agriculture & Environment,* Vol.6, No.3&4, pp. 15-18, ISSN 1459-0255

House, G. J. (1989). Soil arthropods from weed and crop roots of an agroecosystem in a wheat-soy bean-corn rotation: Impact of tillage and herbicides. *Agriculture, Ecosystems & Environment,* Vol.25, No.2-3, pp. 233- 244, ISSN 0167-8809

Koch, W. & Hurle, K. (1998). *Grundlagen der Unkrautbekämpfung.* 2nd edition. Verlag Eugen Ulmer Suttgart, 207 p., ISBN-13: 978-3825205133

Lazauskas, P. (1990). *Agrotechnics against weeds*. Monography. 214 p., Mokslas, ISBN 5-420-00206-X, Vilnius, Lithuania

Lazauskas, P. (1993). The law of crop performance as a basis of weed control. *8ᵗʰ EWRS symposium quantitative approaches in weed and herbicide research and their practical application*, pp. 71-77, Braunschweig

Lazauskas, P. & Pilipavičius, V. (2004). Weed control in organic agriculture by two layer plough. *Journal of Plant Diseases and Protection*, Sp.iss.XIX, pp. 573-580, ISSN 0938-9938

Lithuanian national list of plant varieties (2011). 02.08.2011, Available from <http://www3.lrs.lt/pls/inter3/dokpaieska.dok_priedas?p_id=44875>

Market research / Rinkotyra (2007). *Agricultural and food products / Žemės ūkio ir maisto produktai*. Eds. Šilaikienė, V. Lithuanian institute of agrarian economics, Vilnius, Vol.1, No.35, 112 p. ISSN 1392-6101

Niemann, P. (1981). *Schadschwellen in der Unkrautbekämpfung*. Schriftenreihe des BML, Reihe A: Angewandte Wissenschaft, Heft 257. ISBN-13: 978-3784302577

Pilipavičius, V. & Lazauskas, P. (2000). Regulierung der Unkrautsamenverbreitung im Getreide. *Journal of Plant Diseases and Protection*, Sp.iss.XVII, pp. 469-472, ISSN 0938-9938

Pilipavičius, V. (2002). Preventive weed control in lower input farming system. *5ᵗʰ EWRS Workshop on Physical Weed Control*, Pisa, Italy, pp. 46-56

Pilipavičius, V. (2004). Changes in soil weed seed bank according to spring barley maturity stages. *Agronomy Research*, Vol.2, No.2, pp. 217-226, ISSN 1406-894X

Pilipavičius, V. (2005). Competition of weeds and spring barley in organic and conventional agriculture. *Vagos*, Vol.68, No.21, pp. 30-43, ISSN 1648-116X

Pilipavičius, V. (2006). Three- year assessment of weed dynamics in herbicide-free barley crop: a field study. *Žemdirbystė-Agriculture*, Vol.93, No.3, pp. 89-98, ISSN 1392-3196

Pilipavičius V. (2007a). Whole plant silage nutritive value from spring barley of different maturity. *Veterinarija ir Zootechnika*. Vol.37, No.59, pp. 61-66, ISSN 1392-2130

Pilipavičius, V. (2007b). *Weed spreading regularity and adaptivity to abiotical factors*. Summary of the review of scientific works presented for dr. habil. procedure: Biomedical sciences, agronomy. Akademija, Lithuanian university of agriculture, 30 p.

Pilipavičius, V.; Lazauskas, P. & Jasinskaitė, S. (2009). Weed control by two layer ploughing and post-emergence crop tillage in spring wheat and buckwheat. *Agronomy Research*, Vol.7, Special issue 1, p. 444-450, ISSN 1406-894X

Pilipavičius, V.; Aliukonienė, I. & Romaneckas, K. (2010a). Chemical weed control in winter wheat (*Triticum aestivum* L.) crop of early stages of development: I. Crop weediness. *Journal of Food, Agriculture & Environment*, Vol.8, No.1, pp. 206-209, ISSN 1459-0255

Pilipavičius, V.; Aliukonienė, I.; Romaneckas, K. & Šarauskis, E. (2010b). Chemical weed control in the winter wheat (*Triticum aestivum* L.) crop of early stages of development: II. Crop productivity. *Journal of Food, Agriculture & Environment*, Vol.8, No.2, pp. 456-459, ISSN 1459-0255

Pilipavičius, V. (2012). Spring barley over-ground biomass digestibility in vitro. *Veterinarija ir Zootechnika*, Vol.57, No.79, (*in press*) ISSN 1392-2130

Rabotnov, T.A. (1958). Methods of botanical investigations. *Journal of Botany*, Vol.XLIII(43), No.11, pp. 1572-1581

Rew, L.J. & Cousens, R.D. (2001). Spatial distribution of weeds in arable crops: are current sampling and analytical methods appropriate? *Weed Research*, Vol.41, pp. 1-18, ISSN 0043-1737

Rimavičienė, G. (2005). Žemės ir miškų ūkio augalų pesticidų katalogas / The catalogue of agricultural and forest pesticides. pp.173-232, ISBN 9955597143, Akademija, Kėdainiai region, Lithuania

Sikkemaa, P. H.; Browna, L.; Shropshirea, C. & Soltani, N. (2007). Responses of three types of winter wheat (*Triticum aestivum* L.) to spring-applied post-emergence herbicides. *Crop Protection*, Vol.26, No.5, pp. 715- 720, ISSN 0261-2194

Spiridonov, Yu.Ya.; Nikitin, N.V. & Raskin, M.S. (2006). Efficiency of autumn application of herbicide Difezan on winter wheat crops under conditions of the European non-chernozem zone of Russia. *Plant Protection-Защита растений*, Vol.30, No.1. (Strategy and tactics of plant protection). Minsk, pp. 158-160, ISBN 9856471346

SPSS Science (2000). *SigmaPlot® 2000 user's guide. Exact graphs for exact science*. 435 p. ISBN-13: 978-1568272320

Stasinskis, E. (2009). Effect of preceding crop, soil tillage and herbicide application on weed and winter wheat yield. *Agronomy Research*, Vol.7, No.1, pp. 103-112, ISSN 1406-894X

Statistics Lithuania (2011). Statistikos departamentas prie LRV (2011), M5010302: Agricultural crops in country / Žemės ūkio augalai šalyje, 02.08.2011, Available from
<http://db1.stat.gov.lt/statbank/SelectVarVal/saveselections.asp>

Summary for policymakers, Climate change (2007). The physical science basis. Contribution of working group I to the fourth assessment report of the intergovernmental panel on climate change. Intergovernmental panel on climate change. 05.02.2007, Available from
<http://ipcc-wg1.ucar.edu/wg1/Report/AR4WG1_Print_SPM.pdf>

Tarakanovas, P. (1997). A new version of the computer programme for trial data processing by the method of analysis of variance. *Žemdirbystė-Agriculture*, Vol.60, pp. 197-213, ISSN 1392-3196

Vanaga, I.; Mintale, Z. & Smirnova, O. (2010). Control possibilities of *Apera spica-venti* (L.) P.Beauv. in winter wheat with autumn and spring applications of herbicides in Latvia. *Agronomy Research*, Vol.8, Special Issue II, pp. 493–498, ISSN 1406-894X

Warwick, A.M. (1984). Buried seeds in arable soils in Scotland. *Weed Research*, Vol.24, No.4, pp. 261–268, ISSN 0043-1737

14

Influence of Degree Infestation with *Echinochloa crus–galli* Species on Crop Production in Corn

Teodor Rusu and Ileana Bogdan
University of Agricultural Sciences and Veterinary Medicine Cluj-Napoca
Romania

1. Introduction

Corn continues to be globally one of the main crops, ranking third, after wheat and rice. In Romania it is the main agricultural plant, whose economic importance, especially in the private sector is growing. Given the particularities of this culture, with particular reference to the high sensitivity at infestation with weeds, especially in the early stages of vegetation, corn crop is feasible only if weeds are controlled through various methods. The damage caused by weeds in maize crop are mostly of 30-70% (Sarpe, 1975; Budoi and Penescu, 1996; Oancea, 1998; Bilteanu, 2001; Berca, 2004; Gus et al., 2004; Bogdan et al., 2005; Rusu, 2008) and when the infestation is strong, culture can be fully compromised. The presence in a culture of a small number of weeds is not harmful, but damage caused by weeds grow along with increasing the degree of infestation, depending on the species and age of occurrence of weeds, soil and climate conditions and the moment when weeds are combat (Paunescu, 1996; Bosnic and Swanton, 1997; Perron and Legere, 2000; Bogdan et al., 2001; Fukao et al., 2003; Clay et al., 2005; Rusu et al., 2009). Therefore, specifying the economic threshold of pest is difficult to establish considering the fact that the number of researches in this field until now, is reduced.

Echinochloa crus-galli is one species with a large requirement to water being able to behave as mesophita, mesohygrophita, hygrophita and hygrohelophita (Anghel et al., 1972; Bogdan et al., 2007). It is especially met on the luvosoils, fertile and wet soils, being wide-spread in all the country but in the north sides it has a lower abundance and general dominance than in the south ones. *Echinochloa crus-galli* is met growing on a large variety of soils and grains, from clay sand or sandy clay soils to medium hard soils. The soils with a relative big capacity of water holding and large fertility insure an ideal sublayer.

Echinochloa crus-galli is a weed with a fascicular, powerful root which is hardly drawn by weeding and it easily sprouts after mowing or while weeding. The seeds get to maturity progressively and they can keep the germinal sufficiency till 8-9 years, germinating by installment. They do not support the flooding (Dimancea, 1967).

The success of this weed can be imputable to a very low number of seeds generation, easily dispersed from the plant, owning a latent state of the seeds, a fast development and capacity of blooming in a large range of photoperiods (Păunescu, 1997). The number of seeds made by a plant varies between 200 and 10,000. Chirilă (1967) establishes as limits of seeds number/plant from 150 to 10,000. The medium mass of 1,000 bobsleighs is 2.48 g. (Anghel et al., 1972; Berca, 1996). The reserve of *Echinochloa crus-galli* seeds that can be found in the soil can reach impressive values, correlative with the production potential of the species and the vegetative conditions specific to the infestation areas. The number of seeds found by Kott (1953) reported to the surface of one hectare gets to 1-2.5 billion *Echinochloa crus-galli* seeds. Berca (1996) referring to the seeds of this species and their germination, shows that germination happens after one year of seeds forming, by instalment both as structure and life. The germination happens all over the year, especially in the spring time, 1-2 cm depth when the temperature is over 10 °C. The *Echinochloa crus-galli* seeds have a post maturation period that happens into the soil, especially the upper side of the soil. The length of seminal rest depends by a lot of internal and external factors (Berca, 1996).

The period of germination-rising starts for *Echinochloa crus-galli* in April, depending by the temperature provided by the soil, the minimum germination temperature is 8°C, and it ends in September. The maximum germination is between May and June, after this period there comes an attenuation of germination proportion and plants rising, so that in October it is accomplished to a very low level.

The elongation of the *Echinochloa crus-galli* plants is in a strong connection with the temperature. In the spring time when the temperature is low the elongation is a slow one but in the summer time when the temperature is high the plants grow very fast (Păunescu, 1997; Rusu et al., 2010). At the beginning the plants grow slowly, after 2-3 weeks after their rising starts the tillering period after that the plants start a very fast growing if the conditions of light, humidity and nutritive substances are assured (Berca, 1996). After the floral branches cutting of this weed, or after the first fructification they sprout again during the same year and fructify for the second time (Staicu, 1969).

Echinochloa crus-galli belongs to the yearly monocotiledonate weeds class with late spring germination very harmful for the corn cultures. Growing very fast it asphyxiates the corn crop and infamies the crop.

2. Material and methods

Our researches highlight in terms of Transylvania, the influence of *Echinochloa crus-galli* species (L.) Pal. Beauv. and other weeds on corn production, according to the degree of infestation. Researches have been conducted on Experimental Teaching Resort of University of Agricultural Sciences and Veterinary Medicine Cluj-Napoca. The experiments were located on the northern slope, weak to moderately sloping land, with soil type preluvosol (SRTS, 2003), medium fertile, humus content 3%, texture loam-clay, 42-45% clay. Experience was held between 2004 and 2009 and had more objectives:

2.1 The vegetative cycle and the productivity characters of *Echinochloa crus-galli* depending on the rise period, at Cluj area

The researches were made between 2004 and 2007, respecting the climatic conditions of the soil appropriate to every year. The researches were made outdoors onto 1 m² plots, where we sowed 20 caryopsis of *Echinochloa crus-galli* per plot, at the beginning of April, May, June,

July and August. The observations were made every 10 days between rising and maturity. We analyzed the rising period according to the sowing date; the leaves appearance, the sprouting and the stem elongation, the panicle appearance and the flourishing beginning, the plants' maturation, the caryopsis' maturation and the dissemination and the productivity's characters variation.

In each lot the plants were rarefied, the observations regarding the productivity elements being done upon a number of 3 plants on surface unit. We chose this density for the reason of the necessary space protection for the *Echinochloa crus-galli* growing in order to touch the maximum values of the productivity parametres according to the biological potential of the plant.

The years of experiment with climatic specific (May 1 – August 31)
2004: 405.7 mm (excessively wet climate) and 17.6 °C (normal temperatures conditions);
2005: 349.6 mm (excessively wet climate) and 18.05 °C (normal temperatures conditions);
2006: 455.4 mm (excessively wet climate) and 18.5 °C (warm temperatures conditions);
2007: 167.5 mm (excessively dry climate) and 18.8 °C (warm temperatures conditions).

Year	Specification	The daily average temperature, °C	Rainfall, mm
2004	Value	14.88	562.4
	Deviation	- 0.75	+ 204.4
2005	Value	16.1	505
	Deviation	+ 0.47	+ 147
2006	Value	16.83	572
	Deviation	+ 1.2	+ 214
2007	Value	16.98	250.6
	Deviation	+ 1.35	- 107.4
The normal values		15.63	358

Table 1. The climate conditions during the 1st of April and the 30th of September in Cluj.

2.2 Productivity elements variation of *Echinochloa crus-galli* in accordance to density

In order to follow the variations of productivity parameters to *Echinochloa crus-galli*, the experiences were fixed on the field, in 4 random repetitions, after blocks method, on 1m² lot surface. In the last decade of April there were seeded 200 caryopsis of *Echinochloa crus-galli* on each lot so that the rising of the plants to be assured for the beginning of May.

After the plants rising and the forming of two first leaves, there was done their spacing in order to achieve the density of 50, 20, 10, 5, 3, 2 plants per m². The rating of the plants growing parameters and plants productivity was done in the last decade of July – first decade of August – when the plants were mature enough having as goals: plants height, tillering, panicles length, number of seeds (production).

The results interpretation was done by means, percents, statistical elaboration (variance analyzes). In order to analyze the values that were obtained there were used control data, medium values of the parameters obtained in the variant of density 3 plants/m².

2.3 Influence of degree infestation with *Echinochloa crus–galli* species on crop production in corn

Experience was held between 2008 and 2009. Biological material was the hybrid Turda 201, recommended for this area of culture. The research was done on two agrofonds: unfertilized and mineral fertilized (MF) with NPK 100 kg/ha.

In the unfertilized maize crop were made four variants (I-IV) with different degrees of infestation with *Echinochloa crus-galli*, from about 40 to 100 plants/m² and witness – 2 holings.

In fertilized plots were used the next herbicides for weed control: V_1 - dimetenamid 900 g/l – 2 l/ha applied p.p.i. (pre plant incorporated). V_2 - acetoclor 860 g/l – 2 l/ha applied preemergent. V_3 - isoxaflutol 750 g/l – 0.15 g/ha, applied p.p.i. + (bentazon 320 g/l + dicamba 90 g/l) – 2 l/ha applied postemergent.

Herbicide application was made with the pomp for experience, applying 300 l solution/ha. The experience was organized after randomized blocks method, in four repetitions and area of a plot is 25 m². Competition between corn plants and weeds present was studied in natural density infestation, in unfertilized plots and in those fertilized in which the process of herbicides took place. Weed biomass, corn plants and grain production was measured in the ripening stage. Samples of plants and weeds were harvested using metric frame of 50/50 cm.

3. Results and discussion

3.1 The vegetative cycle and the productivity characters of *Echinochloa crus-galli* in accordance to rising period

The biological particularities of weeds make them be superior to the cultivated plants, as they use more effective the vegetation conditions and the afferent inputs of an agricultural area. *Echinochloa crus-galli* is an annual monocotyledonous species, which germinates late in spring. This species is spread onto extensive areas in the world, covering all continents between 50⁰ northern latitude and 40° southern latitude. In Romania there is plenty of it in all regions, prevailing in the south western part of the country (90%) and in the eastern part (75%). In the other areas, the species varies between 9% (Dobrogea) and 57% (Transylvania). It is very harmful for hoed cultures. In the Cluj County, *Echinochloa crus-galli* represents between 36% and 52% of the weedy rate of the hoed cultures. *Echinochloa crus-galli* produces big damages in Romania's agriculture: in maize – over 70%, in rice – 60 – 65%; in sunflower – 30%; in soybean – 15 – 20%, in sugar beet 25%, in wheat – 10%, in flax – 10%.

In autumn or early spring sowed cultures (that cover the soil to a large extent) *Echinochloa crus-galli* hardly forms a small stem, but when the cultures are harvested off the field, the weed heavily sprouts and it produces a large amount of seeds as it has more space, light, nutrition and moisture.

The high ecological plasticity and adaptability of this species, completed by the possibility of flourishing in a wide range of photoperiods are biological particularities of *Echinochloa crus-galli*.

The plants rising takes place monthly in different percentages until September, when the rising is reduced. The rising period is of 8 – 16 days since the sowing depending on the temperatures. The correlations established between the soil temperature conditions and the *Echinochloa crus-galli* plants rising are very significantly positive. From the specific equations for the experiment years result that the percentage of the plants that are rising is increasing by 5.75 – 6.87 per 1 ^0C of the soil temperature - beginning with 8^0C, the minimum germination temperature.

The plants' growth varies according to the rising period. So, the plants that rise up during the second half of April pass through each specific vegetative stage for a longer period comparing to the ones that rise up during the next months – when the temperatures increase (Table 2).

The rising period/The vegetative stage	April	May	June	July	August
1-3 leaves phase	8 – 10	6 - 8	6 – 7	6	5 – 6
Tillering beginning	19 – 23	17 - 20	13 – 17	12 – 15	12 – 14
The intensive tillering, the adventitious roots rising	32 – 38	28 – 34	20 – 21	18 - 20	16 – 19
The end of the tillering, the culm elongation	50 – 56	48 – 50	30 – 34	28 – 30	26 – 30
The skin stage*	70 – 80	64 – 68	43 – 46	38 – 43	36 – 41
The panicle apparition, flowering*	83 – 90	74 – 79	58 – 61	49 – 54	48 – 50
The grains filling *	94 – 105	80 – 88	64 – 70	60 – 63	59 – 63
The seeds' maturation; dissemination*	110 - 115	95 – 99	79 - 82	70 - 79	65 - 72

*This information is specific for the main stem. The shoots pass progressively these stages after the main stem.

Table 2. The period in days passed by a *Echinochloa crus-galli* plant from it's rising to each vegetative stage (the average period for the years 2004 – 2007 on Cluj-Napoca conditions).

The daytime influences the flourishing period so that the plants that rise later (July, August) reach the flourishing phase in a much shorter period (48 -54 days), comparing to the plants that rise in April (83 days). The shorter days of late August and early September stimulate precocious flourishing and ageing.

The caryopses are maturating in a 20 – 30 days period, after heading (depending on the rising period).

The first panicle dissemination is taking place during 10 -16 August for the plant that rose in April; 15 – 19 August for those that rose in May (first decade); 28-30 August for those that rose in June; 25 – 28 September for the plants that rose in July and 15 – 18 October for those that rose in August.

At the beginning of September the first plants dry out; they are those that rose in April, while those that rose in August dry out at the end of October.

The vegetative cycle of *Echinochloa crus-galli* plants is taking place in summer (Fig. 1). It begins in April for the plants sowed in April and it ends in August. But the cycle for the plants sowed in August, it ends in October.

The length of the vegetation period for a plant and the productivity characters (the height, the shoots number, the panicles number, the panicles' length, the caryopsis number of a plant, the bio weight) are reduced as the plants' rising is late (Table 3).

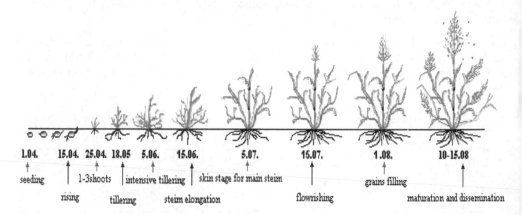

Fig. 1. The growth cycle of *Echinochloa crus-galli* (L.) P.B. on Cluj-Napoca conditions.

The variation of the vegetation period of this species (85 – 140 days) reflects the extraordinary flexibility and the excellent adaptability of *Echinochloa crus-galli* in different environment conditions.

The plants height decreases from 170 cm (the plants that rose in April) to 55 cm (the plants that rose in August). The vegetative growths are reduced as the vegetative period is decreasing and the daily average temperatures are increasing. The plants that rise in spring (April, May) grow and tiller very intensely, they reach considerable heights as a consequence of late flourishing, as the daytime is longer.

The roots of the early plants grow more intensely.

The whole weight reaches impressive values for the long vegetation period plants; it decreases drastically for the plants that rose late. The seeds production decreases as the vegetative periods diminish. It is notable that *Echinochloa crus-galli* seeds production is important (8435 caryopsis/plant) even for the plants that rose in August.

The rising period/The features	April	May	June	July	August
The vegetative period (days)	135-140	125-130	115-120	95-100	85-90
The plants height (cm)	150-170	140-150	120-130	90-105	55,75
Number of shoots	23-25	19-25	16-18	12-14	13-15
Number of panicles	27-31	20-24	16-18	10-16	12-14
The length of the panicles (cm)	12.5-20.4	11.2-18.5	8.5-10.6	7.5-9.4	5.6-7.2
Number of seeds (average/plant)	15794	13406	10898	8762	8435
The length of the roots (cm)	39-45	35-38	30-34	24-27	13-16
Bioweight (gr./plant – herb weight)	895	794	586	338	212

Table 3. The vegetation period and the productivity features of *Echinochloa crus-galli* depending on the rising period on Cluj-Napoca conditions.

There have been variations of the productivity characters among the 4 experimentation years, according to the temperatures and the pluviometric quantities.

The climate of 2006 and 2007 significantly influenced the productivity characters of this species. So, during 2006 – the wealthiest in precipitation, the vegetative growth of the *Echinochloa crus-galli* plants was impressive: the maximum height was 218 cm, the shoots number was 49, and the developed panicles number was 45. During 2007 – when it was drought, there were the lowest values for the vegetative growth, but there were a lot of seeds comparing to the plants height.

3.2 The productivity features variation of *Echinochloa crus-galli* in accordance to density

The productivity features variations of *Echinochloa crus-galli* depend by the plants density to surface unit and climatic conditions specific to experimental years.

The height of the plants is strongly influenced by density, increasing significantly to a low density (2 plants/m² than 3 plants/m²) and a high one (50 plants/m² than 3 plants/m²) due to a strong shading and lack of light (Table 4). This feature (height) did not present a constant variation regarding the climatic conditions specific the experiment years, the single year when the medium difference of height was significantly negative was 2007.

Density / Year	2004	2005	2006	2007	Average
50 plants/m²	172.6***	135.6***	184.5*	65.8^{ooo}	139.6***
10 plants/m²	159.8***	120.1***	158.6^{ooo}	76.3^{oo}	128.7^{ns}
5 plants/m²	142.2^{oo}	101.3^{oo}	180.2^{o}	81.4**	126.3^{ns}
2 plants/m²	152.6***	118.9***	205.6***	84.2***	140.3***
Control 3 pl/m²	145.6	103.5	182.4	78.5	127.5
LSD 5%	1.85	1.19	1. 58	1.41	1.75
1%	2.63	1.70	2.24	2.01	2.48
0.1%	3.80	2.46	3.25	2.91	3.56

Note: ns – not significant, * signification positives, ^{0} signification negatives

Table 4. The height of *Echinochloa crus-galli* (cm) in accordance to density.

The other characters of productivity are strongly influenced by the plants density on the surface unit. Between 3 and 5 plants /m² it is achieved a close tillering, panicles number, panicles length and seeds number, with no significant differences while the density growing is 5 plants bigger on the surface unit all these conditions are decreased to limits between signification negative to very signification negative. The density attenuation under 3 plants on square metre has as effect the growing of the species productivity potential.

The danger that this plant represents even to a reduced infestation of the cultures comes from the possibility of achievement both a high biomass through the vegetative growing elongation, and a very high production of seeds that will represent the source of weeding for the next cultures.

The influence of the nutritional space size and development is very strong upon this species *Echinochloa crus-galli*. The tillering, the panicles number and panicles length are very significantly reduced to densities bigger than 5 plants per surface unit (Table 5).

The differences that appear among the years represent a consequence of species adaptability for adjustment conditions of the productivity in accordance to climatic conditions. If during the rainy years the tillering is influenced from distinctive significant to very negative significant by plants density growing with 2 samples per surface unit, during the very dry year – 2007 - this condition does not suffer any adjustment having a significant growing by density increasing with 2 plants per surface unit. The explanation of this fact is found into the high capacity of tillering of this species when the height growing is diminuated.

Density/Year	2004	2005	2006	2007	Average
50 plants/m^2	3.2^{000}	2.9^{000}	4.2^{000}	2.4^{000}	3.2^{000}
10 plants/m^2	12.9^{000}	12.1^{000}	14.3^{000}	9.3^{000}	12.2^{000}
5 plants/m^2	19.3^{00}	18.8^{00}	20.4^{000}	16.9^{*}	18.9^{ns}
2 plants/m^2	23.4^{ns}	20.1^{ns}	26.2^{*}	18.5^{***}	22.1^{*}
Control 3 pl/m^2	22	20	25	16	20.5
LSD 5%	1.59	0.73	0.97	0.88	1.35
1%	2.26	1.04	1.38	1.25	2.13
.1%	3.27	1.5	2.00	1.81	2.97

Table 5. The tillering per plant on *Echinochloa crus-galli* in accordance to density.

The number of panicles per plant (table 6) follows, in general, the same tendency of a very significant decreasing to densities bigger than 5 plants per surface unit and increasing or decreasing from insignificant to very significant in the situation of a density increasing with only 2 plants, in accordance to the climatic conditions of the year. The weed density decreasing under 3 plants /m^2 has as effect in both character situations (tillering and panicles number) increasing from insignificant to very positive significant. The number of panicles is, especially, the most influenced positively character by the density decreasing.

Density/Year	2004	2005	2006	2007	Average
50 plants/m^2	2.4^{000}	2.6^{000}	3.1^{000}	3.9^{000}	3.0^{000}
10 plants/m^2	13.1^{000}	14.5^{000}	16.6^{000}	9.6^{000}	13.5^{000}
5 plants/m^2	20.8^{000}	21.2^{**}	26.8^{ns}	18.2^{ns}	21.7^{ns}
2 plants/m^2	25.2^{***}	26.1^{***}	32.4^{***}	20.8^{***}	26.1^{***}
Control 3 pl/m^2	23	20	27	18	22
LSD 5%	0.96	0.83	0.95	1.22	1.12
1%	1.36	1.19	1.35	1.73	1.49
0.1%	1.97	1.72	1.95	2.50	2.38

Table 6. The number of panicles on *Echinochloa crus-galli* in accordance to density.

The panicles length is very significant reduced to plants densities of *Echinochloa crus-galli* bigger than 10 plants /m^2 while density of 5 plants /m^2 does not make significant differentiations (Table 7).

The production of caryopsis per plant is very significant reduced to increasing of plants density per surface unit starting with density of 5 plants /m^2. On this density, where the other productive features are less influenced compared with witness density (3 plants/m^2), the seeds production suffers major decreasing especially during the years that are rich in precipitations, when the productive potential of the species is directed to vegetative features (Table 8).

Density/Year	2004	2005	2006	2007	Average
50 plants/m^2	11.4^{ooo}	11.3^{ooo}	12.2^{ooo}	10.3^{ooo}	11.3^{ooo}
10 plants/m^2	14.1^{ooo}	13.4^{o}	14.3^{ooo}	12.1^{ooo}	13.5^{ooo}
5 plants/m^2	16.1^{ns}	13.6^{ns}	16.2^{ns}	13.4^{ns}	14.8^{ns}
2 plants/m^2	16.4^{ns}	15.1***	17.1*	14.9*	15.9*
Control 3 pl/m^2	16.1	13.9	16.4	13.9	14.85
LSD 5%	0.56	0.43	0.63	0.73	0.65
1%	0.79	0.61	0.83	1.04	1.25
0.1%	1.15	0.89	1.29	1.52	1.56

Table 7. The length of panicles on *Echinochloa crus-galli* (cm) in accordance to density.

Density/Year	2004	2005	2006	2007	Average
50 plants/m^2	1,289^{ooo}	1,216^{ooo}	1,482^{ooo}	2,105^{ooo}	1,523^{ooo}
10 plants/m^2	9,462^{ooo}	8,324^{ooo}	9,874^{ooo}	7,304^{ooo}	8741^{ooo}
5 plants/m^2	13,821^{ooo}	10,918^{ns}	12,956^{ooo}	10,021^{oo}	12,179^{oo}
2 plants/m^2	15,659^{ns}	12,164*	16,102^{ns}	11,434*	13,840^{ns}
Control 3 pl/m^2	15,208	11,303	16,018	10,795	13,406
LSD 5%	457.05	667.1	669.8	511.3	678.6
1%	649.71	948.3	952.1	731.08	973.4
0.1%	940.74	1,373.1	1,378.7	1,058.5	1,354.8

Table 8. The number of seeds/plant on *Echinochloa crus-galli* in accordance to density.

The productivity features of *Echinochloa crus-galli* suffer changes in accordance to the weed density per unit surface, to high densities the increasing in high are very visible while the tillering, panicles number, panicles length and the number of seeds produced by a plant are reduced very significant. Between 2 and 5 plants of *Echinochloa crus-galli* /m^2, the productivity parameters vary in more reduced limits, being in the most cases the consequence of the climatic conditions of the experimentation years. The inter specific concurrency is felt even when speaking about the increasing with one plant per surface unit, but this one becomes hypercriticalism in case of density growing with more than 5 plants/m^2.

3.3 Influence of degree infestation with *Echinochloa crus–galli* species upon the maize crop

Echinochloa crus-galli is known as a weed which germinate in late spring, invades especially weeding crops on wetlands, fattened with manure, grows very quickly, suppress and compromise the culture. Precipitation in April - May 2009 (102 mm in April compared to 47 mm multiannual average and 105 mm in May compared to 76 mm) have delayed corn seeding until the end of the optimal period and promoted the accumulation of moisture in the soil of 30% on average depth from 0 to 50 cm and a reserve of water on the same depth of 977 m^3/ha. Under these conditions, sown late, high humidity, fertilization in the last year with manure, favoured an excessive infestation of the culture, with species that germinate in late spring and especially *Echinocloa crus-galli*. At the same time, shortcomings on internal drainage of the soil aggravate the maintenance of crops in critical periods. Under these circumstances, competition for factors of vegetation was quickly won by *Echinochloa crus-galli* which influenced the subsequent development of maize and other weeds (Fig. 2).

In the unfertilized variant, corn invaded by weeds grows anemic and has a yellowish green color, develops storied, on the upper *Echinochloa crus-galli* dominate, in the middle floor develops *Setaria glauca* (L.) Beauv. and in the lower floor a number of dicotyledonous: *Galinsoga, Convolvulus, Matricaria, Lapsana, Hibiscus, Plantago* etc. (Table 9). The amount of weeds, obviously influenced production levels of maize grain and green mass (Fig. 3). Thus, it is found that on unfertilized agrofond with 22,113 kg/ha weed, maize green mass production is 2,100 kg/ha and with 200 kg/ha weed, maize green mass production is 29,790 kg/ha. The total amount of green mass (weed + maize/ha) varies in very close limits between 24,213 kg/ha to 31,740 kg/ha. On fertilized variant, the competition between weeds and maize, on the one hand and between monocotyledonous and dicotyledonous on the other hand, is more balanced, as dicotyledonous come from 1,700 kg/ha in unfertelized variants (Table 9), to 4,159 kg/ha in untreated, mineral fertilized variant (Table 10). On fertilized agrofond in untreated plot, the whole plant corn production was 27,600 kg/ha, and the grain production was 1,965 kg/ha, while the total mass of monocotyledonous weeds weighed 19,560 kg/ha and dicotyledonous weeds 4,159 kg/ha. In the variant treated with dimethenamid the whole plant corn production increased to 48,500 kg/ha, and the grain at 5,070 kg/ha, while total weed mass was 9,671 kg/ha. Similar results were obtained in the variant treated with acetochlor.

The highest production of whole plant corn 53,600 kg/ha and 7,020 kg/ha grain were obtained in the variant treated with isoxaflutol + (bentazon + dicamba). In this variant, because of the high efficiency of herbicides, the total amount of weeds was the smallest, only 950 kg/ha. In this experience, on fertilized background, in variant treated with herbicides, the amount of corn (27,600 kg/ha) + weeds (23,719 kg/ha) totals 51,393 kg/ha, which is practically equal to the best variants treated with isoxaflutol + (bentazon + dicamba), where were obtained 53,600 kg/ha maize and 950 kg/ha weeds, thus in total 54,550 kg/ha.

The reserve of *Echinochloa crus-galli* seeds in the 0-10 cm soil layer determinates at maize harvest shows the danger constituted by late infestations of maize crops with weeds, in maintaining the cultural hygiene of exploitation. From a valuable point of view this reserve of seeds is about 22,264 seeds/m² (average on the three years) in the variant of no disproof, 3,512 seeds/m² in the variant of a classical disproof, 5,394 seeds/m² in a chemical disproof variant through a pre emergent treatment, 6,042 seeds/m² in a chemical disproof variant through a post emergent treatment and 3,816 seeds/m² in a chemical disproof variant through two treatments (p.p.i. + postem.). We can state that the *Echinochloa crus-galli* seed reserve accumulated in the superficial soil layer is tightly related to the biomass of the weeds present in the culture before maize harvest.

3.4 The influence of climatic and technological factors upon the weed characteristics

The variable characteristics of the clime in the hilly area in the spring time, especially in April-May, completed with the particularities of soils workability from this area build for many times one impediment to assure the optimal conditions for corn seeding and establish the optimal time for seeding. The repercussions of these deficiencies can be found for the most times in: culture late rising, culture irregularity, a bigger number of weeds, the passing of some phonological phases by corn plants during inappropriate periods, the differentiation of productivity organs during dryness periods, reduced productions.

Field	Group	Species	Plants/ m²	Mass, kg/ha		
				Species	Group	Total
Plot I	Corn whole plant grains		4	-	-	2,100 288
	Mono	Echinochloa crus – galli	104	18,052	20,364	
		Setaria glauca	10	2,312		
	Dico	Galinsoga parviflora	25	1,516		22,113
		Convolvulus arvensis	4	72	1,749	
		Matricaria, Lapsana, Hibiscus	7	161		
Plot II	Corn whole plant grains		4	-	-	4,630 1,116
	Mono	Echinochloa crus – galli	95	12,633	12,804	
		Setaria glauca	8	171		
	Dico	Galinsoga parviflora	12	1,341		14,587
		Convolvulus arvensis	3	81	1,783	
		Matricaria,Lapsana, Hibiscus	7	361		
Plot III	Corn whole plant grains		4	-	-	130,000 2,526
	Mono	Echinochloa crus – galli	58	8,323	8,433	
		Setaria glauca	4	110		
	Dico	Galinsoga parviflora	22	1,293		10,230
		Convolvulus arvensis	4	102	1,797	
		Plantago,Matricaria, Lapsana	9	402		
Plot IV	Corn whole plant grains		4	-	-	19,720 3,866
	Mono	Echinochloa crus – galli	47	7,080	7,283	
		Setaria glauca	6	203		
	Dico	Galinsoga parviflora	14	1,012		8,464
		Convolvulus arvensis	2	90	1,185	
		Shymphytium, Lapsana	3	83		
Witness (2 holings)	Corn whole plant grains		4	-	-	29,790 5,157
	Mono	Echinochloa crus – galli	3	990	990	
	Dico	Convolvulus arvensis	2	240	960	1,950
		Shymphytium officinalis	2	720		

Mono – Monocotyledonous; Dico – Dicotyledonous.

Table 9. Influence of the density of Echinochloa crus – galli species and of other weed species upon the maize crop in the case of unmineral fertilized soil and without any measure of chemical weed control.

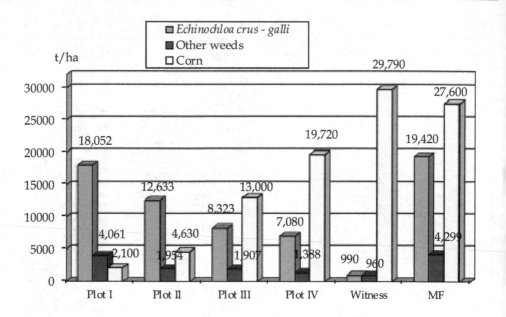

Fig. 2. Influence of *Echinochloa crus – galli* species on the development of other weeds and the green mass corn yield (t/ha).

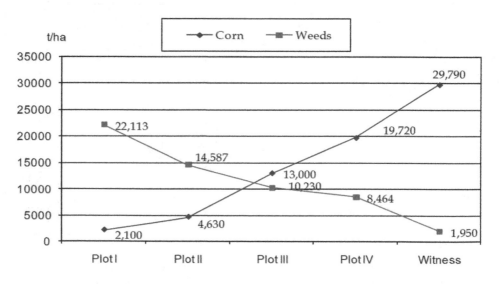

Fig. 3. Correlation between the weeds quantity and the greenery corn (t/ha).

Herbicides kg/ha	Group	Species	Plants /m²	Mass, kg/ha Species	Mass, kg/ha Group	Total
Untrated	Corn	whole plant	4	-	-	27,600
		grains				1,965
	Mono	Echinochloa crus – galli	34	19,420	19,560	
		Setaria glauca	3	140		
	Dico	Galinsoga parviflora	9	891		
		Chenopodium album	8	1,040		23,719
		Polygonum convolvulus	8	1,241	4,159	
		Matricaria, Cirsium				
		Euphorbia helioscopia	14	987		
V₁ - dimetenamid	Corn	whole plant	4	-	-	48,500
		grains				5,070
	Mono	Echinochloa crus – galli	2	2,060	3,100	
		Setaria glauca	4	1,040		
	Dico	Amarantus retroflexus	5	3,230		9,671
		Cirsium arvense	3	940	6,571	
		Chenopodium,Gallinsoga	7	2,401		
V₂ - acetoclor	Corn	whole plant	4	-	-	49,120
		grains				5,421
	Mono	Echinochloa crus – galli	7	3,880	4,890	
		Setaria glauca	6	1,010		7,092
	Dico	Amaranthus, Cirsium	10	2,202	2,202	
V₃ - isoxaflutol +	Corn	whole plant	4	-	-	53,600
(bentazon +		grains				7,020
dicamba)	Dico	Amaranthus retroflexus	1	950		950

Mono – Monocotyledonous; Dico – Dicotyledonous.

Table 10. Influence of density *Echinochloa crus – galli* species upon the maize crop in the case of a mineral fertilized soil and measure of chemical weed control.

The determined correlations confirmed a very strong connection between the climatic conditions and weed amount inclusively with *Echinochloa crus-galli*. There are also very significant direct relations between the overtaking of the optimal seeding date (April 15) and weed of the culture (Table 11).

One significant correlation exists between the quantity of precipitations and *Echinochloa crus-galli* ($r = 0.875$), but this species has a lower dependence to humidity, at least in the first periods of growing comparatively to other weeds, fact that explains the big number of exemplaries, even in the years with a low amount of precipitations and soils with a low reserve of humidity. The air temperature has a lower influence upon the weeding ($r=0.571$) especially during the first period of corn vegetation in conditions in that there were not significant variations of this climatic parameter.

The overtaking of seeding optimal date determinates the increasing range of weed inclusively the amount of *Echinochloa crus-galli* per surface unit. The relation is very significant, the correlation coefficient has values between 0.766 and 0.840 (very significantly) and the regression equation y=2.5148 x + 288.96 shows that every day of seeding delay conduce to weed increase with more than 2 weeds/square metre.

The explanation of the identified correlations is found in the climatic characteristics of the experimental years. The dry periods influence negatively the corn germination and rising taking in consideration the spent period from seeding rising, culture density and its homogeneity. The weeds are also influenced less as frequency and more as rising and development period. During the years that are rich in precipitations the weeds succeeded in germination, rising and assurance of a high infestation of the culture. The plus of humidity and temperature from May and June favoured the weeding both as frequency and phonological development especially between May 30 and June 30. The weeds concurrence to the corn plants in this period it was an acerbic one.

The existent weeds mass in the corn crop before harvesting reflects on one side the climatic specific of the agrarian year, but mostly the effectiveness of each applied method to combat the weeds and not lastly the capacity of weeds concurrence.

The correlations established between the biomass achieved at harvesting moment of *Ecinochloa crus-galli* and yield (Fig. 4, Fig. 5 and Fig. 6) are – very strong, proving once again the fact that this species is a majoritary one both as frequency in corn crops from Cluj area but also as a corn concurrency potential bringing to significant production reductions. The correlation coefficient (r) is very negative significant having values between 0.861 and 0.952.

Characteristic	1	2	3	4	5	6
1. Weed number/m²	1	0.65**	0.85**		0.92***	0.84***
2. Covering range, %		1		0.571°		0.90°°°
3. Number of *Echinochloa crus-galli* /m²			1		0.578*	0.766***
4. Medium rising temperature – 15 days after seeding (°C)				1		
5. The amount of precipitations rising - 15 days after seeding (mm)					1	0.859***
6. Number of days behind seeding						1

r / p 5% = 0.497; 1% = 0.623; 0.1% = 0.742

Table 11. The existent correlations between the weed characteristics, climatic and technological conditions from the corn crop.

The assessment of each combating method both under efficiency in corn crop weeds control aspect and achieved productions level after weeds combating (Fig. 7) it is compulsory and objective. The combating range of *Echinochloa crus-galli* accomplishes with the production a strong positive relation r = 0.959***. Therefore, in the case of a 10% increasing of combating range, the production rises with 48.65 kg/ha.

4. Conclusion

Significant particularity of *Echinochloa crus-galli* species is its growing plasticity according to the rising period and the respective climate during the vegetation period. The vegetative phases and vegetative parameters are adjusted so that the plant would completely pass the generative phase and would assure the species perpetuation.

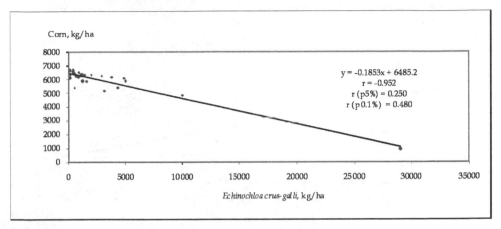

Fig. 4. The relation between biomass of *Echinochloa crus-galli* and corn yield during the dry years.

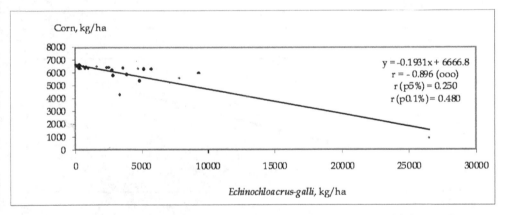

Fig. 5. The relation between biomass of *Echinochloa crus-galli* and corn yield during the rainy years.

The problem of the influence of different species of weeds on the production of agricultural plants has been studies by many researchers. The damage caused by weeds in maize crop is mostly of 30-70%, and when the infestation is strong culture can be fully compromised. Our researches highlight in terms of Transylvania, the influence of *Echinochloa crus-galli* species (L.) Pal. Beauv. and other weeds on corn production, according to the degree of infestation. Researches have been conducted at the University of Agricultural Sciences and Veterinary Medicine Cluj-Napoca, Romania.

The researches was done on two agrofunds: unfertilized and mineral fertilized with NPK 100 kg s.a./ha. The unfertilized maize crop has been made in four variants with different degrees of infestation of *Echinochloa crus-galli*, from about 40 to 100 plants/m² and witness – 2 holdings. In fertilized plots were used 4 herbicides for weed control (isoxaflutol 750 g/l; acetoclor 860 g/l; dimetenamid 900 g/l; bentazon 320 g/l + dicamba 90 g/l). Weed biomass, corn plant and grain production was measured in the ripening stage.

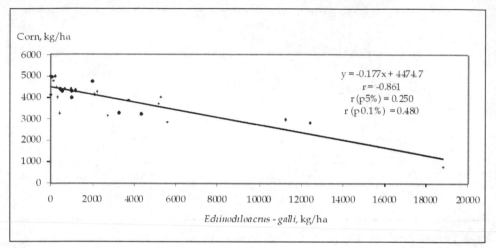

Fig. 6. The relation between biomass of *Echinochloa crus-galli* and corn yield during normal climatic conditions.

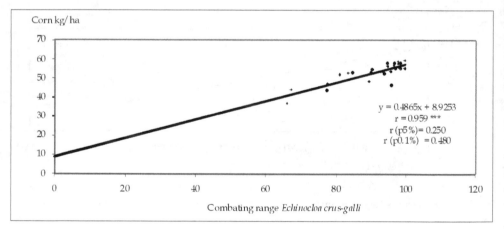

Fig. 7. The relation established between the *Echinochloa crus-galli* combating range assured in the tested variants and production level.

Corn invaded by weeds grows anaemic and has a yellowish green colour, in the unfertilized variant the corn develops storied, on the upper *Echinochloa crus-galli* dominates, in the middle floor develops *Setaria glauca* (L.) Beauv. and in the lower floor a number of dicotyledonous. On fertilized variant, the competition between weeds and maize, on the one hand and between monocotyledonous and dicotyledonous on the other hand, is more balanced, as dicotyledonous come from 1,700 kg/ha in mineral fertilized variants, to 4,100 kg/ha in mineral fertilized variant. The amount of weeds, obviously influenced production levels of maize grain and green mass. *Echinochloa crus - galli* had favorable conditions for maize crop infestation; the losses of production are depending on the degree of weed infestation and can reach 5,000 kg/ha maize grain, compared with those obtained in conditions of weeds control. Production losses in terms of green mass per hectare can be considered equal to the weight of green weeds. At a density of 104 plants/m² of *Echinochloa crus-galli* with green mass of 18,052 kg/ha corn crop is fully compromised.

The prevention of maize crops infestations with weeds and weed control must be adjusted to topoclimate conditions. Along with agro technical, physiomechanical, biological and control means against weeds to share an equal importance for maize crops. All these must be so established as to succeed in efficiently controlling weeding all through the vegetation period of maize.

In central Transylvania, abundant rainfalls in July, August and even September and high temperatures favours late infestation of maize crops with annual species, very plastic as concerns the springing period and the bio-mass accumulated in the time period, especially *Echinochloa crus-galli*. Thus, when maize is harvested we can observe a high weeding level and the weed seeds reserve accumulating in the soil increases.

The protection of maize crops in the centre of Transylvania against weeds must to be into consideration some factors that are specific for that area. These factors are: large weed seeds reserve in soil, which, every year, provides a high weeding degree of crops; whimsical rainfalls; alternative springing of dominant weed species and their biological specific features, in order to reduce specific maize weeding under the economic deleterious level.

5. References

Anghel, G., Chirilă, C., Ciocârlan, V., Ulinici, A. (1972). *The weeds from agricultural cultures and their combating*. Ed. Ceres, Bucharest.

Berca, M. (1996). *Actual issues of weeds combating in Romania.The 10ᵗʰ Herbology National Simposium* , Sinaia, p. 7-16.

Berca, M. (2004). *Weeds integrated management*. Ed. Ceres Bucharest.

Bîlteanu, Gh. (2001). *Crop production*. Ed. Ceres, Bucharest.

Bogdan, I., Gus, P., Rusu, T. (2001). *The green polution – an priority for the Cluj's area agriculture*. The economic increase, development, progress – Scientific communications session –Babes-Bolyai University, Cluj, Romania, XXX, 1703-1713.

Bogdan, I., Gus, P., Rusu, T. (2005). *The vegetative cycle and the productivity characters of the Echinochloa crus-galli (L.) P.B. depending on the rise period, in the Cluj county*. Bulletin of USAMV-CN, 61, p. 405. Ed. AcademicPres®, Cluj-Napoca.

Bogdan, I., Guş, P., Rusu, T., Moraru, P.I., Pop, A.I. (2007). *Research concerning the weeding level of autumn wheat – potato – maize – and soybean crop rotation, in Cluj county*. Bulletin of USAMV-CN, 63-64, p. 283-290. Ed. AcademicPres®, Cluj-Napoca.

Bosnic, C. A. and Swanton, C.J. (1997). *Influence of barnyardgrass (Echinochloa crus-galli) time of rising and density on corn (Zea mays)*. Weed Science Journal, n. 45, pp. 276-282.

Budoi, Gh. and Penescu, A. (1996). *Agrotehnica*. Ed. Ceres, Bucharest.

Chirilă, C. (1967). Contributions to knowing the weeds seeds from maize crops. Scientific Papers IANB, Serial A X.

Clay, S.A., Kleinjan, J., Clay, D.E., Forcella, F., Batchelor, W. (2005). *Growth and fecundity of several weed species in corn and soybean*. Agronomy Journal, vol. 97, n. 1, pp. 294-302.

Dimancea, S.,(1967). *Agrofitotehnica*. Ed. Didactica si Pedagogică, Bucharest.

Dincă, D. (1957). *Agrotehnica*. Ed. Agrosilvică de Stat, Bucharest.

Fukao, T., Kennedy, R.A., Yamasue Y. and Rumpho, M. E. (2003). *Genetic and biochemical analysis of anaerobically-induced enzymes during seed germination of Echinochloa crus-galli varieties tolerant and intolerant of anoxia.* Journal of Experimental Botany, vol. 54, n. 368, pp. 1421-1429. Oxford University Press.

Gus, P., Rusu, T., Bogdan, I. (2004). *Agrotehnica.* Ed. Risoprint, Cluj-Napoca.

Ionescu-Sişeşti, G. (1957). General Soil Cultivation Course, weeds and their combat Workhouses of didactic material Inst. Agro. N. Bălcescu, Bucharest.

Kott, S.A. (1953). *Korantinnâie sorta i boriba s nimi.* Selihoz. Izdat, Moscva

Oancea, I. (1998), *Treatise of agricultural technology.* Ed. Ceres, Bucharest.

Păunescu, G. (1996). *Contributions to the Setaria glauca and Echinochloa crus-galli species biology.* The 10th EWRS Symposium, 213-218.

Păunescu, G. (1997). Researches regarding the depth influence and durability upon caryopsis germination of Echinocloa crus-galli. Proplant 97, p.155-164.

Perron, F. and Legere, A. (2000). *Effects of crop management practices on Echinochloa crus – galli and Chenopodium album seed production in a maize/soyabean rotation.* Weed research Journal, vol. 40, n. 6, pp. 535-547, Blackwell Science, Edinburgh.

Rusu, T. (2008). *Agrotehnica.* Ed. Risoprint, Cluj-Napoca.

Rusu, T., Gus, P., Bogdan, I., Moraru, P. I., Pop, A. I., Clapa, D., Doru, I. M., Oroian I. and Pop, L.I. (2009). *Implications of Minimum Tillage Systems on Sustainability of Agricultural Production and Soil Conservation.* Journal of Food, Agriculture & Environment, vol. 7(2/2009), p. 335-338, Helsinki, Finlanda.

Rusu, T., Gus, P., Bogdan, I., Moraru, P.I., Pop, A.I., Sopterean, M.L., Pop, L.I. (2010). *Influence of infestation with Echinochloa crus-galli species on crop production in corn.* Journal of Food, Agriculture & Environment, vol. 8(2/2010), p. 760-764, WFL Publisher Science and Technology, Helsinki, Finlanda.

Sarpe, N. (1987). *Integrated control of weeds in agricultural crops.* Ed. Ceres, Bucharest.

Staicu, I. (1969). *Agrotehnica.* Ed. Agrosilvică, Bucharest.

SRTS. (2003). *Romanian System of Soil Taxonomy.* Ed. Estfalia, Bucharest, 182 pp.

Permissions

The contributors of this book come from diverse backgrounds, making this book a truly international effort. This book will bring forth new frontiers with its revolutionizing research information and detailed analysis of the nascent developments around the world.

We would like to thank Prof. Dr. Mohammed Naguib Abd El-Ghany Hasaneen, for lending his expertise to make the book truly unique. He has played a crucial role in the development of this book. Without his invaluable contribution this book wouldn't have been possible. He has made vital efforts to compile up to date information on the varied aspects of this subject to make this book a valuable addition to the collection of many professionals and students.

This book was conceptualized with the vision of imparting up-to-date information and advanced data in this field. To ensure the same, a matchless editorial board was set up. Every individual on the board went through rigorous rounds of assessment to prove their worth. After which they invested a large part of their time researching and compiling the most relevant data for our readers. Conferences and sessions were held from time to time between the editorial board and the contributing authors to present the data in the most comprehensible form. The editorial team has worked tirelessly to provide valuable and valid information to help people across the globe.

Every chapter published in this book has been scrutinized by our experts. Their significance has been extensively debated. The topics covered herein carry significant findings which will fuel the growth of the discipline. They may even be implemented as practical applications or may be referred to as a beginning point for another development. Chapters in this book were first published by InTech; hereby published with permission under the Creative Commons Attribution License or equivalent.

The editorial board has been involved in producing this book since its inception. They have spent rigorous hours researching and exploring the diverse topics which have resulted in the successful publishing of this book. They have passed on their knowledge of decades through this book. To expedite this challenging task, the publisher supported the team at every step. A small team of assistant editors was also appointed to further simplify the editing procedure and attain best results for the readers.

Our editorial team has been hand-picked from every corner of the world. Their multi-ethnicity adds dynamic inputs to the discussions which result in innovative outcomes. These outcomes are then further discussed with the researchers and contributors who give their valuable feedback and opinion regarding the same. The feedback is then collaborated with the researches and they are edited in a comprehensive manner to aid the understanding of the subject.

Apart from the editorial board, the designing team has also invested a significant amount of their time in understanding the subject and creating the most relevant covers. They scrutinized every image to scout for the most suitable representation of the subject and create an appropriate cover for the book.

The publishing team has been involved in this book since its early stages. They were actively engaged in every process, be it collecting the data, connecting with the contributors or procuring relevant information. The team has been an ardent support to the editorial, designing and production team. Their endless efforts to recruit the best for this project, has resulted in the accomplishment of this book. They are a veteran in the field of academics and their pool of knowledge is as vast as their experience in printing. Their expertise and guidance has proved useful at every step. Their uncompromising quality standards have made this book an exceptional effort. Their encouragement from time to time has been an inspiration for everyone.

The publisher and the editorial board hope that this book will prove to be a valuable piece of knowledge for researchers, students, practitioners and scholars across the globe.

List of Contributors

Antonio Ruiz-Medina and Eulogio J. Llorent-Martínez
University of Jaén, Spain

András Székács and Béla Darvas
Department of Ecotoxicology and Environmental Analysis, Plant Protection Institute, Hungarian Academy of Sciences, Hungary

Raj Mohan Singh
Department of Civil Engineering, MNNIT Allahabad, India

Šárka Klementová
Faculty of Science University of South Bohemia, Czech Republic

Bettina Bongiovanni, Cintia Konjuh, Arístides Pochettino
Laboratorio de Toxicología Experimental, Departamento de Ciencias de los Alimentos y Medio Ambiente, Argentina

Alejandro Ferri
Departamento de Química Analítica, Facultad de Ciencias Bioquímicas y Farmacéuticas, Universidad Nacional de Rosario, Rosario, Argentina

Elaine C.M. Silva-Zacarin
Laboratório de Biologia Estrutural e Funcional (LABEF), Universidade Federal de São Carlos – UFSCAR, Sorocaba, São Paulo, Brazil

Grasiela D.C. Severi-Aguiar
Programa de Pós-Graduação em Ciências Biomédicas, Centro Universitário Hermínio Ometto, UNIARARAS, Araras, São Paulo, Brazil

Aurélio Vaz De Melo, Rubens Ribeiro da Silva and Hélio Bandeira Barros
Federal University of Tocantins, Brazil

Cíntia Ribeiro de Souza
Federal Institute of Education, Science and Technology of Pará, Brazil

Wiesław Moszczyński and Arkadiusz Białek
The Institute of Industrial Organic Chemistry, Poland

Roberto Rico-Martínez and Jesús Alvarado-Flores
Departamento de Química, Centro de Ciencias Básicas, Universidad Autónoma de Aguascalientes, Aguascalientes, México

Juan Carlos Arias-Almeida
Limnología Básica y Experimental, Instituto de Biología, Universidad de Antioquia, Medellin, Colombia

Ignacio Alejandro Pérez-Legaspi
División de Estudios de Posgrado e Investigación. Instituto Tecnológico de Boca del Río, Boca del Rio, Veracruz, México

José Luis Retes-Pruneda
Departamento de Ingeniería Bioquímica, Centro de Ciencias Básicas, Universidad Autónoma de Aguascalientes, Aguascalientes, México

N.P. Singh
Indian Institute of Pulses Research, Kanpur, India

Indu Singh Yadav
Indian Institute of Pulses Research, Kanpur, India
National Research Centre on Plant Biotechnology, New Delhi, India

John-Pascal Berrill and Christa M. Dagley
Humboldt State University, California, USA

Y. Baye
Centre Régional de la Recherche Agronomique de Tadla, Beni Mellal, Morocco

A. Taleb and M. Bouhache
Institut Agronomique et Vétérinaire Hassan ll, Rabat, Morocco

Vytautas Pilipavičius
Aleksandras Stulginskis University, Lithuania

Teodor Rusu and Ileana Bogdan
University of Agricultural Sciences and Veterinary Medicine Cluj-Napoca, Romania